THE SILENT LANGUAGE

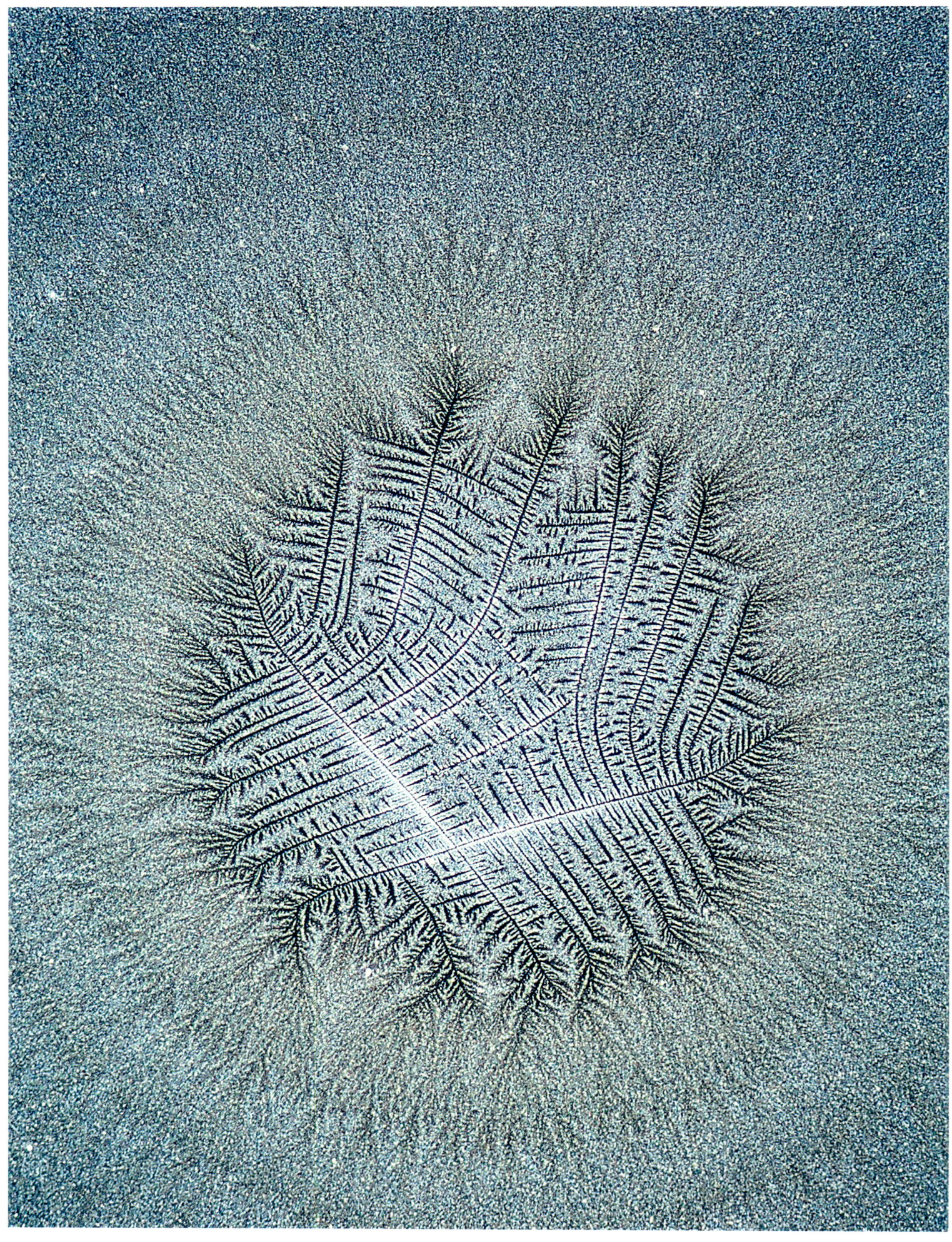

Inge Just-Nastansky

THE SILENT LANGUAGE OF LIFE

Research into Formative Forces in Water Drops

Portalbooks ≈ 2024

*There is nothing great but truth,
and the smallest truth is great.*

J. W. v. Goethe

Portal Books

An imprint of SteinerBooks/Anthroposophic Press, Inc.
834 Main Street, PO Box 358, Spencertown, NY 12165
www.steinerbooks.org

Copyright © 2024 by SchneiderEditionen.
All rights reserved. No part of this publication may be reproduced, stored in a retrieval system, or transmitted, in any form or by any means, electronic, mechanical, photocopying, recording, or otherwise, without the prior written permission of the publisher. This book was originally published in German as *Die stille Sprache des Lebens: Bildekräfteforschung im Wassertropfen* (SchneiderEditionen, Stuttgart, 2018).

Design: Walter Schneider, www.schneiderdesign.net

Library of Congress Control Number: 2024932920

ISBN: 978-1-938685-47-7

Printed in the United States of America
by Integrated Books International

Contents

Foreword by Prof. Dr. Bernd-Helmut Kröplin … 9

Preface by Armin Husemann, MD … 11

Introduction to the Research … 13

Material and Method … 19

Chapter I: Water and External Influences … 23

 1. The Drop 23
 2. Sunlight Effect 25
 3. Change of the Base 25
 4. Air Effect 26
 5. Water Turbulence 27
 6. Circumferential Effects on Swirled Water 28
 7. Transference 30
 8. Color Effects 30
 9. Thermal Effects 32
 10. Cold Effects 34
 11. The Dripper Changes: Water Drops after Eurythmy 36
 12. Reflections on Chapter I 43

Chapter II: Salt between Substance and Form … 45

 1. Salt 45
 2. Salt Process: Movement between Substance and Form 46
 3. Turbulence and Dilution of a Salt Solution 47
 4. Salt—Heat Drying 48
 5. Salt and the Sun 51
 6. Salt and "Plant-water" 52
 7. Swirled Plant Salt in Water (Yam Salt) 56
 8. Salt and "Mineral Water" 58
 9. Salt Ash Water 60
 10. Reflections on Chapter II 64

Chapter III: Minerals, Gemstones, Metals in Water — 73

Chapter IV: Plants, Fruits, Seeds, Barks, and Ashes in Water — 85

1. *Plant Stem and Plant Root, Flowering Plant, Larva, Butterfly, Bee 86*
2. *Berries in Water, the Dark Core in the Center 95*
3. *Seeds — Kernels of the Plant in Water 112*
4. *Seeds of the Plant in Water: The Importance of the Seed Shell 119*
5. *Barks of Young Twigs and Trees in Water, the Cambium 124*
6. *Reflections on the Seed in the Center and the Cambium 134*
7. *Charcoal and Ash of the Plant in Water:
   Mistletoe Charcoal (Viscum carbo) in Water 135*
8. *Exposed and Unexposed Plant 146*
9. *Four Preparations from Rudolf Steiner's Agricultural Course 151*
10. *Peat from Ireland and the Swabian Alps, Refined Peat 154*
11. *Reflections on Chapter IV, Sub-chapters 8–10 161*

Chapter V: Human Fluids — 163

1. *Fluids from the Sensory Area 163*
2. *Cerebrospinal Fluid (CSF) 165*
3. *Pleural Exudate and Breath 171*
4. *Blood 172*
5. *Drop Communication 180*
6. *Human Serum 181*
7. *Patient Serum before and after Treatment 184*
8. *Reflections on Chapter V 186*

Three Phenomena — 188

Conclusion and Acknowledgments — 193

References — 195

About the Author — 197

Foreword

Darkfield photography is a fascinating thing in itself. Known and used in chemistry, physics, and medicine, it provides insight into transparent bodies that cannot be examined under a brightfield microscope.

But it gets really exciting when you examine transparent water drops and find that after drying, systematic, reproducible structures remain in the droplet pattern, which apparently depend on the ingredients and effects on the water and are often of breathtaking beauty. This has resulted in our long-term water research *"Welt im Tropfen"* (World in a Drop), which pictorially shows and phenomenologically compares various effects on water and the associated changes in its structure. These photos often touch the viewer deeply. Since ingredients are dissolved in the water and seem to have lost their original material form, the images in the water are all the more astonishing, because they provide a glimpse into an "in-between world" that apparently exists beyond the material manifestations. This world speaks a different language, which must first be learned in order to be understood—and this hidden world contributes to our recognition of the connections in the living.

This brings us to the new dimension of this book. Inge Just-Nastansky's work begins with the influence of the environment on the drop images (heat, cold, observer), moves on to substance and form (salt) and then questions living nature. Plant branches, leaves, barks, flowers, seeds, fruits, roots—everything that belongs to life is placed in the water and examined. The climax is the "bodily fluids" of the human being such as blood, serum, spinal fluid, etc. in their manifestations under the microscope.

How can one now understand this pictorial language and the emergence of the images in the dried drop? The photos themselves in their pictorial expressiveness touch the reader deeply. Just-Nastansky gains even deeper insight through her method of comparative consideration of the different phases of life. Again and again, the "completed" images are connected with those in the process of "becoming." From rows of drops under the darkfield microscope, coherent natural phenomena emerge against the background of years of study of the formative forces of nature, which can often be explained surprisingly conclusively with the help of an anthroposophical understanding of the world.

It is a glimpse into another, normally hidden world, in which the cosmos and the person of the observer meet. Water is the mediator between heaven and earth. All earthly phenomena of life come to reality through water, it is the basis of all life processes. One is continually astonished by the enigmatic and yet so familiar design language in the water droplets.

It looks as if a path is opening up here that, parallel to physical-chemical structural research in water, could quietly track down the essence of life from a completely different side. Further research and contemplation is therefore required!

Prof. Dr. Bernd-Helmut Kröplin

Image 1: Crataegus fructus (hawthorn berries)

Preface

Water carries movements of all living processes. When a drop dries, a life process coagulates into a form. Inge Just-Nastansky opens herself up to the forms she observes under the microscope as a scientist, artist, and doctor, and thus as a whole person.

> "Man in himself, in so far as he makes use of his healthy senses, is the greatest and most exact physical apparatus there can be."
> (Goethe, *Proverbs in Prose,* ed. by Rudolf Steiner, Stuttgart 1961, p. 21)

She brings to these hieroglyphics of life the life experiences with her patients. As a eurythmist, she has awakened to the life processes in human and speech gestures. As a thinker, she tries to make unbiased judgments in the balance between outer and the inner experiences.

The Goethean researcher has the courage not only to offer a prefabricated reductionist reflection to external phenomena, but to open his soul to them in an unbiased way with the question: What does the phenomenon do to my sensation, what does it do to me when I immerse myself in the dynamics of its gesture in the sense of a trained exact imagination?

The goal of such research paths is a further developed imaginative thinking that does justice to life processes.

> "It is...an experience of knowledge that nature, in order to follow it in its creation, demands the transition of logically formed ideas into artistic pictorial forms. For example, one will be able to express the human physique up to a certain point through logical thinking. But from this point onward, one will have to let the comprehension enter into artistic forms, if one does not want a shadow, a kind of ghost of man, but man in his living reality. And one will be able to feel that in the soul, in that it experiences the bodily form in itself artistically and pictorially, the reality of the world reveals itself just as it does in the logically formed ideas."
> (Rudolf Steiner, *The Goetheanum in Its Ten Years,* pp. 22–23)

A fertilization of scientific thinking through experiences of the arts is the prerequisite for an organic science.

The work of Inge Just-Nastansky joins the ranks of the so-called "image-creating methods" that have emerged in anthroposophical natural science. One thinks of the pioneer Lili Kolisko, who studied the coming to rest of liquids in rising images. For example, at full moon and at new moon radically different rising patterns of the silver salt solution are formed. Inge Just-Nastansky also shows the polarity of the droplet image at full moon and new moon.

Macroscopy brings balance to microscopy.

Inge Just-Nastansky does not want to deliver fixed results. She wants to draw attention to novel phenomena. She presents her way of dealing with them in an open-minded and free way.

In this way, the reader can take the path to the living through the gate of wonder.

Armin Husemann, MD

Image 2: Belladonna, late harvest

Introduction to the Research

The inspiration for this research work goes back to an exhibition—"Welt im Tropfen" (world in a drop)—at Stuttgart Central Station in 2001.

Professor Dr. Bernd Kröplin, head of the Institute for Statics and Dynamics of Aerospace Structures at the University of Stuttgart, showed enlarged, dried water drops in images on numerous display walls. Various waters, essences, potentized plant juices, saliva drops before and after mobile phone radiation, drop communications, etc. were displayed.

He had taken over the idea and practice of *this* "drop-image method" with representation in the darkfield microscope from the Stuttgart artist Ruth Kübler with amazement and enthusiasm and worked it out scientifically.[1]

It involves water drops of various origins, dripped onto a microscope slide, showing an image after drying. These are looked at in a darkfield microscope and photographed at various magnifications.

This quite straightforward, actually simple method of finding out something about water from its drop fascinated me instantly. The question came to me whether plants, minerals, precious stones...objects from the realms of nature leave an image in the dried drop of water when you put them in water for a longer period of time?

I immediately started putting plant stems in water for a week, as well as minerals and gemstones. The water was taken out in a 1 ml disposable syringe and dripped onto a microscope slide with a fine needle. In the brightfield microscope of the practice, after drying, faintly visible, but meaningful, partly plant-like structures actually appeared. The other objects also produced different image formations.

This was astonishing to me beyond measure and aroused my interest. So, in 2002 I began researching these phenomena in a darkfield microscope with a camera attached.

The physical processes are described in detail in "Material & Method."

While walking through nature, everything was now questioned: What might it show in the drop image if it is put into water for a certain time?

Plant branches, leaves, barks, flowers, seeds, fruits, roots, minerals, etc. were "set" in water glasses on the laboratory table.

Astonishing images emerged, characterized by order, rhythm and harmony, which soon grew into a great abundance.

How could one understand their *formation* in water on the one hand, and their *pictorial message* on the other? Minerals do not dissolve in water, neither does the plant stem made of cellulose—so what does the image produce?

Comparative observation became the main task in order to discover a certain order in these rather specific results. An exciting journey of discovery began with obstacles, errors, and astonishing insights into coherent natural phenomena. This is reported.

In the background were the years of preoccupation with the *formative forces* of nature and their visualization in the realm of substance. The drop pattern method, which Theodor Schwenk (1910–1986), engineer and water flow researcher, had developed until 1967 and which is still used today at the Institute of Flow Sciences in Herrischried to test water quality, is another method. The seminal publication of his water research, *Sensitive Chaos: The Creation of Flowing Forms in Water and Air*, has been a companion to me for many years.[2] In his droplet imaging method, distilled water drips into a glass dish with a homogeneous mixture of sample liquid and the purest glycerin (17.5 ml sample liquid with 2.5 ml of the purest glycerin) every 5 seconds. Due to the increased viscosity of the sample liquid, the droplet pattern grows from the center. In symbol drawings, seven drop image types can be distinguished.[3] I had also dealt with the ice crystallizations of Masaru Emoto, MD (1943–2014)[4] and the water-sound paintings of Alexander Lauterwasser (b. 1951)[5].

Out of medical-humanistic interest, the "image-creating methods" were a concern for me. This should be mentioned briefly: Lili Kolisko (1889–1976) used capillary analysis (called capillary dynamolysis by her, later the rising image method) to prove cosmic forces in metal salt solutions with immersing flow paper. "Sternenwirken in Erdenstoffen" (Star work in earth substances) is, among many other publications with pictorial plates, the title of a small paper published by her in Stuttgart on March 30, 1927.[6] In it she documents a fraction of her many years of work at the Biological Institute of the Goetheanum and the Clinical Therapeutic Institute in Stuttgart.

The "capillary-dynamic blood examination" (rise and fall method) according to Werner Kaelin, MD (1888–1973) gave me valuable diagnostic indications for patients over many years through the research laboratory of Geerd Seelig (1932–2015) in Winterbach/Gütersloh.

Out of diagnostic interest, I worked for a while with the "sensitive blood crystallization" (small image method) of Dr. Ehrenfried Pfeiffer (1899–1961) and Alla Selawry, MD (1913–1992). Here it is the copper chloride which, in solution as a reagent on living forces of a drop of blood, forms various organ-specific crystallization shapes in the drying process—this in a circular image area, viewed with a magnifying glass.

These different ways of visualizing living forces are described in numerous publications by the researchers.

The drop image method presented here had primarily nothing to do with medical activity. Only the many astonishing events in water droplets and the deciphering of their enigmatic formal language were of interest.

In the three following years after the beginning of my investigations, a loose cooperation with Prof. Bernd Kröplin resulted in the form of joint lectures with his and my images—for example in the Urania/Berlin, in Munich and Stuttgart.

In 2006, the book *Vom Wesen des Wassers* (The Essence of Water) was published by Frederking & Thaler.[9]

Among the contributions of the seventeen water researchers there is also a contribution with some of my images and text.

Over the years, I presented my research results to a wider audience by means of slide lectures. Since the lecture participants repeatedly expressed the wish to see these images in a book, I have decided to present part of the research in this publication with all its expressiveness and also "mysteriousness."

Let us first familiarize ourselves with this research method.

Although the drop of water is no longer liquid when we look at it in the image, we immediately lose the "safe ground." We find no foothold for our understanding, because the three spatial directions are no longer given for orientation.

The three-dimensional world of experience that surrounds us gives us support for perception, conceptualization, and cognition. Thus, natural science starts from what is. The origin of what exists must also be found in this itself. For this purpose, the senses—above all, the sense of sight—are expanded by extraordinarily refined methods of investigation and technical devices. Theories and models are designed to find explanations and causal connections in the world of the reality of being.

Are we also dealing with a reality in this water drop research and its results—the images? If so, which one?

In the microscope we do not see the enlargement of objects of the world of existence—e.g., a liver cell or an insect eye. In the drop image we see something that cannot be classified in known forms of existence. So it is a mirage?

The images, however, speak to us instantly through their beauty, harmony, and order. They arouse the greatest astonishment and excite our curiosity. What is it that I see there? How can I understand it? Where can I place it? Why do these images leave a deep impression on my soul for years?

The images are always reproducible in the same or similar way with the same experimental set-up and are often stable in the water for years!

This gives reason to look for a way to decipher them beyond pure amazement. The images themselves led us down this path.

First of all, it must be established that we are dealing with a reality. Something works in water when a sprig of lavender, a chalcedony or grape seeds are put into it. Over a certain period of time they are together—a communication develops between water and object. You could also say, "A conversation begins." The result of this conversation is the image—in the circle of the surface.

This is now something newly created. You can follow this emergence in the still liquid drop under the microscope. These are often breathtaking events.

In the course of the research, the impression grew that a creative willpower is at work here—it creates something new in the water and wants to express itself in the design of the image.

Every design requires ideas and forces so that the form can be built up and appear. In short, one could say that spirit and substance enter into a creative relationship in water.

The painter Alexej von Jawlensky perhaps spoke of this process when he wrote: "Often one paints and paints and paints, and then one feels—an angel leads the hand."

One of the questions in this research is: from which areas of force do the creative impulses come? For the painter, this area can be experienced in artistic creation as a spiritual-soul world, as an angelic being or also as intuition.

We know that water is the mediator between heaven and earth. All earthly phenomena, especially those of life, come to reality through water.

Thus, we know this water in all its possibilities of transformation from ice over the oceans, from the springs up to mist, rain, snowflakes, and cirrus.

What inexhaustible encounters this *mercurial* element experiences with light and air, warmth and cold, earth and rock, with plants, animals, and humans, all of which it permeates. It selflessly surrenders to these impressions, allows itself to be transformed and transforms everything in ceaselessly changing movement—it becomes the basis of all life processes.

In the following, we want to accompany water on these paths

Introduction to the Research

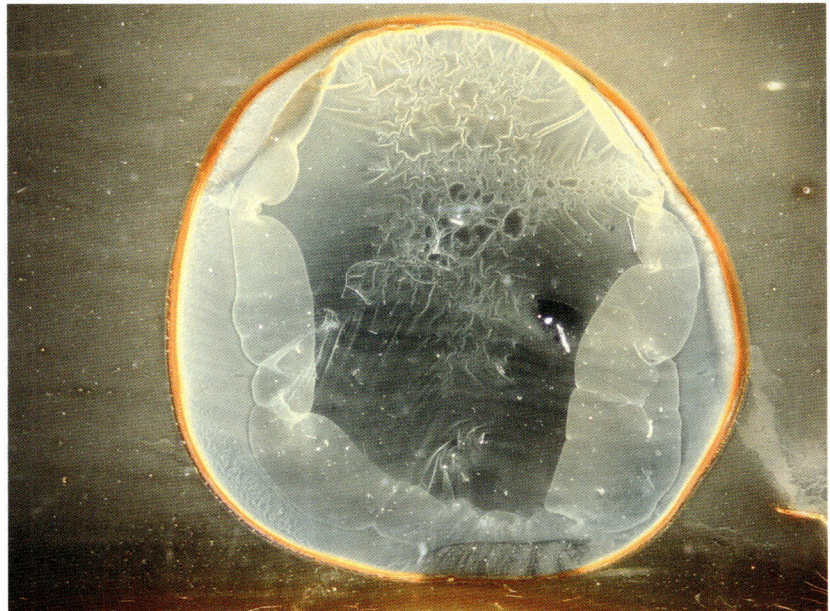

Image 3: Rosehip deposit smear/salt drop 0.9%, liquid (40x)

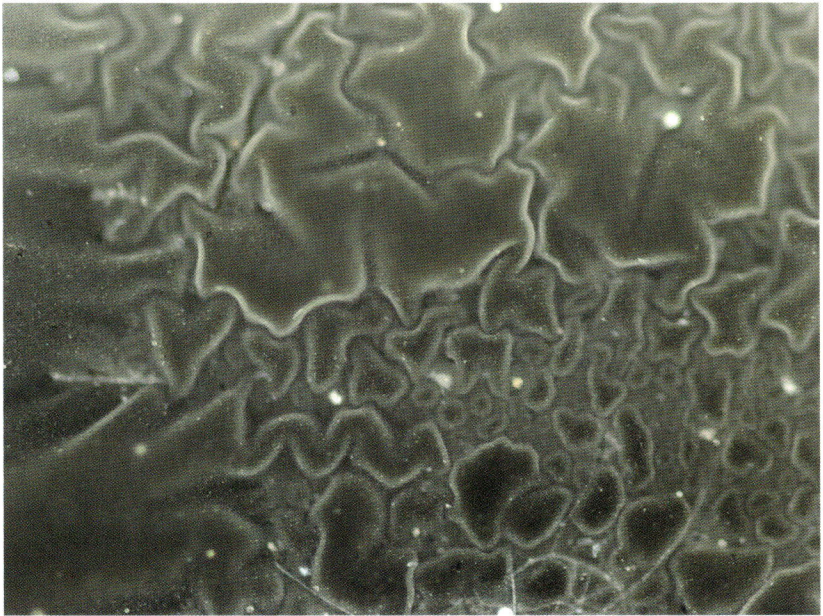

Image 4: Rose-hip smear/salt drop 0.9%, liquid (200x)

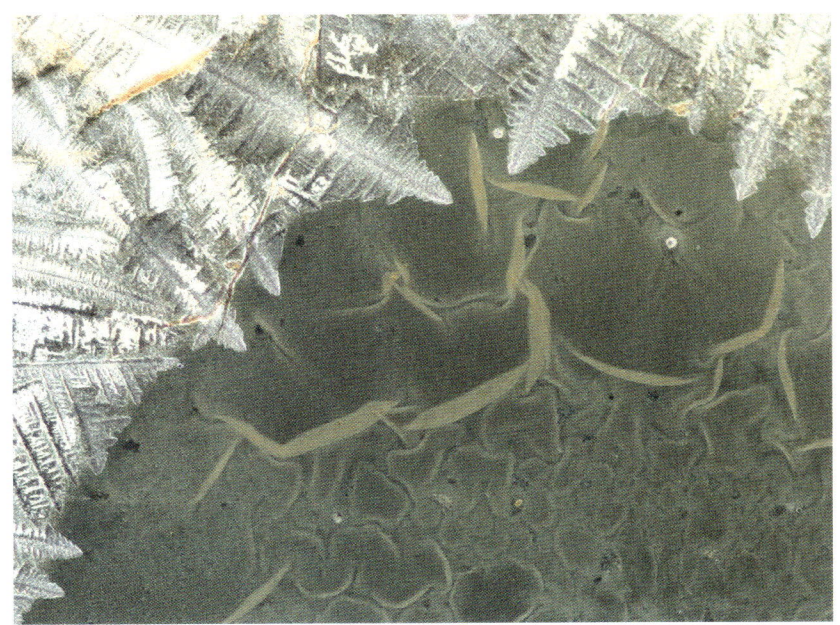

Image 5: Rose hip deposit smear/salt drop 0.9%, beginning salt crystallization (200x)

Image 6: Rose hip deposit smear/salt drop 0.9%, progressive salt crystallization (200x)

of transformation, because it can tell us something about its encounters in the drop image. The order of the five chapters is the result of observation alone. For example, when I see the image of a lavender branch in the water, I need to know what part the surrounding circle could have played in its creation? Thus, the path goes from the circumcircle to the earth substances to the plants and finally to the human being, as will be briefly described in the following.

With regard to the reality and truthfulness of the image statements, it is often objected that the effects of the surrounding circle decisively change the drop image; the dripper also gives the drop image a subjective character. These objections are quite justified, and they had to be subjected to close scrutiny.

This is found in chapter I. Here, we see only the effects of the circumcircle on the water, into which nothing has been added. The influence of the dripper is also shown in this chapter, e.g., dripping before and after eurythmy, as well as in the following chapters, which lead us into the realms of nature. These images show structures imprinted by objects in water. The imprint that the water has experienced through its long-term encounter with the object is stronger than the influence through the dripper and also stronger than the circumcircle effects. A salt solution, a chalcedony or seeds in water lead to the respective typical image, independent of the individuality of the dripper. In this way, these justified objections can be clarified.

The suggestion for the salt experiments (chapter II) came from the need to gain "a solid ground," because one is dealing with an earth substance and its centripetal force orientation, even if it dissolves in water. They lead to a concrete insight into centripetal and circumferential structural unfolding, as will be seen.

In chapter III we then see images of the non-soluble earth substances which, after many months of their encounter with water, leave their very own imprints in the drop.

This is followed in chapter IV by an insight into the plant kingdom. The plant in itself is completely dependent on water and is permeated by it. The question was what comes to light when its various organs are put into the water.

Finally, chapter V is devoted to the fluids of the human being. There is no need for a period of exposure to water, for the images appear immediately after the liquid has been dripped on and dried.

Briefly, we will return to the preconditions for the creation of the images in chapters II, III and IV: The water glasses with the objects from the natural kingdoms are on the laboratory table. Inside, the water is in constant motion: It evaporates upward, pushes cooler substance from the bottom, flows on fine paths, circles in delicate eddies, constantly pushing its gossamer lamellae past everything present, grasping its mineral materiality. Some of this can be observed under the microscope.

Essentially, it can be said from experience: When the water has finished its "conversation" with the natural object after weeks and months—it matures, so to speak—the drop image does not change essentially in its image statement; except as a result of comprehensible events, such as in the case of seed maturation and seed germination in the water. The image often remains stable for years. If I remove the object from the water, this imprint remains—also for a very long time—in the water, and each new drop shows the familiar image.

Water with memory?

This is how *certainty* can be gained in this research. All images created in water have a relationship to the object—they are reproducible in an object-typical way and can be easily assigned. However, we have not yet *understood* what is trying to speak to us through the images. We are looking for a path to knowledge.

The background to this is anthroposophy and the spiritual science of Dr. Rudolf Steiner (1861–1925). From this background, a clue to understanding emerged in this new territory, as well as suggestions for the most diverse experiments. Some of this will be seen.

At the beginning of the evaluation now stands the ever more insistent observation and comparison of the image statements. This is always preceded by amazement. Then questions upon questions arise, and one notices that the mode of observation changes in comparison to that given by the normal sensory world.

The *sensory* aspect of the drop image is not found in this way in the world of appearances. First of all, it originates in all the conditions and procedures of the experimental set-up, which is precisely known to me. In order to come to an understanding of the image, the soul now tries to retrace the creative or formative process that preceded the formation of the image. On this conscious and deliberate path, it encounters the fundamental question of how substance and form meet in water. This is then less a question than an experience of various impulses of movement. The images will be used again and again to talk about the groping search for understanding.

The abundance of the images, of which only a small selection can be shown here, appeals to the viewer, as already mentioned above, immediately and without prior knowledge. This leaves one free to reach an understanding through looking and inner re-experiencing.

Words by Novalis have accompanied me throughout this work. In *Die Lehrlinge zu Sais* (The apprentices of Sais), the chapter entitled "Die Natur" (Nature), Novalis wrote about the special *sense* that the student should try to develop in order to gain a true understanding of nature: "Does he only learn to feel once? This heavenly sense, this most natural of all senses, he still knows little...." And further:

> To everything that man undertakes he must direct his undivided attention, or his "I"..., and when he has done this, thoughts, or a new kind of perceptions, which seem to be nothing but delicate movements of a coloring or rattling pen, or whimsical contractions and figurations of an

Introduction to the Research

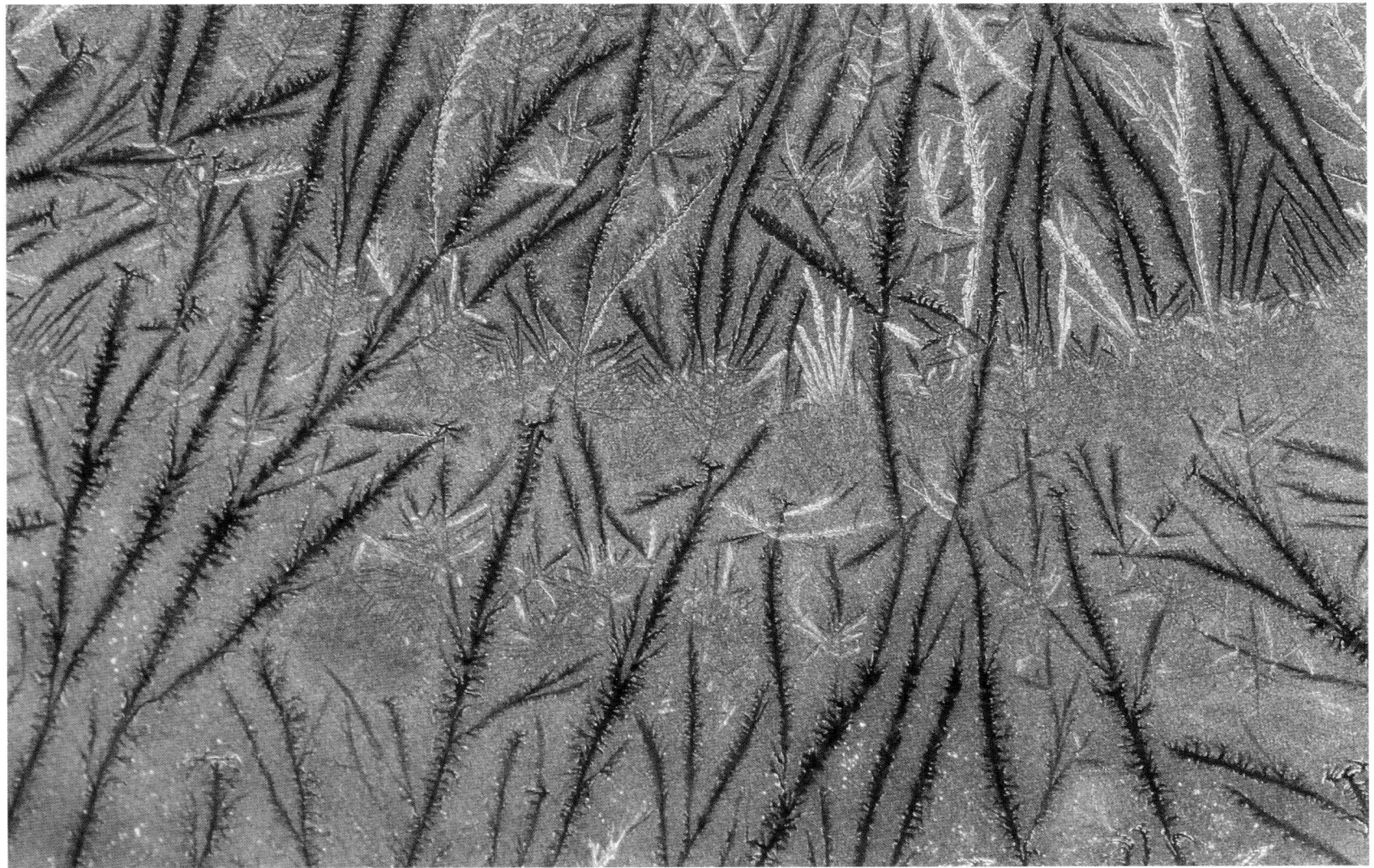
Image 7: Pollen / mistletoe charcoal (1:1 preparation, 200x)

elastic fluid, soon arise in him in a wonderful way. They spread from the point where he firmly pricked the impression in all directions with lively mobility, and take his I away with them.[10]

This book does not attempt to give ready-made explanations for the manifold events in the drop of water. These ready-made explanations do not seem to exist at all, because the sense for the process would be lost. Just as one has to walk a eurythmy form again and again in order to understand it through movement, so it is—on another level—with these images.

"The Silent Language of Life" asks for our thinking, comprehending co-experience in looking at what is given. Nothing is to be put into the images—however, the search for understanding expands again and again into more comprehensive contexts in order to be able to encounter the phenomena cognitively. These contexts are closely linked to the phenomena—i.e., the phenomena emerge from them, as one can experience with an unbiased approach. Only brief hints can be given about the scientific elaboration of the contexts. The images may inspire further scientific research.

What has emerged from activity in the water, what has emerged from movement in the water, wants to be deciphered by feeling into it, by feeling and moving thinking. This requires artistic sense—then we look through a "keyhole" into another living world that appears through the drop of water. More and more we can become at home in what this research can inspire. In this way, we can eavesdrop on nature and its manifold phenomena, which it otherwise conceals from us.

Image 8: Rose thorn water (200x)

Material and Method

Image 9: Laboratory

Image 9 shows the laboratory. Minerals, precious stones, salts, metals, plants, roots, fruits, seeds, ashes, etc. from the kingdoms of nature are *prepared* in water in various water glasses (image 10). The jars are covered with glass to prevent rapid evaporation. They are labelled and dated, and their size is adapted to the respective objects; they must not be too small so that the water can circulate; they must also be placed at a certain distance from each other, as otherwise "transfers" (see below) can occur. Minerals remain in the water for months, fruits and seeds for many weeks, plant stems for a few weeks or days until the image is completed.

From these *preparations*, water is taken at different intervals—the drop image matures—in a 1 ml syringe (B. Braun Melsungen Omnifix-F, 1 ml) and dripped onto a well-cleaned microscope slide (OTMM microscope slide with matted rim, Fa. A. Hartenstein) with a fine needle (Sterican, 0.40 x 20 mm, size 20, B. Braun). Three rows of approx. 6 to 8 drops are produced (image 11).

Image 10: Glasses of the preparations

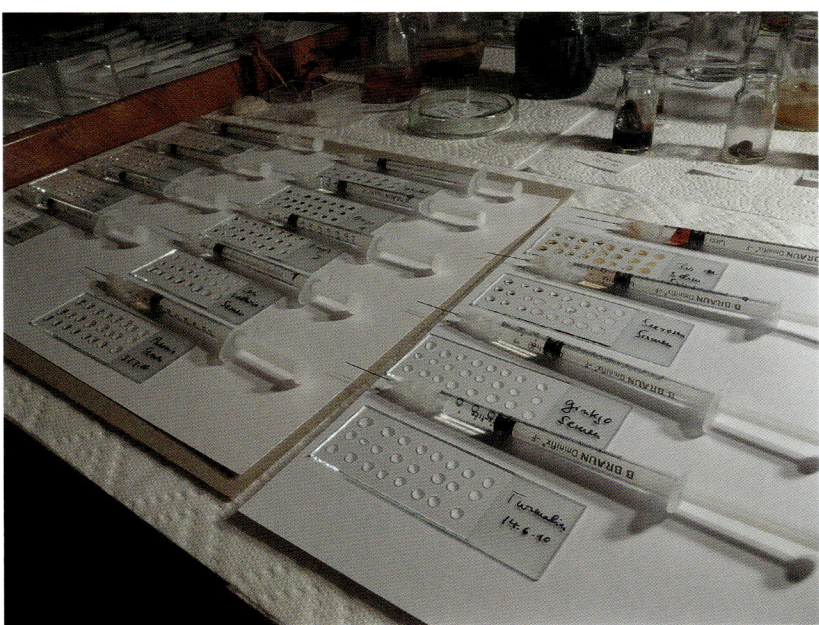

Image 11: Dripped-on drops

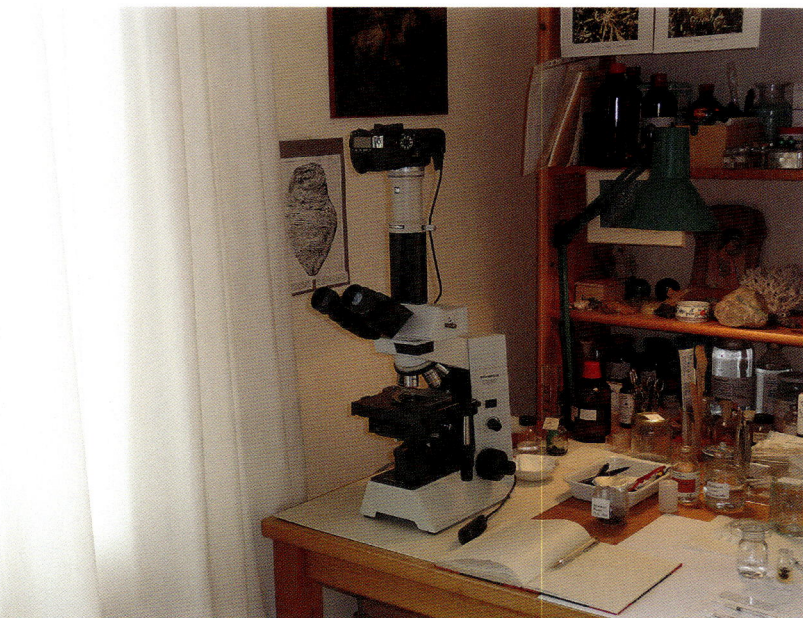
Image 12: Darkfield microscope

Image 13: Hand swirler

The drying process is completed within 30 to 50 minutes, depending on the spherical shape of the deposited drop. Short-term preparations with weak imprints result in very flat drops that dry faster. The resulting drop images are viewed and photographed in a darkfield microscope (Olympus OM System, CX40 with universal condenser, image 12) at various magnifications (40x, 100x, 200x, 400x). A camera with remote release is mounted on the microscope tube for this purpose (Olympus camera OM-4Ti).

The beam path of the darkfield microscope comes from the periphery and is not direct light. The objects appear through the light rays bent on them in such a way that the finest distortions in the structures become visible. Dark materiality illuminates as if from within. Light materiality, as in the salts, remains light. The background remains black.

All procedures and processes, as well as the sequence of images, are documented in the laboratory book. Since the beginning of the research in 2002, 5 laboratory books and 12 epistemological books have been created. From the latter, however, only individual suggestions, conjectures, and questions can be put on paper here.

Slides were produced in a photo lab, as were prints in various enlargements.

Until 2014, the images were taken with the analogue Olympus camera and are available as slides for lectures. Unfortunately, it was then necessary to switch to a digital camera (Canon EOS 6D), as there were no more films for microscope images.

Since 2015, work has been done with the digital camera. With a few exceptions, the images in the book were taken with the analogue Olympus camera. This is noted in the text.

The drops dry at room temperature—but drying has also been done at high heat or in the refrigerator, which is noted in the text and images.

The water used over the years is *Volvic*, because this one, of the many different waters tried for the preparations, shows a conciseness in the image design like none of the others. These others produce the same, but only very weak image statements (see below). The following were tested: distilled water (only produces images under special circumstances; see below), various spring waters, Lourdes water, Lauretana, St. Leonard, Plose, Hirschquelle, Stuttgart water and others.

One reason for this different behavior of the waters can be assumed to be their mineral composition.

Volvic has a very high SiO_2 content (32 mg/l). Other waters have no silicic acid at all or only a very low proportion, e.g., Plose Naturale from the Dolomites: SiO_2 6.5 mg/l.

"Volvic's natural mineral water is filtered through six layers of volcanic rock in one of Europe's largest nature reserves.... It comes from 90 m below the 'Puy de Dôme' in the French Auvergne."[11]

The most important element in magmas and lavas is silicon. When combined with oxygen (SiO_2, silicon dioxide), it forms various tetrahedral shapes (see below). The silicates are built according to the principle of the six-number and the three-number.

"The same formative forces as in quartz are hidden in the forms of honeycombs."[12]

"The silica process is a shaping process."[13]

Of course, it was also necessary for the comparative research to work with one and the same water. There was no significant difference between Volvic in glass or plastic bottles. As long as it was available, Volvic from glass bottles was used.

Understandably, only a few of the many images (well over 15,000) can be shown. But the viewer must know that he is only seeing individual examples from a large number of the same species.

Material and Method

The hand swirler (Ralf Rössner, IMTON, image 13) was often used to increase the sensitivity of the water and make it more responsive, especially for circumferential forces. Salt solutions were also frequently swirled. Two egg-shaped glass vessels are flexibly connected to each other so that after filling with water, the lower part can be brought upward with a swing. The water flowing into a vortex immediately peripheralizes itself in the lower vessel by flowing down the glass wall. Suction and centrifugal forces increase the membrane formation (see below) of the water and "lighten" its mineral materiality. Therefore, one can speak of "revitalization." In most cases, this reversal was carried out about 20 to 24 times during the swirling process.

The colors of the images usually correspond to what one sees in the microscope. In water, the objects often produce a strong color tinging. However, slight color changes sometimes occur due to the recording technique and reproduction. Unfortunately, the photographs taken with the digital camera sometimes show a brownish undertone that does not correspond to what is seen in the microscope.

The details of the individual magnifications of the images are given in brackets. The hundredfold magnification should not be misleading since the droplets are simply different sizes.

The caption "center magnified" does not always refer to the previous drop, but to a center of the many other drops of the same species.

After months and years, a thick layer of mold can form on some plant roots. The water underneath is completely unaffected. The drop images remain the same. (Of course, it is not necessary to work under sterile conditions.)

A vigorous shaking of the preparation water does not destroy the designs.

Of great importance is the experience that the image reappears in clear structures when the materiality of the drop image is scraped off with a scalpel, placed on another slide and fed with a new drop of water!

This experience raises many as yet unanswered questions about the formation and "fixation" of the drop image.

Image 14: B.W. new moon, center magnified (400x)

Chapter I
Water and External Influences

Before we turn to the external influences on water, we need to look at some of the secrets of the water drop.

1. The Drop

It is the small transparent vessel in which—invisible to the eye—all the events in the water are imprinted in order to appear in the image.

The images actually reveal themselves only in the circular form, which has proved true many times.

If I use a brush to apply the liquid onto the slide, the characteristic image statement is completely lost.

Here is an example, images 15 and 16 (image detail, digital camera): Rose thorn water.

According to this, the spherical shape has a significance for the creation of the images. What might be the reason for this?

If we look at the circular drop area, we know that it is incalculable in the finite.

Drop area = circular area: $F = \pi \cdot r^2$

The number Pi (Ludolf's number) is an irrational, a transcendental number; it eludes final mathematical calculation.

It is obtained by dividing the circumference of a circle by its diameter: $\pi = 3,141592\ldots$

For centuries, mathematicians and geometers strove to transform the circular area into a square of equal area. The secret of "squaring the circle" could not be solved by calculation alone. The solution to this problem can be found in the booklet of the same name by Dr. med. Kaspar Appenzeller.[14]

The series of digits of the number Pi after the decimal point was extended immensely, but never brought to an end. It continues into the infinite, the cosmic, the transcendent.

The word *transcendent* comes from the Latin *trans-scendre* = to transcend, to exceed, to climb over. What is perceptible to the senses becomes supersensible, supernatural.

Thus in the drop-sphere ($F = 4 \cdot \pi \cdot r^2$) we have an incalculable vessel.

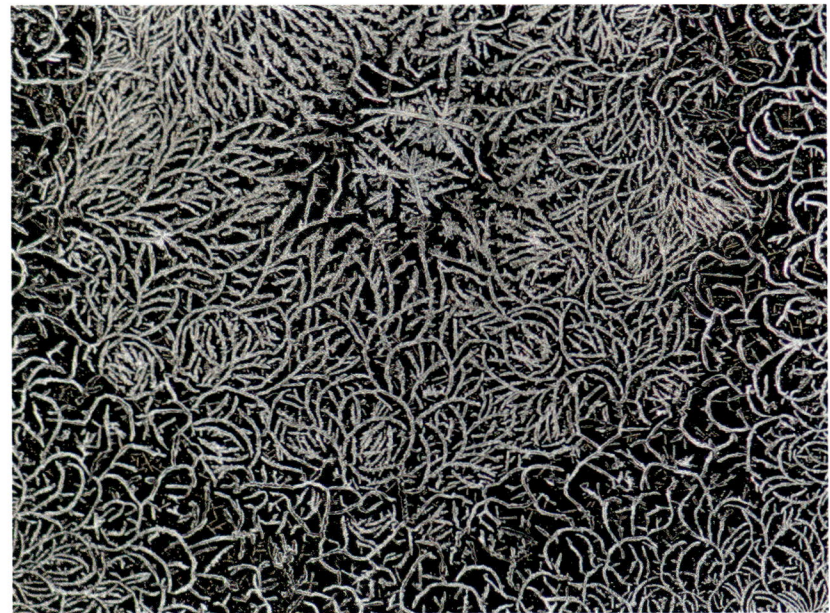
Image 15: Rose thorns with drops (200x)

Image 16: Rose thorns without drops (200x)

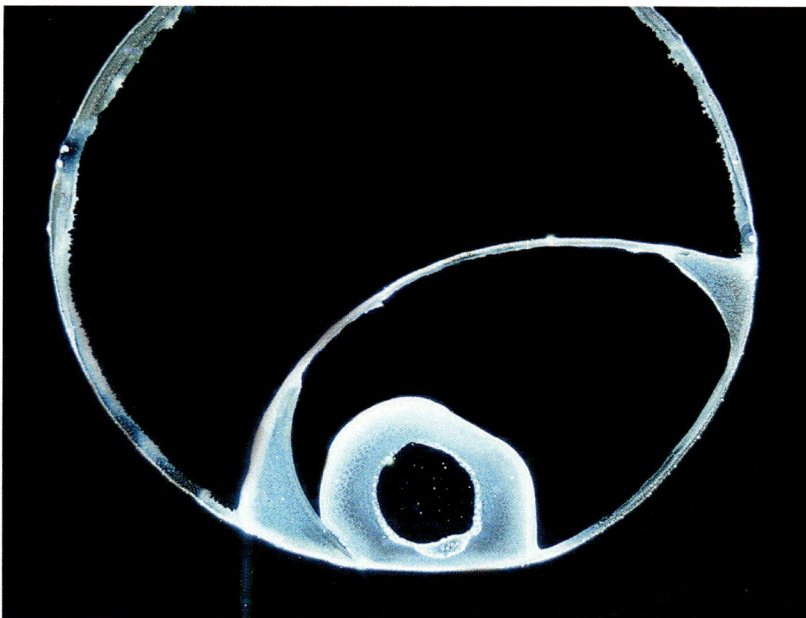
Image 17: Water drop (40x)

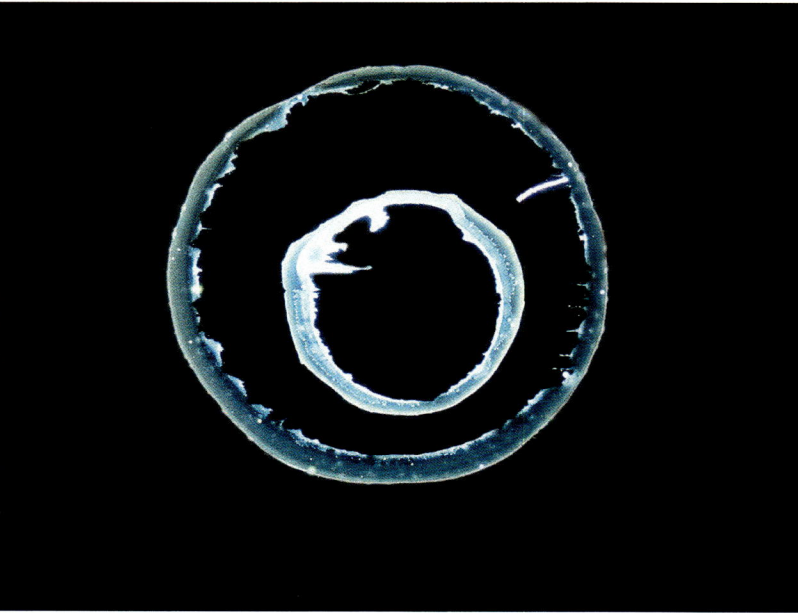
Image 18: Water drop (40x)

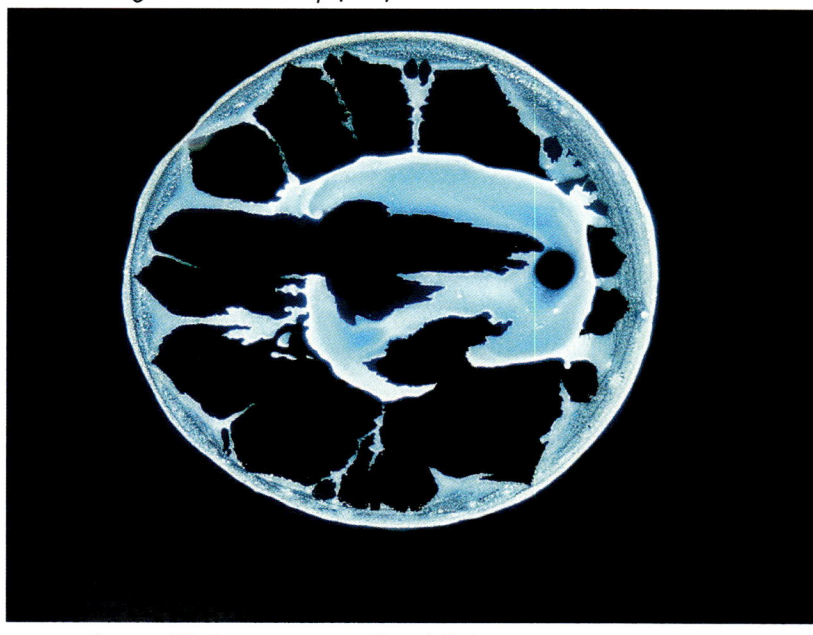
Image 19: Stuttgart water drop (40x)

In the dried surface we actually see something new, unknown, which we do not encounter anywhere else in the natural environment.

Due to its surface tension, the drop sphere is enclosed by "a skin." It mirrors our surroundings and turns them upside down.

Just as the circumference is reflected on the surface of the drop, this drop is also formed by the circumference (see p. 43). We observe this in the rising mist, in the dewdrop, in the rainbow with its colors, etc.

It would be a strange idea to think that a drop of water is formed from the inside by building up its molecules and minerals. It is not like that. The environment "creates" the rising drops from the water through light, air, and heat. The continuity of the water surface is dissolved in mist or water vapor and appears in billions of small droplets that have overcome the earth's gravity. Isn't it always an astonishing experience to watch the "clouds of water weighing tons" as they migrate?

The spherical shape, as an image of the cosmos, is intimately connected with *cosmic forces*. These permeate the water as creative forces and call upon it to bring the world of life into being together with earthly materiality. The whole earth with its oceans is also a sphere.

We must think of the outer skin of the drop as continuing into the interior as fine, layered skins. Water is a surface being - the more it is in motion (e.g., in a vortex), the more numerous are these delicate surfaces (chapter 1, image 69). Like "thin silk sheets" they glide past each other in flowing and swirling; in this way, the outside can be reflected within, and water can become a mediator between heaven and earth.

The stratification of the water can be observed in the liquid drop under the microscope. We observe tiny "light sparks" that criss-cross the black field of observation, but in a targeted manner and at different speeds. It is noticeable that these sparks always remain in one plane and do not change from one layer to the other. This becomes apparent when the depth of field is changed in the microscope.

A poetic question is: How many sheets of paper does the creative spirit have at its disposal to write on...? In this regard, perhaps Novalis speaks of the "clattering pen."

It should also be mentioned that the loss of the three spatial directions in the drop surface described above is replaced by a new orientation.

There are three zones: center, middle, and edge.

These are characterized by the fact that the dropped drop, as a hemisphere, unfolds quite different dynamics. In the black, still liquid field of observation, the edge zone usually settles first and comes to rest. One can then observe an interplay between the center and the middle zone. The center of the drop remains dark and mysterious for the longest time, until it very suddenly becomes distinct or remains dark due to various events. Of course, these three zones create every conceivable variation in the formation of the image.

Water and External Influences

In the first chapter we will look at the circumambient effect on the water in which nothing has been put. We will see what happens in the water under certain external conditions.

We first see three drops of water (images 17 to 19) as they usually appear in variations after dripping on.

They show the circular as a boundary, large dark fields and arbitrary traces of materiality, which will concern us below. The third drop is taken from the flowing stream of water (Stuttgart water). The first two drops are Volvic water and can be considered as control drops for all the following images.

2. Sunlight Effect

We observe the effect of sunlight on the water drop alone.

The slide is loaded with three rows of drops of water. The sunlight, focused by a magnifying glass, is focused on one of the drops at a time until it has dried (images 20 and 21).

A wooden board on the balcony grid serves as a base for the slide. The drops dry relatively quickly in the sunlight.

Under the microscope, a small world has suddenly appeared: Tiny centered shapes are grouped around a dark center and seem to grow peripherally into concentric circles. The dark fields (see control drops) have disappeared in favor of a harmonious texture. All drops illuminated in this way show these blossoming shapes. Is this always the case, or does the water imprint change due to the sun when the slides are placed on a different, mineral support?

3. Change of the Base

Now the slide with its freshly dripped drops is placed on different mineral supports and exposed to the sun in the same way.

Lava base (image 22): It has become quite dark in the drop image - the sunlight seems to have been swallowed up by the lava lump. You can see in the image why wine is grown in such lava cavities on Lanzarote. Sunlight and heat are stored by the volcanic rock.

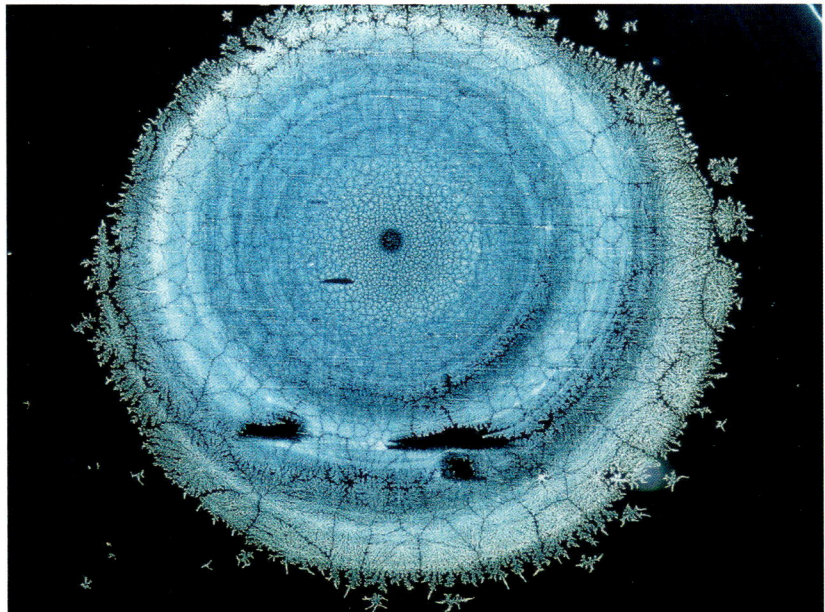
Image 20: Water, sun, magnifying glass (100x)

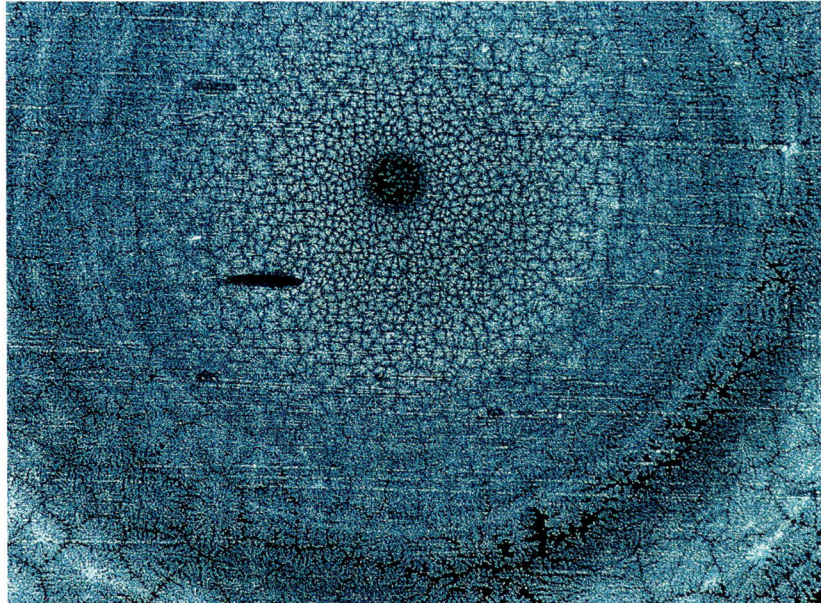
Image 21: Water, sun, magnifying glass (200x)

Image 22: Water, sun, magnifying glass, lava (100x)

THE SILENT LANGUAGE OF LIFE

Image 23: Water, sun, magnifying glass, chalcedony (400x)

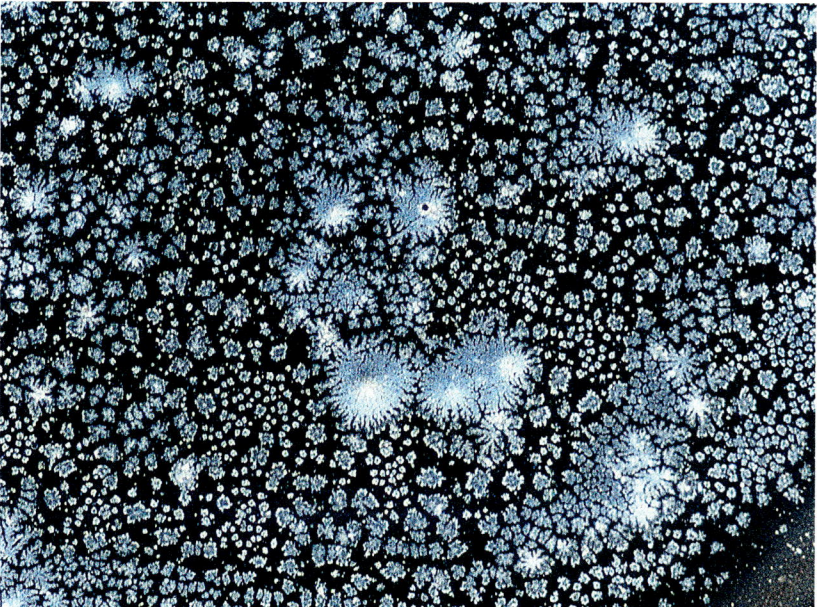

Image 24: Sun through green tourmaline disk (200x)

Image 25: Sun through green tourmaline disk (400x)

Chalcedony base (image 23): Drops dried on this gemstone give a harmonious image with fine, differently colored circles (drop edge zone enlarged).

On these we see the delicate surface layering of the water, because in the drying process the circular surface edges settle from the outside to the inside.

Thus we can see how water can be impressively transformed by its proximity to the physical-mineral world. Many experiments bear witness to this. Sunlight provides the creative impulse.

Another experiment with sunlight:

A green tourmaline disc lies on a water-filled petri dish, which is placed in the sun for two hours (images 24 and 25).

The entire drop surface is then covered with very small flowers, and in between appear larger plant-like, centered clumps that branch out (image 25).

4. Air Effect

What happens in the water droplet when its air environment is intensely changed?

A slide with fresh drops is placed under a square box made of plexiglass.

Tobacco air (cigarette smoke) is blown into this enclosed space through a small opening until the drops have dried (image 26). Circles of strong expressiveness, separated from each other by black rings, are grouped around a strongly pronounced center (materiality appears bright in the darkfield microscope). The circles of the peripheral zone are less rigid and interspersed with fine rays. The water was strongly imprinted by the poisonous effect of tobacco smoke. Further experiments with the poisonous effects of other substances in the surrounding air envelope would be of great interest here.

Image 26: Water, tobacco smoke (100x)

Water and External Influences

5. Water Turbulence

(Turbulence is marked as B. W. = *Belebtes Wasser* [revitalized water].)

From Theodor Schwenk, the engineer and unsurpassed water flow researcher, we learn what a "living water" is: one that is moved. Whether something of this vitalization through turbulence appears in the drop image is to be examined here.

Water is swirled in the hand swirler (see the chapter "Material and Method") and then dripped on.

In the central zone, rosettes sprouting from centers appear, which peripherally change into plant-like growth. They are grouped harmoniously around a crystalline center with many crosses (image 27).

This time the control drops had been forgotten. It was made up from the water bottle that had stood next to it during the swirling.

The astonishment at these "control drops" was very great—they showed very similar images to those of the swirled water, but the crosses were missing in the center! The water bottle had witnessed the intense activity of the swirling (image 28). Here we encounter the phenomenon of "transmission."

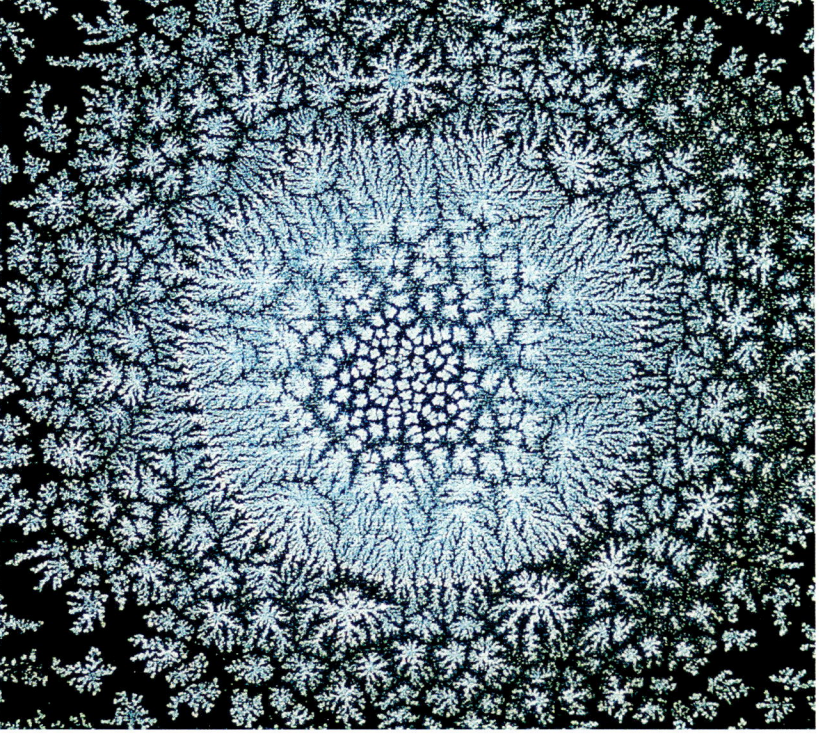

Image 28: Water next to B. W. (200x)

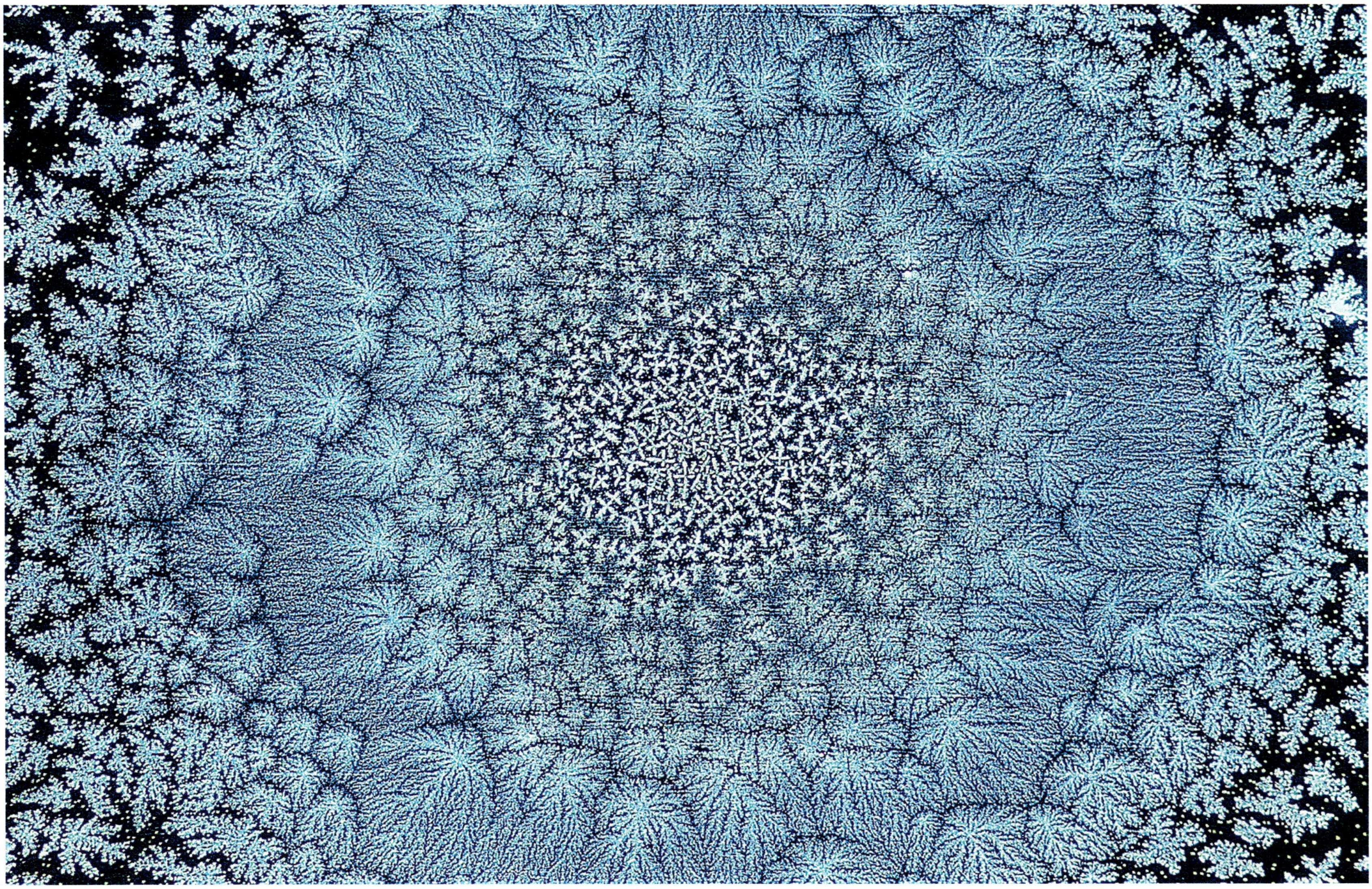

Image 27: Revitalized water (B. W.) (200x)

— THE SILENT LANGUAGE OF LIFE —

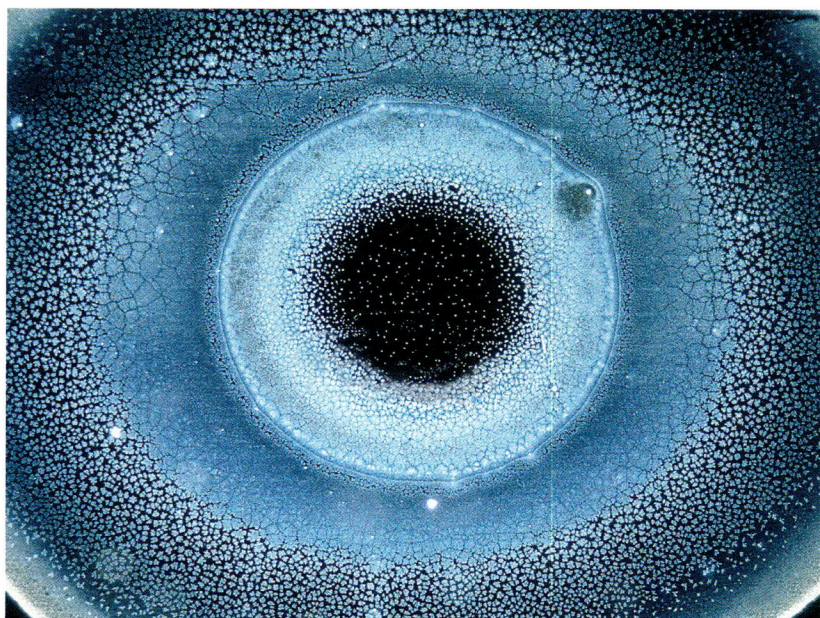

Image 29: B. W. Full moon (100x)

Image 30: B. W. New moon (100x)

6. Circumferential Effects on Swirled Water

Here the question arose as to whether swirled water undergoes the same "revitalization" under different external circumstances?

Image 29: Swirled water at full moon
Image 30: Swirled water at new moon
Image 31: Swirled water at new moon, center magnified
Image 32: Swirled water during lunar eclipse
Image 33 and 34: Acorns in the water
Image 35: Swirled water during lunar eclipse,
    edge zone magnified
Image 36: Swirled water after lunar eclipse,
    edge zone magnified

We see familiar but also very different imprints in the water.

In image 32, a "sun" appears in the still liquid drop, which disappeared after drying. This "sun" is not an isolated phenomenon, but appeared again and again spontaneously in a wide variety of drop images.

Images 33 and 34 show this with acorns in water. Here the "sun with protuberances" has dried. This is inserted here only as an example of the phenomenon "sun in a drop image."

In image 35, we see small "rings of light" in the edge of the drop, which chain together with short arms.

After the night of the lunar eclipse, the same water (image 36) was swirled anew in radiant sunlight. In the drop rim, centered *flower figures* appeared again, as we already know them in a similar way. How quickly water can change its imprint! Where have the small ring formations gone? The sunlight has caused a complete rearrangement of the water.

Image 31: B. W. New Moon, center magnified (400x)

Image 32: B. W. Lunar eclipse (100x)

Image 33: Acorns in the water (100x)

Image 34: Acorns in the water (200x)

The figures speak to us—but how can we express what the soul experiences when looking at them?

"That looks like..." A familiar figure from the world of appearances alone does not lead to an understanding of the pictorial language. It would be a comparison based purely on external points of view and does not lead beyond the level of the intellectual activity of understanding the sense-perceptible world. The phenomenon and its occurrence can be intertwined in the sensation and co-experience of the movements of the water event that have coagulated into forms.

The three images of full moon and new moon (images 29, 30 and 31) are an example of this. In the full moon image (29), do we not experience a calm, full, swelling mood in the soul? In contrast, the dynamic circling in the new moon image gives us structured, actively centering feeling, sucking inward rather than toward the periphery (images 30 and 31). Thus, full moon and new moon effects can be experienced in qualitative differences in the image.

In lectures, it can be experienced again and again that the viewers often describe very similar, unanimous sensations quite spontaneously.

We do not want to prejudice anyone, but the fact is that in the deepened perception and empathy for what has happened—the swirling of the water in the light of the full moon and in the light of the new moon—a path can open up to the harmonization of the formed phenomenon in the image and its process of becoming.

In this way we eavesdrop on nature, the artist, at work. She speaks to us through the drop painting, and trust grows in another, true reality that appears in the painting as a result of living natural processes.

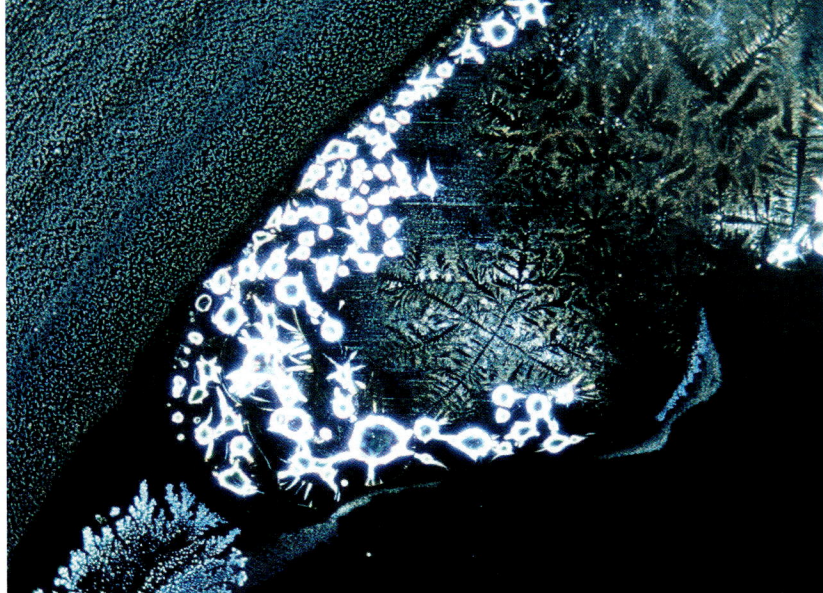
Image 35: B. W. Lunar eclipse (400x)

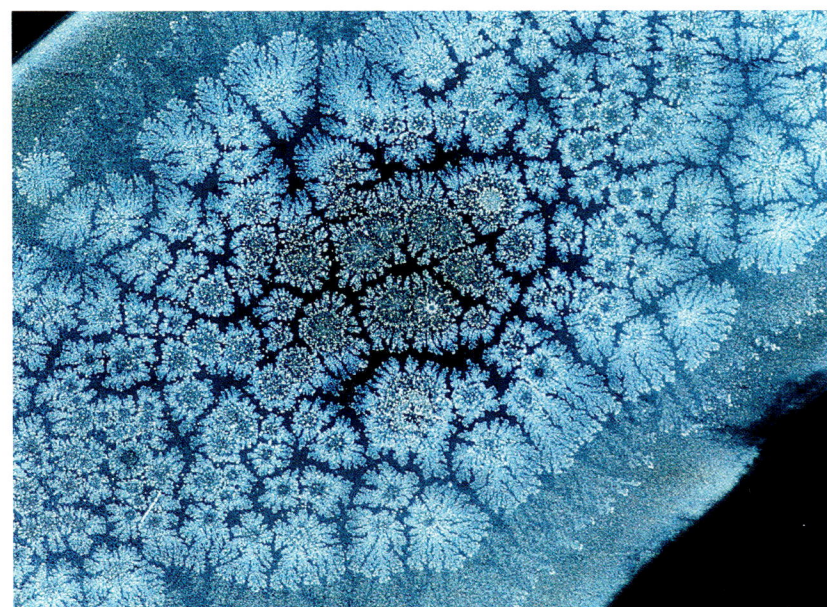
Image 36: B. W. after lunar eclipse (400x)

Image 37: Water spray in the lavender branch (200x)

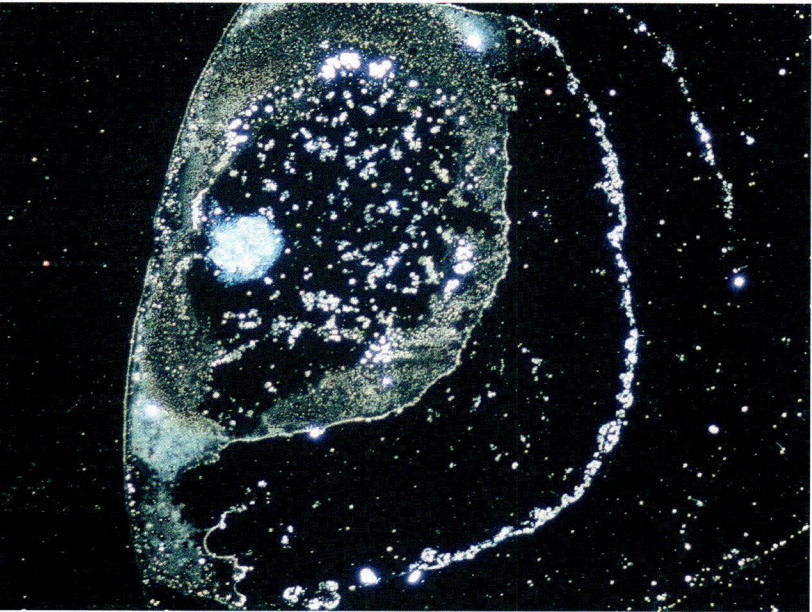

Image 38: Water spray on barcode (200x)

7. Transference

From the far periphery, the path in the following images leads to the more immediate surroundings, with which the water "enters into conversation."

A water syringe was put into the top of a lavender branch — the lavender branch stands in the water. After three weeks, the droplet image from this syringe shows plant-growing structures unfolding from a center to the periphery (image 37).

Another water syringe lies on the bar marks of a barcode during the same time (image 38). Only a section of the drop can be seen, as the rest of the drop field was black.

This image needs no description, but here too the soul experiences something in the comparison of order and chaos. So we see how the same initial water reacts very sensitively to different "neighbors." It responds to what happens in its environment.

8. Color Effects

Can colors also change the water?

Water syringes are placed on the colors red, green, yellow, and blue respectively for three weeks.

The color red already showed an extraordinary activity in the water during the drying process, similar to a bubbling water surface. No structures appear (image 39).

The color green (image 40) shows calm circling, also without structures.

The color yellow (image 41) causes branches radiating from a center point, these are surrounded by circles.

The color blue (image 42) forms a circle-like branch around a black center and then branches out into the periphery. There are concentric circles in the drop border zone.

Here, the different imprints of the image can perhaps reveal something of the creative forces hidden in the colors.

Water and External Influences

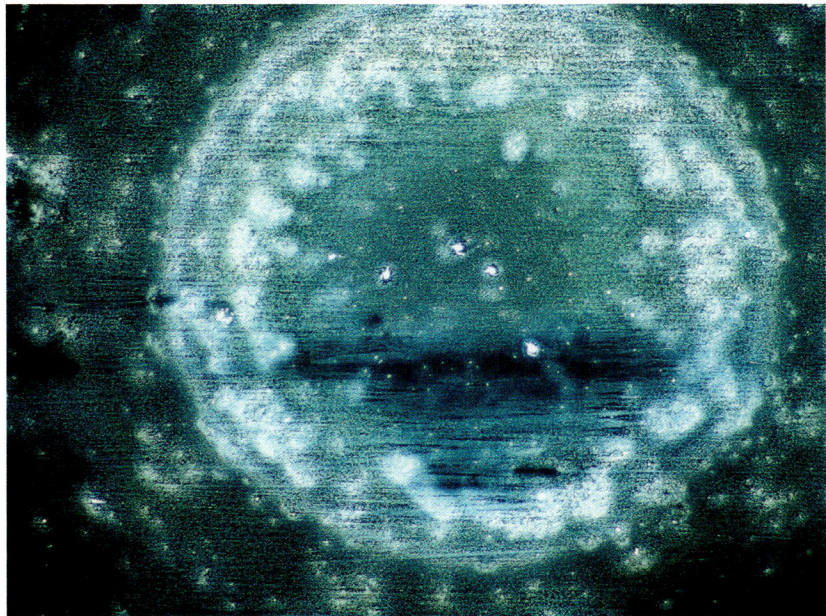
Image 39: Red color (100x)

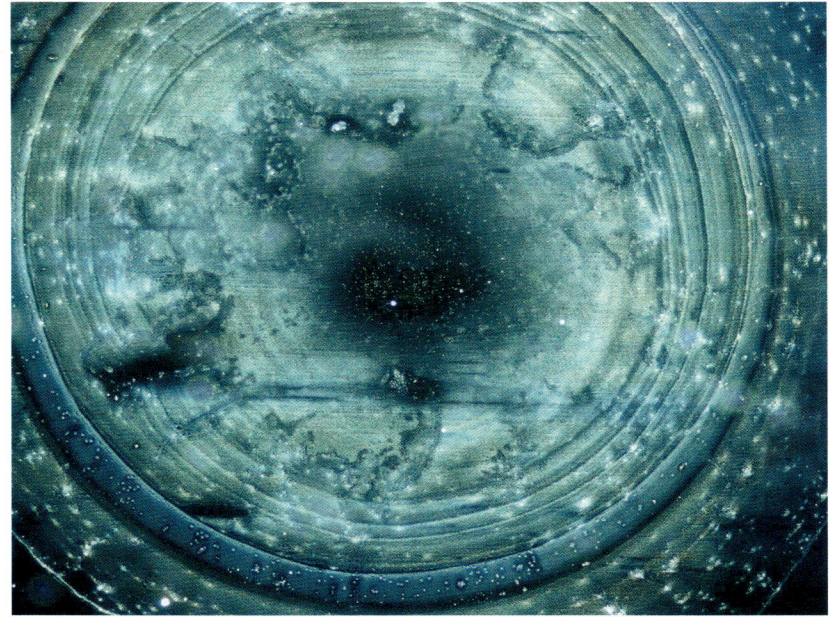
Image 40: Green color (100x)

Image 41: Yellow color (40x)

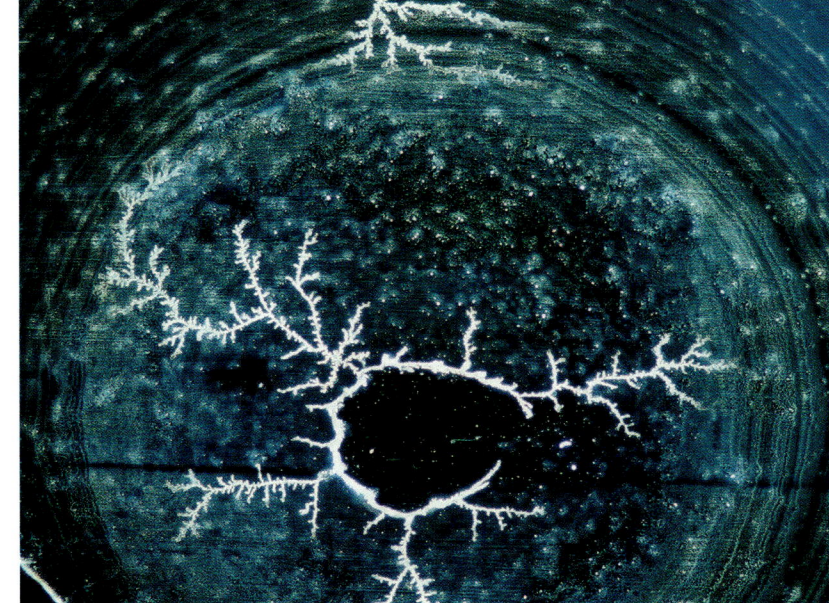
Image 42: Blue color (100x)

THE SILENT LANGUAGE OF LIFE

Image 43: Onyx water (100x)

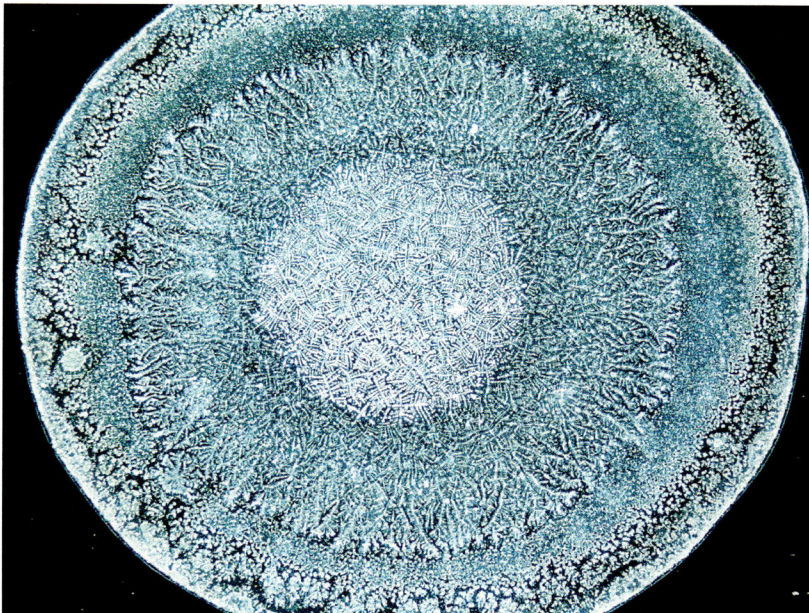

Image 44: Onyx water, heat (100x)

9. Thermal Effects

Here, the influence of heat on the crystallization process in the drying drop was investigated. (See below for the distinction between mineral and "living" crystallization.) Crystallization requires rest and time.

Heating causes the water to move more briskly. Depending on the temperature, the drying process is considerably shortened. Can crystallization occur at all under such circumstances? And if so, what does it look like?

Drop images were compared with others from the same batch water that dried under the influence of heat.

An onyx stone lies in water for several months. The batch of water is called "onyx water." Its typical drop pattern (image 43) shows a crystalline design in the center of the drop extending into the central area. Hexagonal branches are discernible. The drop edge zone remains free. (See also chapter III.)

Now the same water is again dripped on and dried on a hot (approx. 60°C) red sandstone. (It should be noted that heat drying causes clearly distinguishable structural changes on various other mineral substrates.) After the drying process, which was shortened by heat, we see a changed overall appearance (image 44). The rhythmic gradation of the design zones is still more noticeable here, but the much softer, no longer crystalline figurations spread over the entire drop surface to the edge.

An object from the plant kingdom is also tested for heat effect: A bovist (mushroom) lay in water for three weeks (image 45). It shows the typical image for plant fruits (see below) with

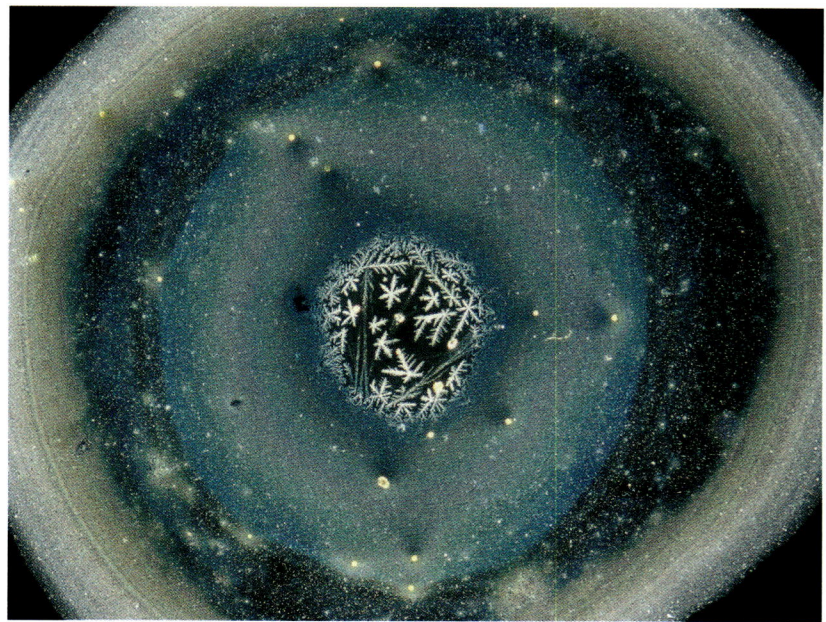

Image 45: Bovist water (100x)

Image 46: Bovist water, heat (100x)

Water and External Influences

Image 47: Bovist water, center magnified (400x)

Image 48: Bovist water, heat, center magnified (400x)

a clearly defined center. In this, crystalline six-pointed stars are discernible.

After drying the same water on hot red sandstone, the drop center has clearly transformed, but without leaving the area of the center (image 46).

In addition, the individual crystalline six-pointed stars have transformed into a coherent star. It is likely a special zone in the drop center, and perhaps also a special materiality if the heat does not cause its usual peripheralization as in onyx water.

The center magnifications of other Bovist drops show the happening more clearly (images 47 and 48): without heat, the typical interior, occupied by single-branched "snowflake figures" (image 47).

Under the influence of heat, the "snowflakes" arrange themselves into a radiant star. On closer observation, however, the star spikes originate in the periphery and radiate inward, forming a dark center (image 48). What a transformation!

In order to get closer to these phenomena, we should refer to fundamental explanations Rudolf Steiner gave to the natural scientists in Stuttgart concerning the heat system.

In the lecture of March 11, 1920, toward the end, Rudolf Steiner drew a small sketch on the blackboard, in which a round red and a round blue color field meet each other in a deeply furrowed wavy line (image 49).

The explanations that accompany this simple drawing lead us to the border between the material and the immaterial, between the physical and the spiritual, between the extensive and the intensive, between pressure and suction.... The blue field must be thought of spatially, as a region of pressure effects, the red field non-spatially, as a region of suction effects. Rudolf Steiner speaks of what occurs in the area of the heat being. In the drawing it is indicated: a continual playing over of the spaceless (red field) into the spatial (blue field). There is talk of a physical-spiritual vortex,"...we have a continual absorption of what is in space by the entity that is outside of space."[15]

These explanations are to be seen together with the heat phenomena in the drop of water. The preceding images serve this purpose. An object from the mineral kingdom (Onyx) and an object from the plant kingdom (Bovist) are considered as examples. It must always be remembered that we are dealing with the water element and, through the drop image, with the second dimension, the surface.

The droplets from the onyx sample (image 43) had "enough time" during the drying process (approx. 40 min.) to form their

Image 49: Sketch by Rudolf Steiner

33

crystallizations in the central area of the droplet.

The effect of heat on the newly dropped droplets (image 44) accelerates this process considerably (approx. 10 min.). In this image we see how the materiality spreads over the entire drop surface. Here, the peripheral circular areas appear almost without design, the middle zone ring is more vegetatively designed, the center is brighter white and shows grid-like crystallization. "Brighter white" means that the microscope light refracts more materiality. In the center, the water remains liquid the longest—i.e., the drying time was prolonged here, and the crystallization was also favored by the slow cooling of the colored sandstone on which the slide rests.

What did the heat in the mineral water cause in the surface?

1. Expansion of the materiality, gradually in different zones up to the periphery, where drying begins and the heat has the strongest effect.
2. Dilution of the materiality in the peripheral area.
3. Dissolution of the hexagonal structures.
4. Indicated vegetative growth tendency in the central zone.

With reference to Rudolf Steiner's blue-red sketch, the onyx-water/heat-drying could initially be located in the blue field—representing spatiality—i.e., in the region of physical phenomena.

However, the peripheralization, dilution of matter, and dissolution of the hexagonal structures into vegetative ramifications indicate a certain tendency toward dematerialization caused by the heat. This would mean that the heat being—in the sense of the "physical-spiritual vortex"—pushes forces (red field of the sketch) into the material (blue field) in order to destroy the physical so that the spiritual can emerge. Especially the vegetative forms could show this transition. However, in the onyx heat images this would only be read out as a certain *tendency* from the image structures.

This tendency can be seen a little more clearly in the Bovist water images. Images 45 to 48 were described above.

Especially the enlargements of the drop center show clear differences without heat (image 47) and with heat (image 48).

The six-pointed stars (image 47), even if they spring from the rather vegetative center edge, partially separate, crystallize out of their own crystallization nucleus and branch out in geometric regularity, as we know it from snowflakes.

We could align the painting with the blue surface of the sketch from this point of view. It has materialized somewhat out of the water after drying.

Image 48: Under the influence of heat we see a star formation emerging from the "crystallization nuclei" of the periphery. Very delicate tufts of rays originate there and radiate inward, leaving out a black center. But it could also be that the black center is created by a suction effect.

The radiating lines are of extraordinarily fine white materiality and are usually accompanied by a black parallel (see also salt chapter; black "lines of force").

It is a moving question whether one can speak here of *dematerialization* to a certain extent—i.e., in relation to the blue-red sketch—of the red surface?

In comparing the two images (47 and 48) we see in any case an impressive transformation of the shapes, in that a center-related, more material crystallization (six-pointed stars in image 47) is transformed by heat into an almost dematerialized, coherent star figure (image 48). This clustered star cannot be found in this way in the world of external phenomena! In the sense of the "physical-spiritual vortex" described by Rudolf Steiner, it could be seen as a spiritual counter-image to the more physical six-star crystallization. "But as the material arises, the immaterial arises on the other side, which slips into the material, destroys its materiality...."[15]

Physicality grows out into space from a crystallization nucleus (snowflake, image 47). The clustered star (image 48) shows the opposite type of formation: ray clusters are sucked in from the periphery.

The heat causes a reversal of the formations, as in a vortex. The explanations of the spiritual scientist lead to a deeper understanding of what one experiences by looking at the images.

10. Cold Effects

Image 50 shows hoarfrost, which in itself crystallizes in needles, here it has been liquefied and dripped on.

Images 51 and 52: Melted hailstones. These are small spheres, and the design in the water seems more lively.

Water and External Influences

Image 50: Hoarfrost (400x)

Image 51: Hail (200x)

Image 52: Hail (400x)

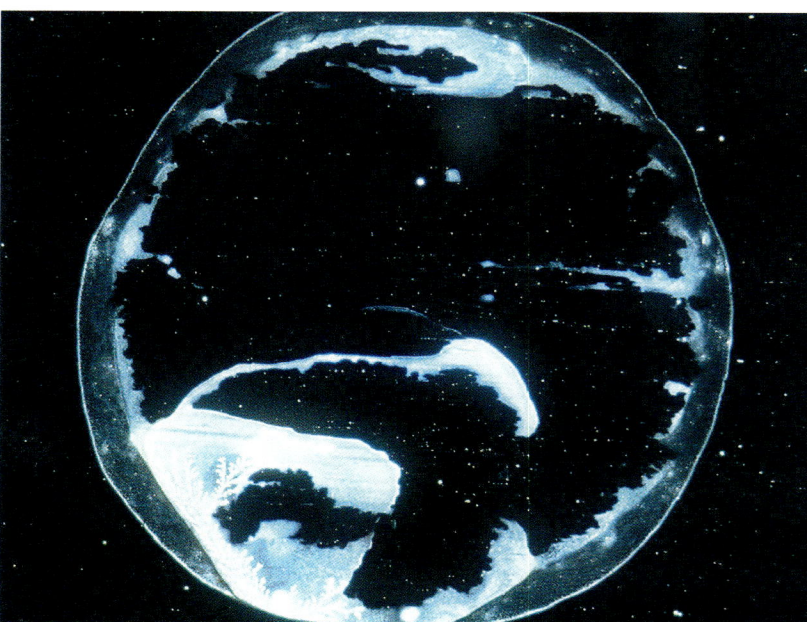

Image 53: Water drop (100x)

Image 54: Water drop after eurythmy (100x)

11. The Dripper Changes: *Water Drops after Eurythmy*

In the following drop images we see that a human being (he is called the "dripper" here), after having done eurythmy, evokes pictorial formations in the dripping of the water, which would not appear without eurythmy. (Eurythmy was performed outside the research rooms).

A small selection from the large number of these "self-experiments" undertaken again and again over the years is shown here. However, the first impression made by the drop image created after a short eurythmic movement was particularly impressive.

A coherent wholeness emerged from random traces of fabric in black dark fields (images 53 and 54). This delicate veil covering the whole drop surface is deeply touching next to the center ring and a harmonious circle around the center. The dripper has therefore called up a force within himself through his previous activity, which works through the microcirculation of the fingertips right into the water syringe. This force calls upon the water in the syringe to reveal itself as a surface being. A delicate skin covers the entire surface of the drop. An all-encompassing connection is created in which each part "knows" about the other because it is connected to it. We see here the subjective imprint of the dripper.

The eurythmy experiments revealed many other secrets besides the coherence-creating component.

If, for example, the water was dripped on after an hour of speech eurythmy, the drop pattern showed mainly material structures (image 55). From the center they grow into delicate ramifications, spreading harmoniously into the middle zone; in this a relaxed round dance of differently shaped figures appears, while the edge zone is covered with calm circles.

The enlarged center of the drop (image 56) shows the filigree of the fabric more clearly. Crosses and small "source stars" appear in the center, which grow peripherally into strong clusters.

In images 57 to 60 we also see drop images after speech eurythmy.

Water and External Influences

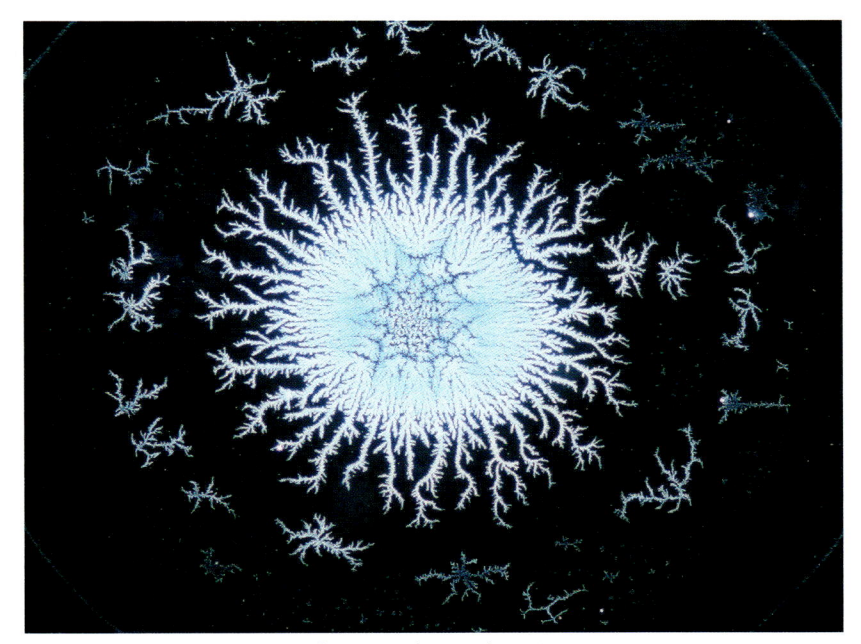

Image 55: Water after speech eurythmy (100x)

Image 56: Water after speech eurythmy (400x)

Image 57: Water after speech eurythmy (100x)

Image 58: Water after speech eurythmy (100x)

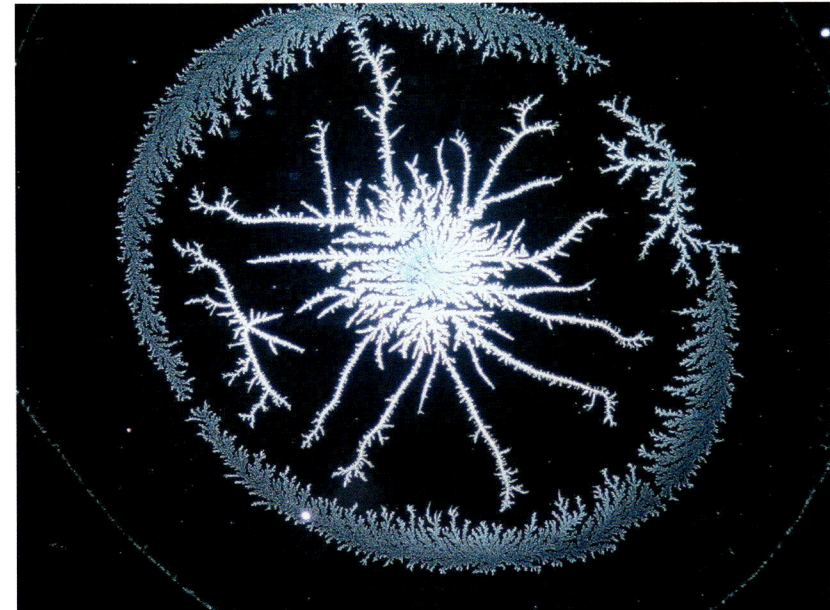

Image 59: Water after speech eurythmy (200x)

Image 60: Water after speech eurythmy (100x)

THE SILENT LANGUAGE OF LIFE

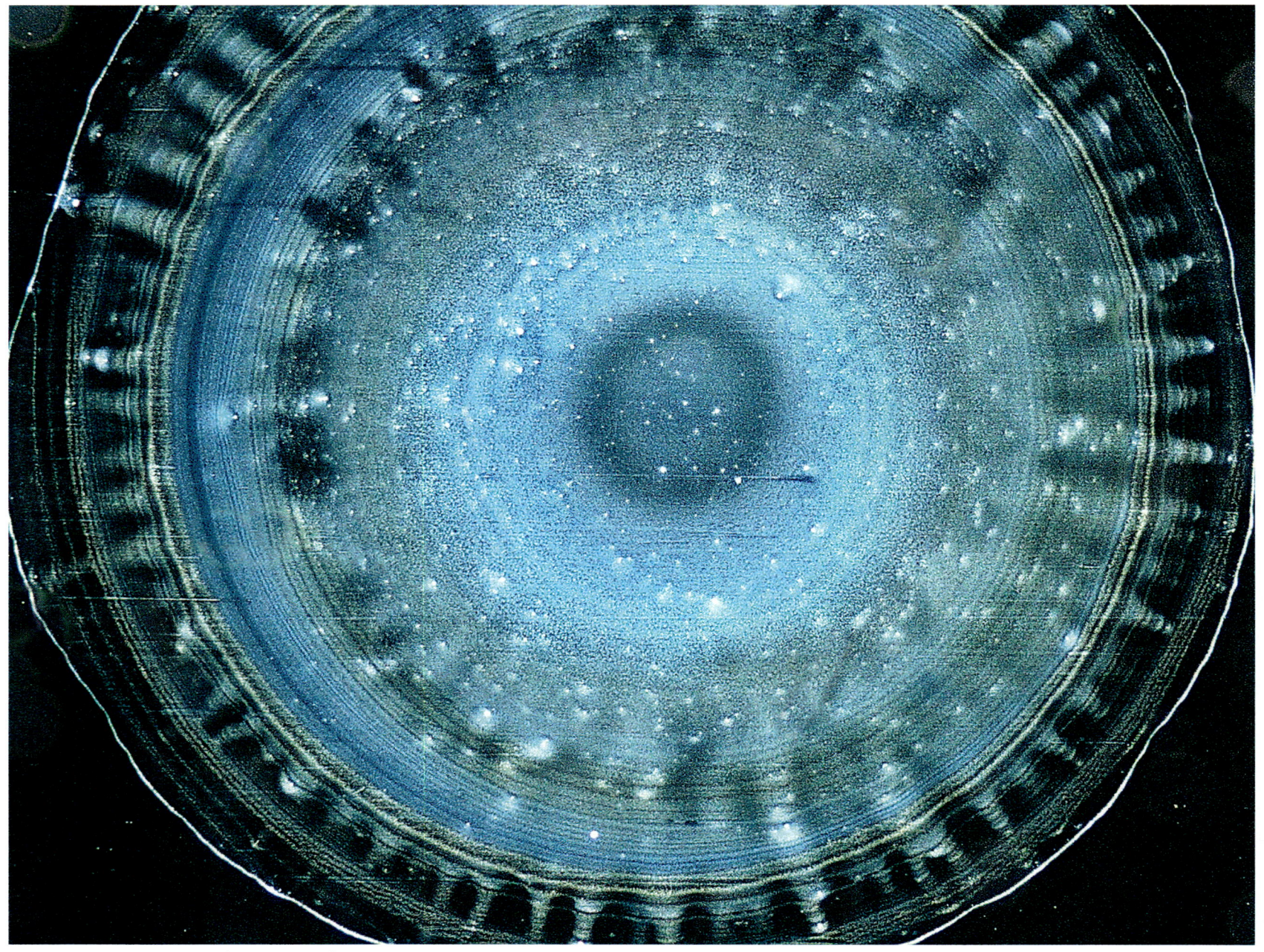

Image 61: Water after tone eurythmy (100x)

When tone eurythmy was performed for over an hour to piano music and then the water was dripped on, it usually seemed to have "forgotten its materiality" (image 61). We do not see any shape drawings, but only calm, very fine rhythmic circling around a dark center! The rhythmic irradiation of the entire drop surface is also striking. The magnification of another section of the drop (image 62), dripped on after tone eurythmy, allows us to see more clearly the delicate rhythmic circling with indicated rays in the periphery.

In images 63 to 66 we also see sections of the droplet after tone eurythmy.

Speech and tone eurythmy seem to influence the dripper in different ways. In speech eurythmy, the water is called upon more strongly for material structuring, which does not happen after tone eurythmy. The deepening of this puzzling question may only be hinted at here.

Water and External Influences

Image 62: Water after tone eurythmy, droplet section (400x)

Image 63: Water after tone eurythmy (200x)

Image 64: Water after tone eurythmy, droplet section (200x)

Image 65: Water after tone eurythmy, droplet section (200x)

Image 66: Water after tone eurythmy, droplet section (200x)

We ask about the material that we have encountered so far in so many different ways in the image designs: Spring stars, clusters, branches, ramifications.... How could they come about?

With the help of an archetype of the water structure, one could say in a strong simplification: Water consists, physically speaking, of H_2O molecules in a tetrahedrally-hexagonally arranged structure (see *Vom Wesen des Wassers*[9] [On the nature of water], Avraham Karltheodor Schmidt: *Das Gedächtnis im Hexagon* [The memory in the hexagon]). An image of this is given by the snowflakes as six-pointed stars and hexagons. These molecules chain together with the minerals contained in the water in the form of ions to form what is called a cluster. The stimulus for this cluster formation comes from the forces acting on the water. Distilled water, in which no minerals are present, only produces images under special circumstances. Due to the manifold forces acting on the water, these material structures are thus created on which the light of the microscope refracts so that they become visible as bright figures. Comparatively speaking, one could speak of a kind of crystallization process, insofar as the mineral materiality of the water comes to rest in structures. In comparison to physical crystallization, as will be seen in the salt chapter, however, the angular laws of the mineral world and the centripetal force orientation are missing here. Instead, an unpredictable, playful material event appears, whose stimulus for its appearance seems to come from the surrounding area.

Now—except after tone eurythmy—many other drop images appear without these material structures. The mineral content of the water is the same, but it does not appear in the form of shaped clusters. So there must be forces of a different kind at work when rhythmic circling and jets appear in the image.

We imagine a moving water surface with waves. A stone dropped into the water forms concentrically growing wave circles. A small paper boat, however, as we know, would not be carried on these waves to the periphery, but would rock up and down in its place. The skins of the water stop it—the silk sheets do not move—the wave travels through them. It is all in motion in this way, but not by change of place. Let us now imagine that the whole thing comes to rest. It slowly dries up until the movement comes to a standstill. Then the center appears surrounded by delicate circles. The continuous rhythmization seems to have prevented the formation of clusters. Crystallization needs rest.

So here the water is characterized in two ways. Through clustering, on the one hand, the mineral substance appears in linear branching or centered clusters. On the other hand, in the imprint of the "skin formation" with rhythmic oscillation in circles, no material design appears.

These abilities of water, which appear here and in other images, express its various possibilities of being the carrier of all living things in nature. However, on closer inspection and reflection, a trinity is revealed. First of all, it is the ability to grasp, transform, and shape the earth substance. The second ability of water lies in its lamellar or surface formation. This membrane formation is an expression of the highest sensitivity and mirrorability. In this way, water becomes a "sensory organ" to absorb the finest external influences. The third ability lies in the fact that water immediately responds to even the most delicate impressions with rhythmic wave movements, as shown by a soft breeze over a water surface.

A trinity emerges: "metabolic activity," "sensory organ formation," and "rhythm"—three basic water elements for all phenomena of life in nature. The rhythmic activity is the mediator between material and sensory activity. In the human nerve-sensory system, membrane or surface formation is expressed in many ways—e.g., in the medullary sheath of the nerves (see below). Metabolic activity and rhythm are also life-bearing basic elements in the physiology of the human organism.

Let us return once again to the different imprints of water after speech and tone eurythmy and venture a possible explanation.

Basically, our life body (etheric body) comes into a more lively activity through eurythmy. We experience how our body is lifted out of heaviness into more lightness. We are more permeated by breath and warmth all the way to the periphery of our organism. This is what can be experienced when we do eurythmy.

In consonantal eurythmizing, the things of the world tell us their "names" through forms. The consonant forces of the zodiac sculpt our shape. In eurythmy therapy, we use consonants to get our metabolism moving.[16] The nutrients are stimulated for better processing. Thus, the consonantal has to do with the substance and with the inhalation of the formative forces.

The vowel and the musical aspect of tone eurythmy are different; here the human soul speaks on the exhale and reveals its innermost feeling. Through the vowels we are connected to the dynamically orbiting planets.

In tone eurythmy we are more deeply permeated by breath, rhythm, and sound. We think of the continuous wave movement of the stone thrown into the water.

Could it be that this prevents material "quasi-crystallization"? (One might be reminded here of the water sound images of Alexander Lauterwasser.[5])

Of course, in speech eurythmy, not only the consonants speak, but also the vowels, so that we see "mixed images" with and without fabric design.

The eurythmy for these water experiments was carried out by one and the same eurythmist. Do other eurythmists show similar drop results? How does the "dripper" change with other types of movement such as tai chi, yoga, qigong, or bodybuilding? Unfortunately, there was no way to see these interesting questions answered by the drop image.

The following two images were made by doing a speech eurythmy gesture with the hands over freshly dripped drops until the drops had dried (about 45 minutes).

Water and External Influences

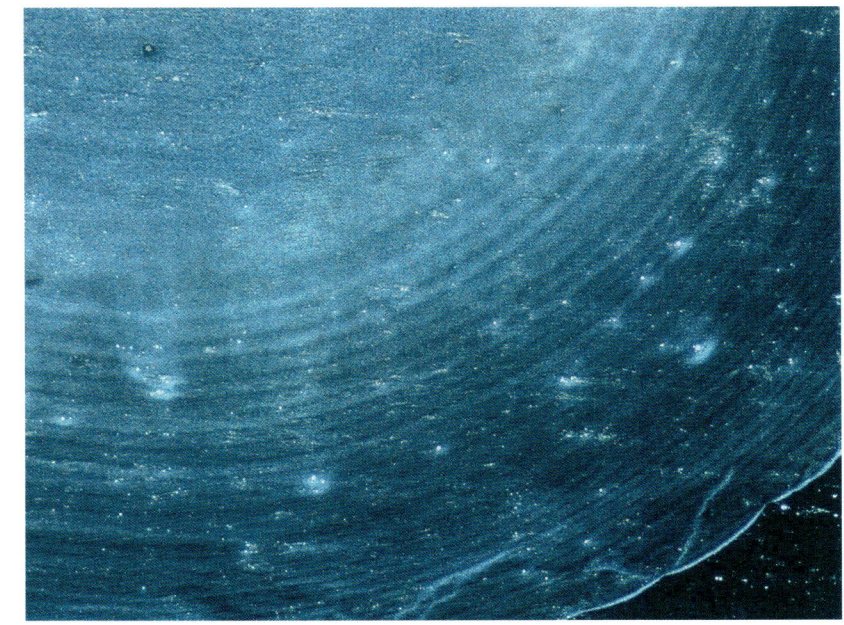

Image 67: Water after Eurythmy-M, drop section (200x)

Image 68: Water after Eurythmy-L, drop section (200x)

Image 69: Water, lamellae (400x)

— THE SILENT LANGUAGE OF LIFE —

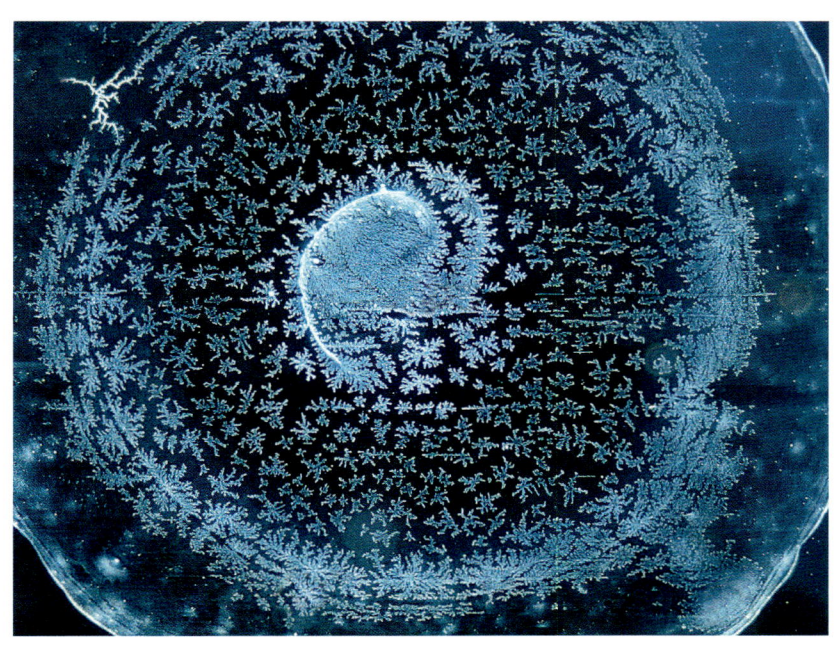

Image 70: Water, bathroom (100x)

Image 67 shows the water imprinted by the sound "M."
Image 68 shows the water imprinted by the sound "L."
Image 69 shows the delicate lamellae of water, which have often been mentioned. This basic ability of water—to form lamellae, skins, membranes—rarely appears in the image. Here it is shown in the water of a withered bouquet of lavender.

At the end of this chapter, which shows us circumferential forces, we see images from two water glasses that stood in different locations for three weeks.

Image 70: A water glass stands in the bathroom for three weeks. The daily activities have caused cluster traces in the drop image.

Image 71: The second water glass stands on a small table for the same period of time, where "mental work" is done. Physical and mental-spiritual activities seem to influence the water in different ways. It takes part in its surroundings!

Image 71: Water, small table (100x)

42

12. Reflections on Chapter I

After drying, the most diverse events appear in images in the small, transparent, delimited space of the drop.

These originate from the water with its minerals, which gives itself up selflessly in lamellar and whirling movements. This water opened itself to the light of the sun and the moon, to the darkness of the earth (lava and chalcedony as a base), to the air interwoven with tobacco smoke, to the living plant (lavender branch), to the colors, the warmth, the cold, and allowed itself to be transformed, also by the dripper after eurythmy. The interplay of the elements (earth, water, air, and warmth) with the light and other formative forces from the surroundings forms the background for the appearance of the images in the drop. Rudolf Steiner speaks about the drop of water:

> I have drawn your attention to the way in which the etheric actually gazes down upon us from the extraterrestrial world, and how everything, whether it is a large or a small drop, is made round, spherical, from the etheric.[17]

Elsewhere he says again in relation to the drop:

> ... we have the drop shape in nature. It is usually imagined in such a way that one thinks of the drop as being held together from the inside. But one does not need to think this way. You can also think of the drop from all sides, formed from the outside.[18]

As already explained above, the spherical shape of the drop has a decisive function in the composition of the image.

Effective forces from the surroundings in the drop of water come to reality on the surface and lead to the concept of the *etheric*. The designed drop image has to be brought into relation with the life-creating forces surrounding the earth.

As the research progresses, we will see how this "circumference around the earth" is shifted inward by the objects placed in the water, so that other structures appear in the drop image.

For the viewer who is even less familiar with the concept of the *etheric*, it should be noted in abbreviated form: Etheric forces are life forces and, at the same time, forces of levity. They counteract the elemental and material forces through peripheral suction. They handle the matter in the form of water. They cause plant growth out of the earth and all growth and regeneration processes in animals and humans. Here one can speak of an etheric body.

The earth is enveloped by these living-expanding forces in four ways: warmth ether, light ether, sound or chemical ether, and life ether. They become effective via the elements earth, water, air, and warmth—in polar relationship to them. Each ether has its own dimension, which will be discussed later. It should also be borne in mind that the ethers of warmth and light have a more expansive effect, while the ethers of sound and life have a more concentrating effect.

However, what has so far only been expressed in individual images (e.g., onyx, bovist) are species-specific creative forces—the actual image forces that radiate in from the wider stellar sphere of the earth. They are the astral and spiritual forces of the planets and fixed stars. They constitute, so to speak, through the etheric forces, the phenomena of the living world that have become physical on earth.

> The world has a transition from earthly lawfulness to cosmic lawfulness, which shines in. The world is a closed whole, and when you come to the end—you have only to imagine it figuratively—then you meet everywhere the interior of a spherical surface. That is where the astral radiates in. The astral begins to work in from the outside by taking possession of the etheric.[19]

The following chapters attempt to indicate which living formative forces are at work in particular instances.

To gain insight into this comprehensive field, we refer to the booklet by Dr. Ernst Marti, MD, *Die vier Äther. Elemente—Äther—Bildkräfte*[20] (The four ethers. Elements—Ether—Image Forces). This booklet gives a summary of Rudolf Steiner's teachings on the elements, ether and image forces.

What has been briefly described here shows, when looking at the events in the drop of water, that these active forces are to be experienced as activities of living beings of the elementary world. Even if we can no longer easily recognize and distinguish these living, creative beings today, we are deeply in awe of them when we see what they bring about.

Rudolf Steiner has given us comprehensive insights into their differentiated work.[21]

Image 72: Salt crystal (400x)

Chapter II:
Salt between Substance and Form

1. Salt

Soon after the research began, in 2002, salt experiments were started in order to gain "solid ground" in this research. Salt, as an earth substance, dissolves in water—what does it show in the drop image? Does it change when it is swirled, for example, or when it comes together with the initial waters of plants?

Thirty-two different salts were dissolved in water and dripped on. After numerous experiments, it turned out that common salt (NaCl) showed a very clear reactivity, for example, to turbulence and to the effects of plants and minerals, which then led to the question of the salt process and to humans.

Initially, droplets that had not produced an image could even be "awakened" by salt. The impression deepened that here, too, common salt can act as a revealer, as we are familiar with from the salting of our food.

The salt of the earth is stored in huge salt domes on all continents: saline salt, common salt (NaCl). Once formed by evaporation and cooling from the world's oceans, they now contain about 3.5% salt. The salt content of human blood is about 0.9 g per 100 ml. The main part is common salt.

The first two images show rock salt (saline salt) from Friedrichshall (image 73) and from the Himalayas (image 74).

Image 75 shows sea salt with two triangular crystals and image 76 Sinai sea water (Red Sea with corals). Here the black structure lines are striking.

The salts were put into solution and dripped on.

In the drop periphery, salt crystals are deposited, which are already macroscopically visible growing out of the surface into space. They lie isolated or interpenetrate in the process of formation, as we also see with the quartz crystals.

The previous images did not show us any such earthly spatial shapes. The crystallization process is completely unaffected by external influences. Striking in all crystals are the black crosses that radiate as diagonals through the square or the fragmentary cube. They are of great importance for the structure of a salt crystal, which we can observe in the microscope—here, of course, largely in the surface (image 77).

Image 73: Rock salt, Friedrichshall (40x)

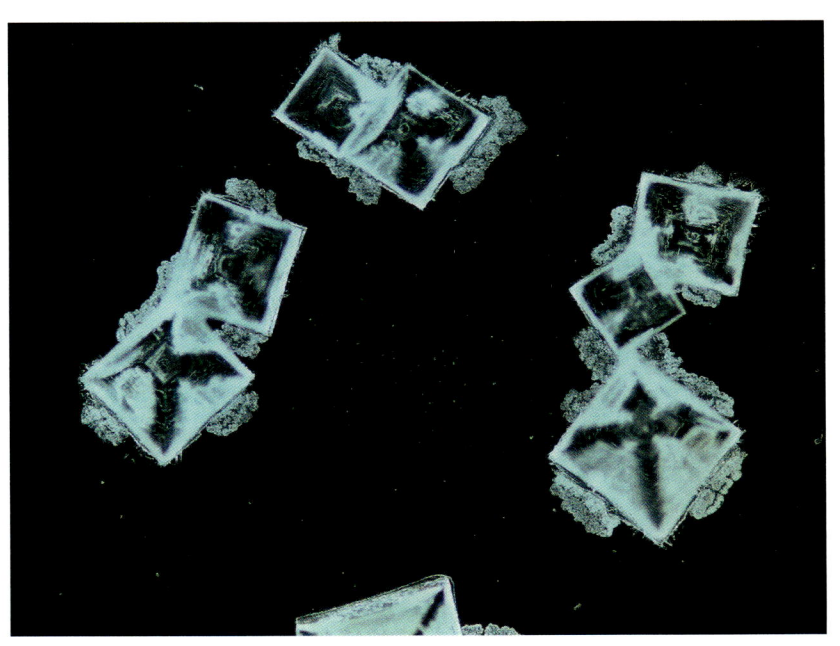

Image 74: Rock salt, Himalaya (40x)

Image 75: Sea salt (40x)

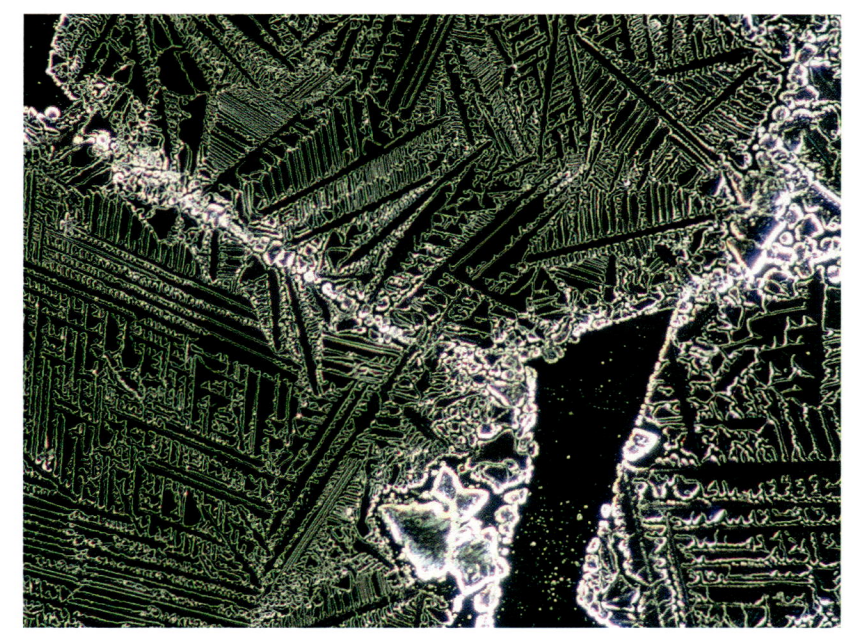

Image 76: Sinai Sea Water (Red Sea) (200x)

Image 77: Salt build-up (400x)

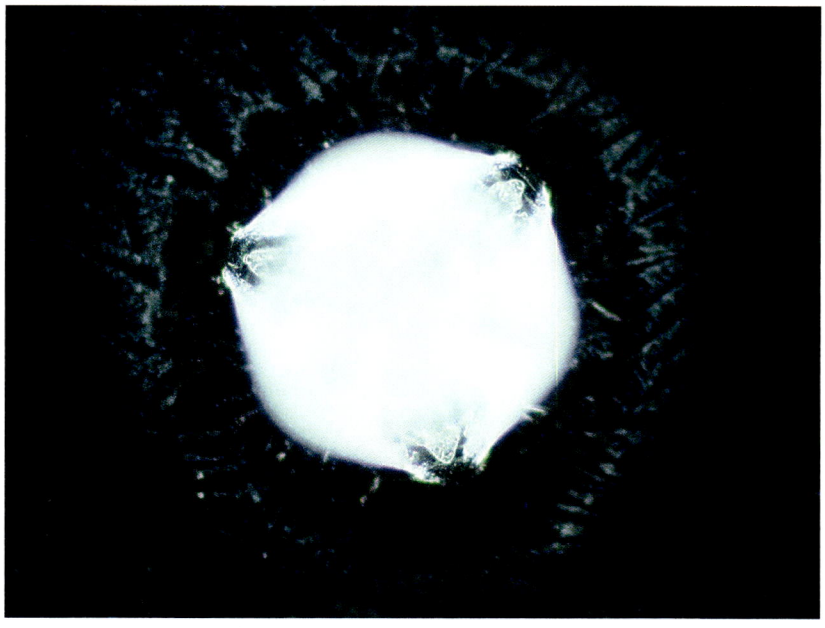

Image 78: Salt crystal, triangular (400x)

The process is as follows: Starting from a crystallization nucleus in the center, tiny white lines are arranged at right angles in the quadrilateral—always in relation to this center. Step by step, new lines are added, which are extended in width in strict accordance with the law, until the closed edge zone of the quadrilateral is completed. The black diagonals are clearly marked, from which the white saline seems to be sucked in by stages. A small "salt pyramid" has been created in the surface.

The interplay of material and mineral forces of form—here the square—occurs with breathtaking precision and calm, purposeful activity. What impresses here as white materiality is transparent, crystal-clear in a salt crystal—which, for example, can be allowed to "grow" on a wool thread in a saturated solution after cooling.

The light guidance in the darkfield microscope makes it appear white as a materiality, much like the salt we encounter every day. The beauty of a single crystal in the plane is shown in image 72, see page 44.

A triangular salt crystal is shown in image 78, which we will deal with later.

2. Salt Process: Movement between Substance and Form

Does the salt given in solution, turbulence, dilution, and heat reveal something of its inherent forces in the drop image? In the physiology of the human organism, we encounter salt (NaCl) *only* in its living processes. In the warmth of the heart, the saline blood is swirled. In the nervous system, the basis of our consciousness, we encounter the inflow and outflow of the sodium ions (Na^+) of the salt in the nerve cell to build up the electrical potentials. In the stomach, it is the chlorine ions (Cl^-) in the hydrochloric acid formation that meet the food

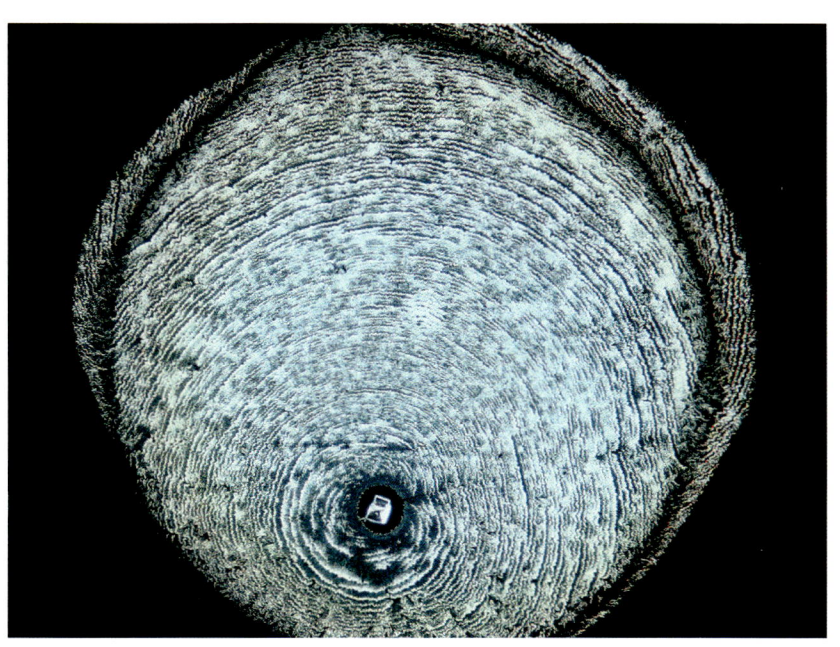

Image 79: B. W. Brine solution (40x)

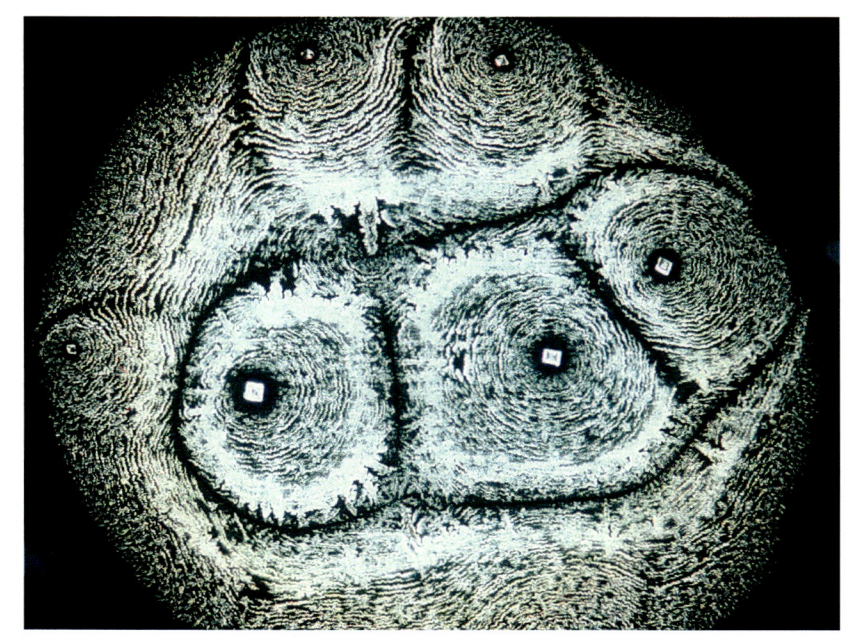

Image 80: B. W. Brine solution (40x)

substances. Thus, this mineral materiality as *substance (substare = to be under)* in the salt process serves all basic functions of the human organism.

In the following experiments we want to see what happens to the salt through various processes.

3. Turbulence and Dilution of a Salt Solution

(Swirling is marked as B. W. = Revitalized Water).

One teaspoon of a salt-brine solution is added to 300 ml of water and swirled in the hand swirler. Images 79 and 80 show how the salt spreads over the entire drop surface, arranging the substance in rhythmic circles around a central crystal.

After dilution I and renewed swirling, a filigree seeding of small radiating salt fields becomes visible (image 81).

In the magnification (image 82) we see how the salt spreads out playfully in branches from a center in orderly white "trails." These do not interpenetrate or cross each other, but meet in opposition. The reason for these phenomena is the "invigoration" of the water caused by turbulence.

With further dilution (image 83) and turbulence, the white material content of the trails is lost—these are now black. In these black, rhythmic paths, the result of formative forces becomes visible. The formative forces themselves are of course invisible, hence appear black. However, they reveal themselves in the handling of the salt, which on the one hand has the spiritual formative power of the mineral inherent in it and which on the other hand is exposed to suction forces through the swirling of the water. The materiality arranges itself at right angles in between. The originally white salt crystal turns into a black cross (image 83). Other drop areas (images 84 and 85) no longer show squares, but only black paths with materiality arranged at right angles.

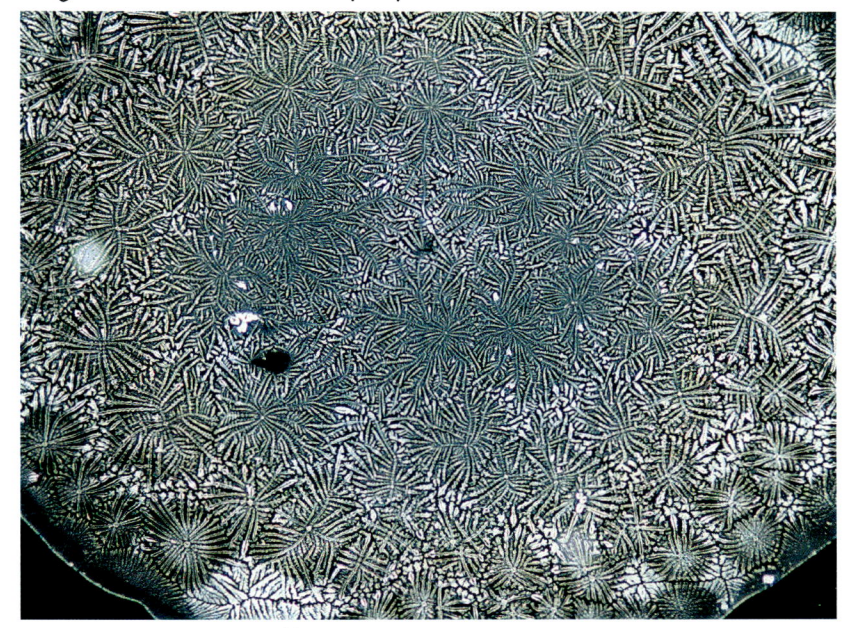

Image 81: B. W. Brine, dilution I (100x)

Image 82: B. W. Brine, dilution I (200x)

THE SILENT LANGUAGE OF LIFE

Image 83: B. W. Brine, dilution II (200x)

Image 84: B. W. Brine, dilution II (200x)

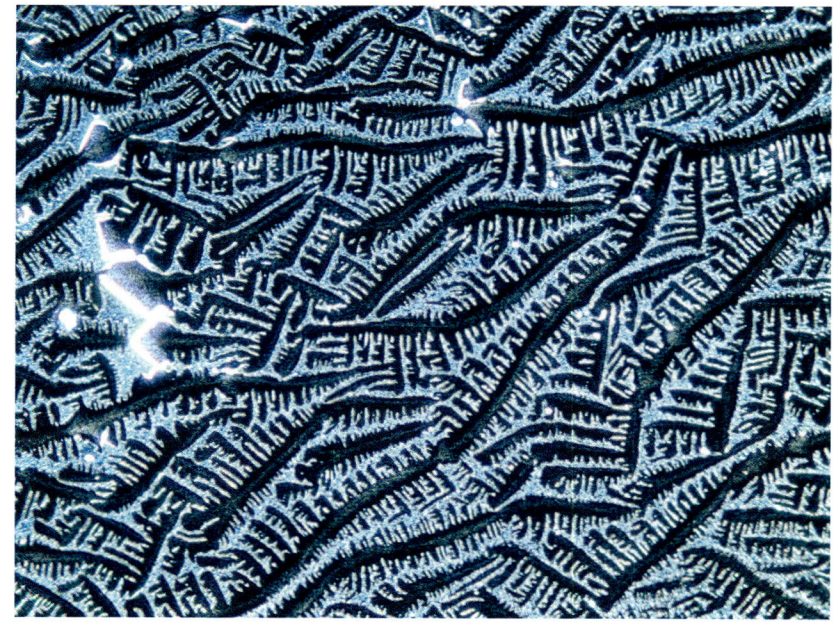

Image 85: B. W. Brine, dilution II (400x)

4. Salt—Heat Drying

It was eagerly awaited to see how the salt would change under the influence of heat.

An unvortexed, concentrated salt solution is dripped on and dried on a hot (approx. 60°C) red sandstone (image 86). Instead of the familiar salt squares, strong salt needles form, radiating inward from the periphery. The crystallization process is reversed by heat (see also chapter I, image 48). What had previously (images 73 and 74) been deposited in the periphery in the quadrangle, now strives from there in inwardly growing needles.

Images 87 to 91 show what happens on a small scale under turbulence, dilution, and the effect of heat. We see the formation of circles and the irradiation of salt into a center.

The salt tends to dematerialize and form "spheres" due to the effect of heat.

In image 92, we see swirling water from the Dead Sea. The vegetation radiating in from a peripheral round testifies to "life processes" through the swirling. They have torn some of the materiality from the salt crystal in the center, four triangles have remained (white).

How might the sun itself transform the salt? This is shown in the following images.

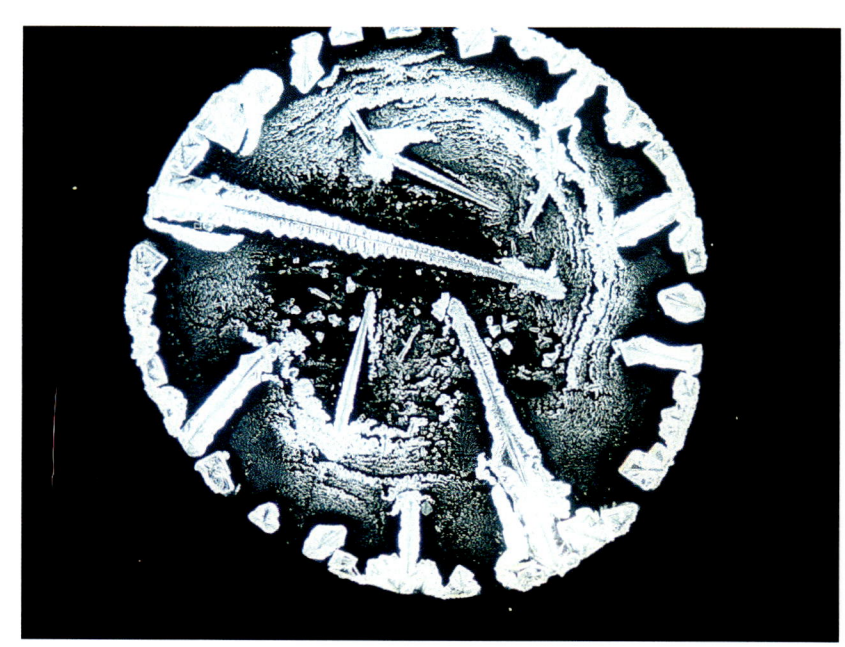

Image 86: Salt - heat drying (40x)

Salt between Substance and Form

Image 87: Brine, B. W. heat drying (100x)

Image 88: Brine, B. W. heat drying (100x)

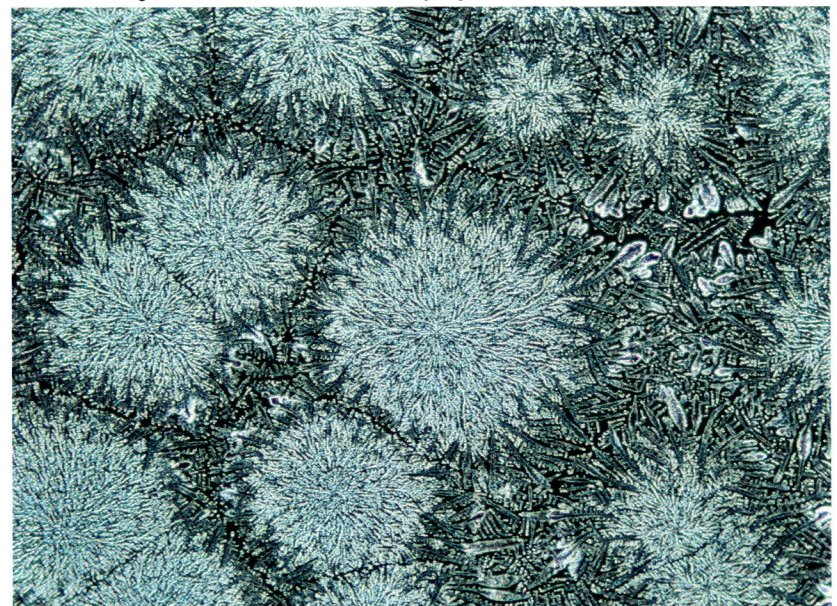

Image 89: Brine, B. W. heat drying (200x)

Image 90: Brine, B. W. heat drying (200x)

Image 91: Brine, B. W. heat drying (400x)

Image 92: B. W. Dead Sea salt (200x)

THE SILENT LANGUAGE OF LIFE

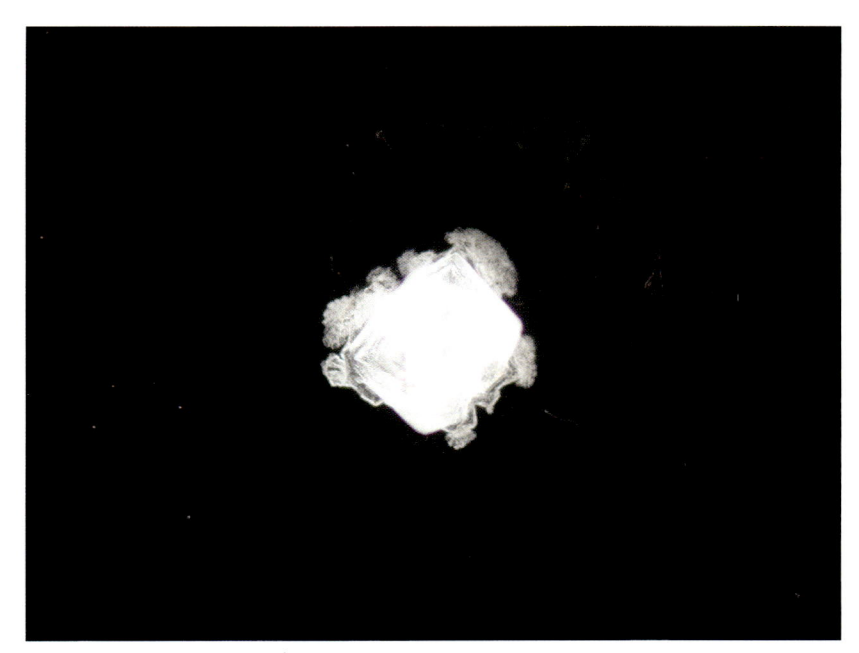

Image 93: NaCl 0.9% (40x)

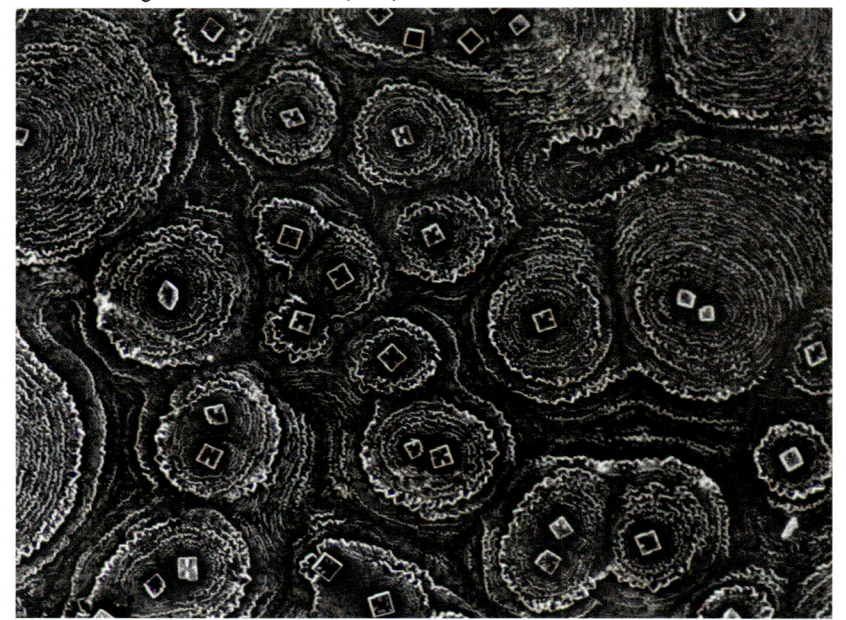

Image 94: Salt, sun (100x)

Image 95: Salt, sun (200x)

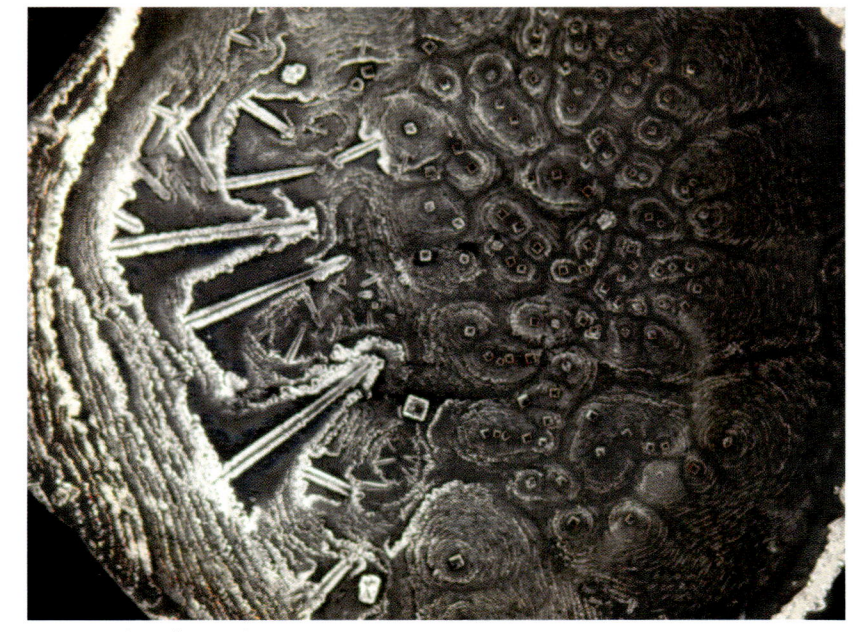

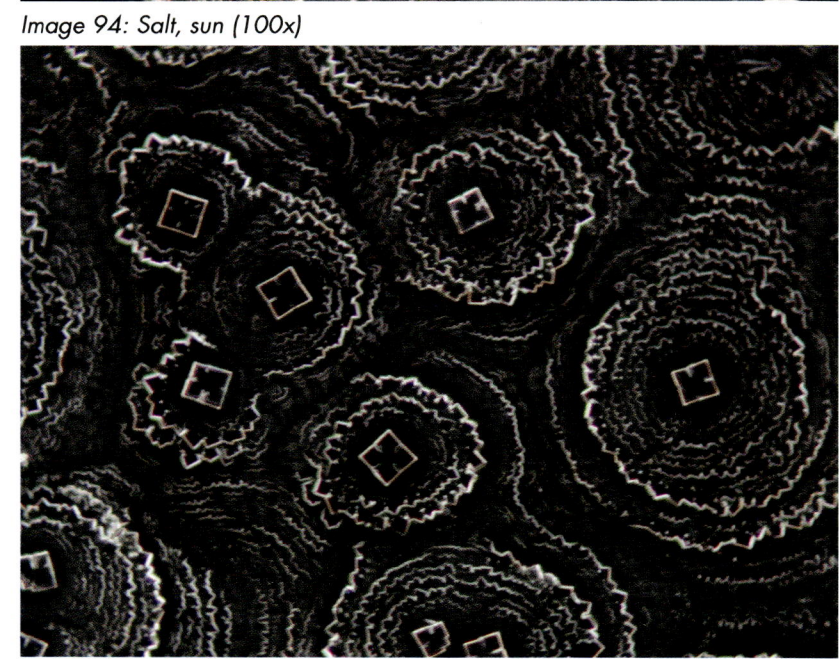

Image 96: Salt, sun (400x)

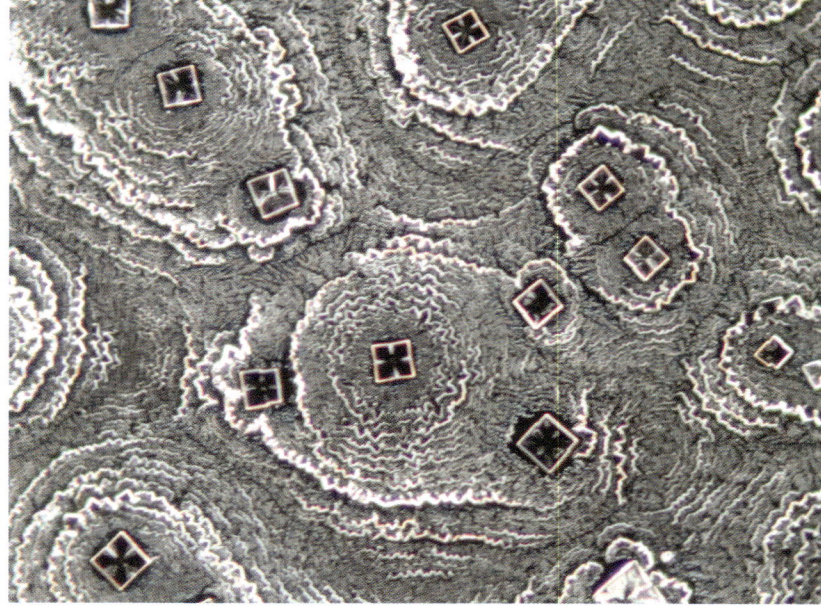

Image 97: Salt, sun (200x)

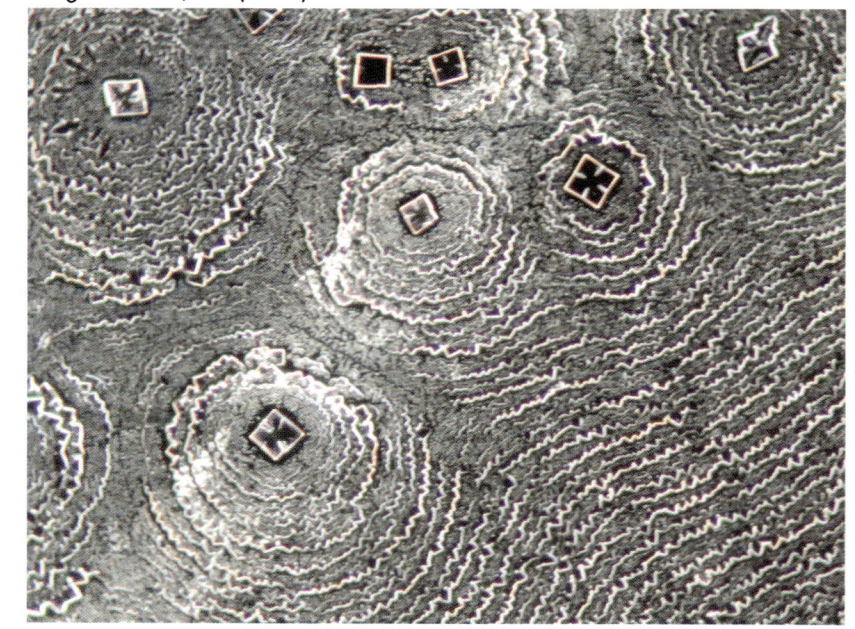

Image 98: Salt, sun (200x)

5. Salt and the Sun

Image 93: Around the time of St. John's Day, a 0.9-percent NaCl solution is dripped on and left to dry on the laboratory bench (digital camera).

Further drops of the same solution are dried in the sun with the magnifying glass (images 94 to 100). The images show us in different magnifications the subtle transformations of the salt in the light of the high sun. Substance and form play with each other in the rhythmic-moving interaction of forces. The numerous black crosses in the very small salt crystals are impressive. Here, too, we encounter an incipient "dematerialization" through the light and warmth of the sun's action.

The last image of this series (image 100) should be especially noted: A delicate "salt case" that no longer contains any materiality is deeply impressive due to its pure form.

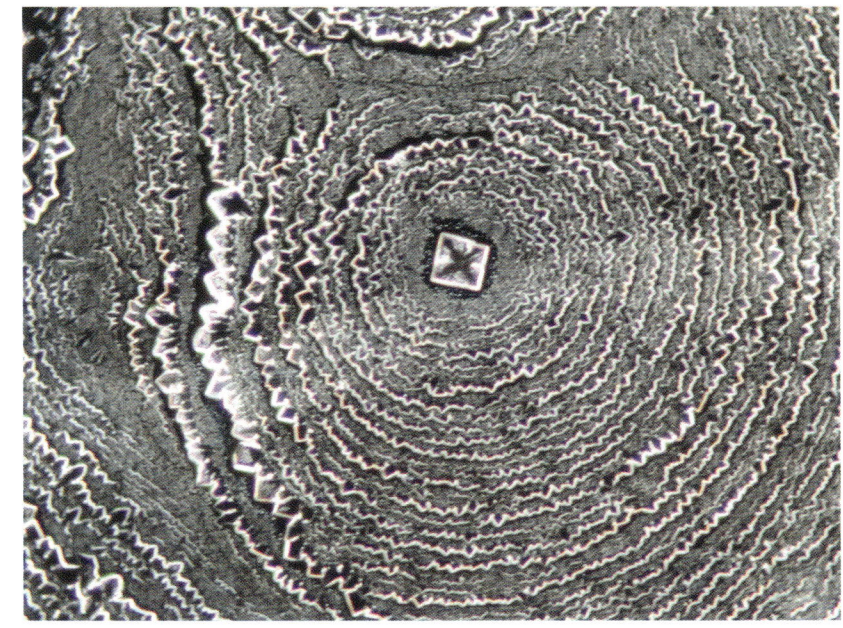

Image 99: Salt, sun (400x)

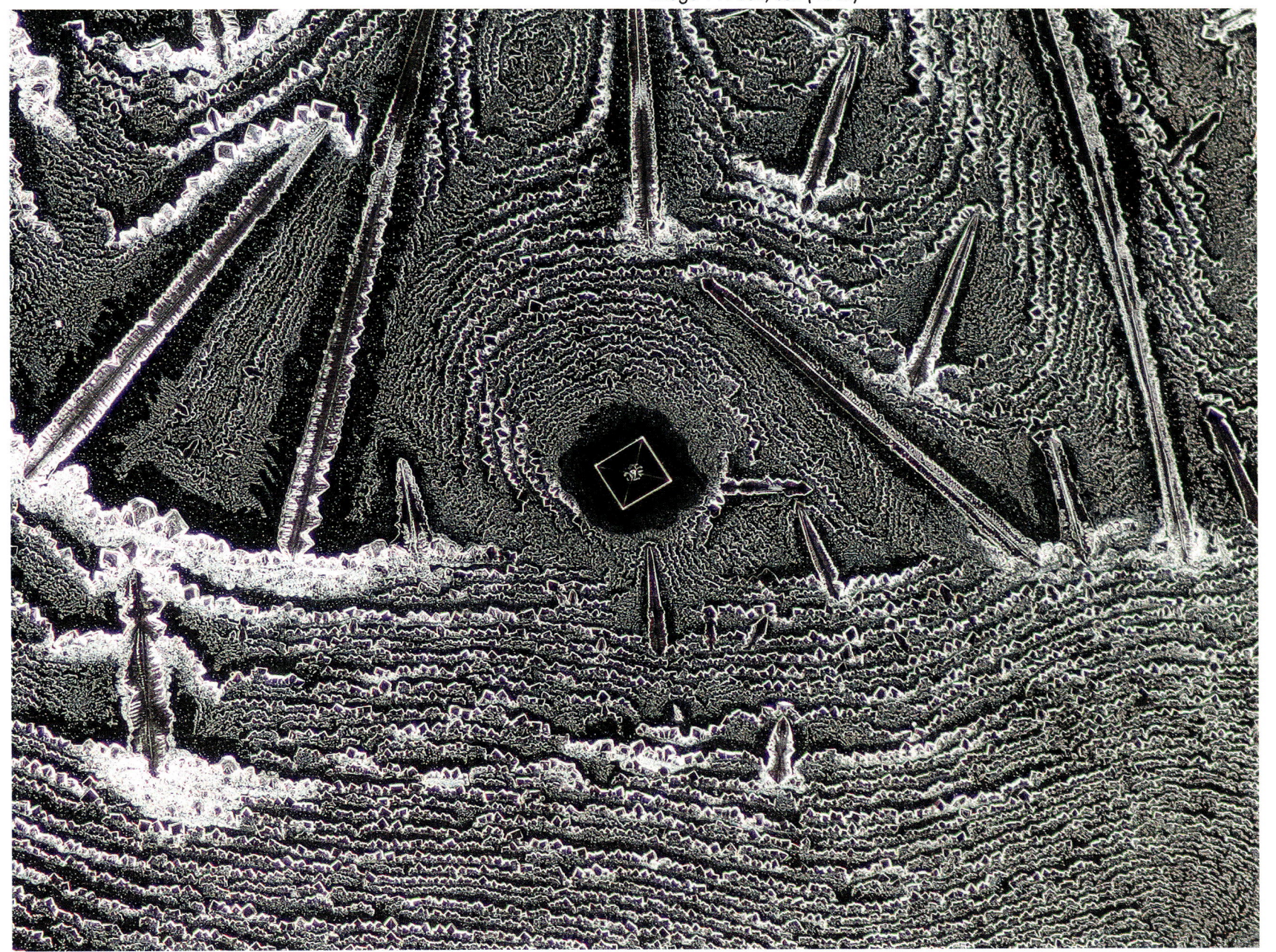

Image 100: Salt, sun (200x)

Image 101: Dead Sea salt, aloe juice, shaken (400x)

Image 102: Wheat water salt (200x)

6. Salt and "Plant-water"

So far we have seen how salt sets out and is transformed by dilution, turbulence, heat, and sunlight.

Now the question was what happens when the salt comes together with the water of the roots of plants and fruits. This water of the roots is called "plant water" here. We are not interested in the actual droplet images of these roots (see plant chapter), but only in their influence on the salt.

In the first image of this series (image 101), however, it is not preparation water, but the juice of the aloe plant, which is shaken with salt water from the Dead Sea in a 10 ml syringe. It grows out of the crystalline substance like a plant. The salt water from the Dead Sea in Israel is always available for experiments because the small white salt balls that cover the shore in wide areas have been collected in jars. Pea-sized, spherical salt crystals were a surprise! They are dissolved in water in various dilutions. Due to the numerous minerals, no really square crystals are usually formed.

The images 102 to 108 show us what happens when the dripped-on plant water (= preparation water) is mixed with a second drop of a 0.9-percent NaCl solution, drop by drop. (For medical use, the ready-made physiological saline solution (NaCl 0.9%) is always available in sufficient quantities).

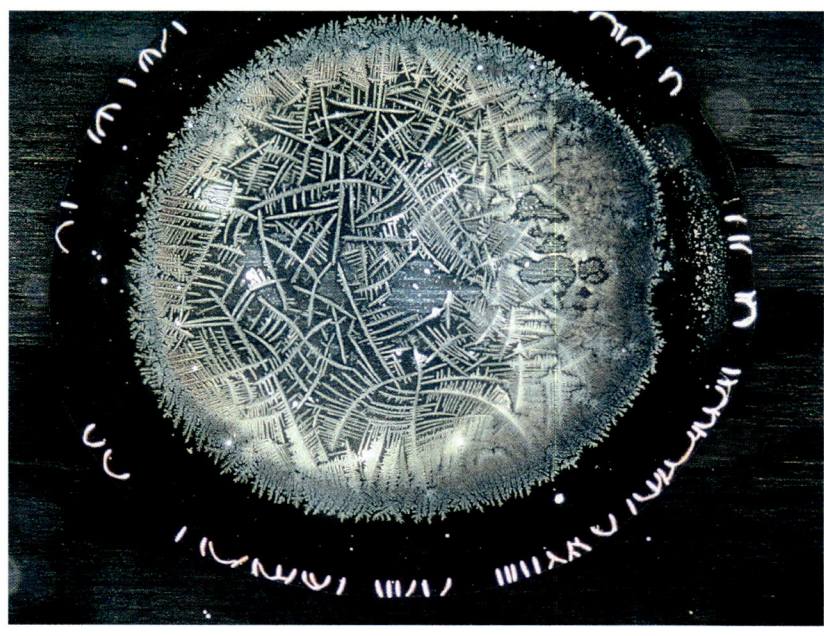

Image 103: Acorn water salt (40x)

Image 104: Physalis water salt (400x)

Salt between Substance and Form

Image 105: Physalis water salt (400x)

Image 106: Rosehip water salt, NaCl 0.9 %, diluted 1:10 (200x)

Image 102: Wheat water salt.

Image 103: Acorn water salt.

Image 104: Physalis water salt.

Image 105: Physalis water salt.

Image 106: Rosehip water salt, diluted 1:10 (digital camera).

Image 107: Lavender branch water salt.

Image 108: Mistletoe seed water salt.

It can be seen how various "plant waters" transform the salt crystallization into moving, partly *plant-like* structures. It must be a superordinate, no longer solely mineral formative force that takes hold of the earthly substance to wrest it from its earthly crystalline nature. This force seems to have formed during the time when the plants were immersed in the water of the preparations. In a specific way it is manifested in modifications by the salt. Central and circumferential forces come into the image: the centripetal force effect of the salt formation recedes in favor of a centrifugal effect which, as a plant force, animates the salt materiality and carries it into the circumference.

Let us take a closer look at what happens on a small scale.

Image 107: Lavender branch water salt (200x)

Image 108: Mistletoe seed water salt (200x)

53

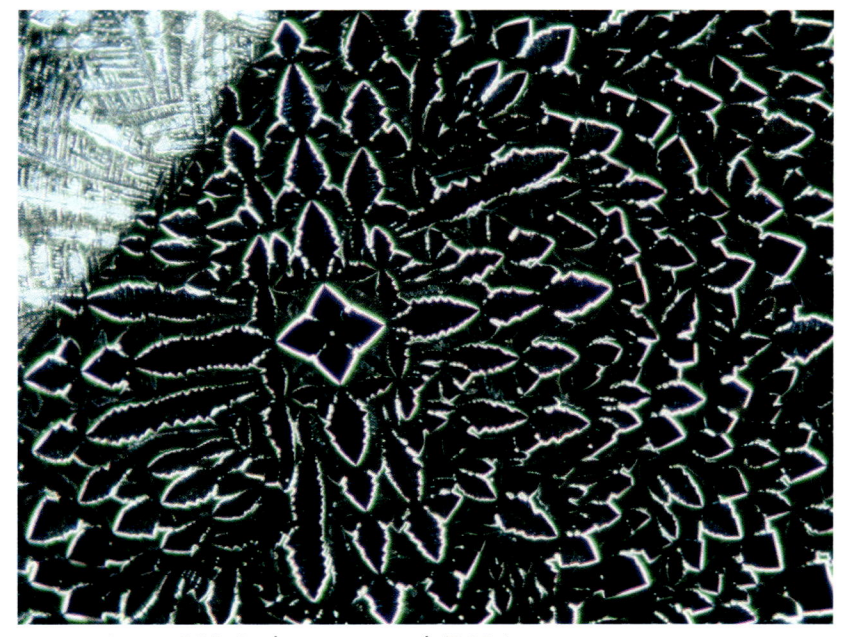

Image 109: Barberry water salt (200x)

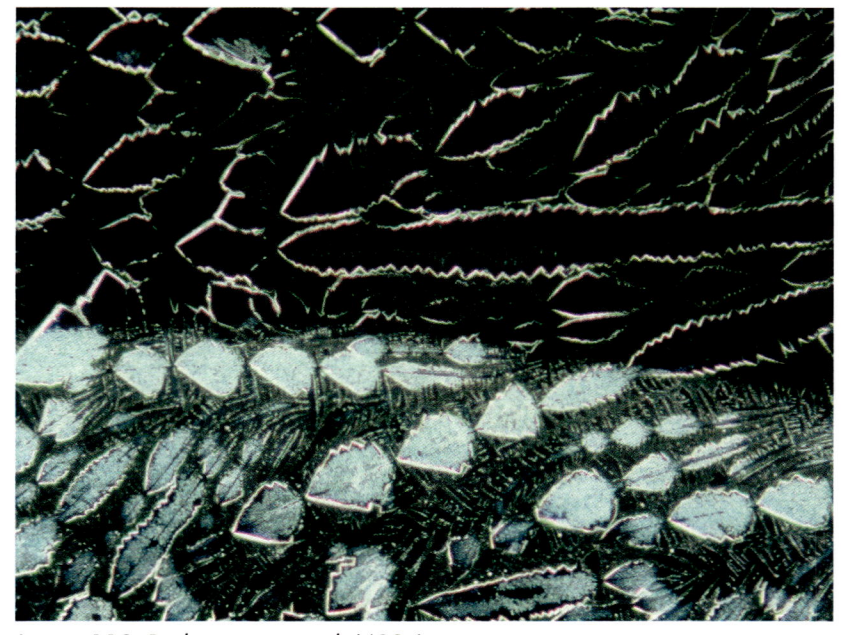

Image 110: Barberry water salt (400x)

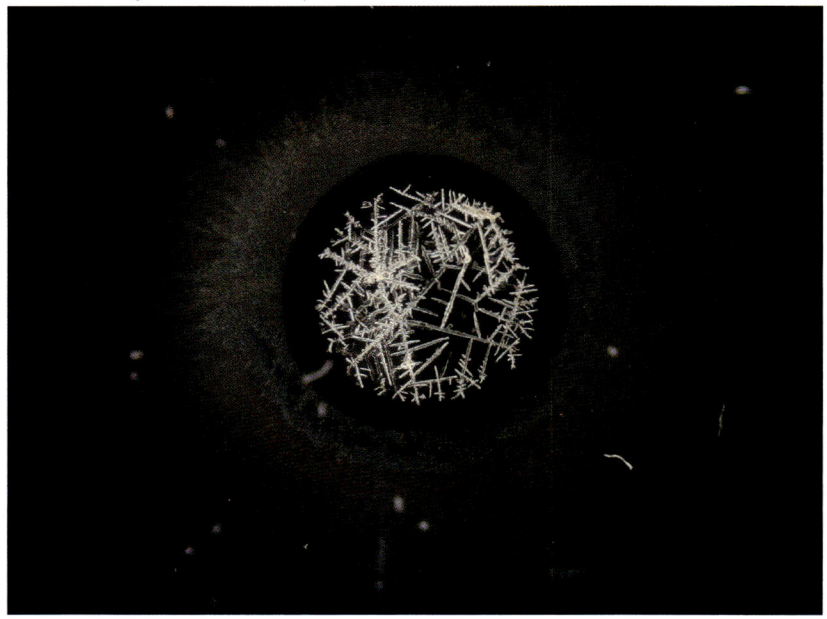

Image 111: Grape seed water (40x)

Image 112: Grape seed water salt (200x)

In images 109 and 110 we can clearly see how the saline substance has left the salt form through the action of barberry berries. Of course, the salt cannot disappear from the water; it is therefore deposited in other areas. In this case, an actual *revitalization* cannot be seen, but a clear linear rhythmization can.

In another experiment we see again the effect of seed-water on salt (images 111 to 113, digital camera).

After fourteen days, grape seeds showed a centered image typical for fruits and seeds. The center shows a crystalline-geometric branching in a black, clearly circular marked area. It is the "early phase" of the "drop image ripening" of fruits and seeds (see below). The centering is also transferred to the salting process.

Here, the water at the base of grape seeds is also shown. Figure 111: Grape seed water, dripped on.

After adding 0.9 percent salt solution, diluted 1:10, to these preparation drops, a center is again formed (image 112).

In the magnification, we see four plant-like shapes branching out from the four corners of a substance-empty crystal into the surrounding circle (image 113)! We experience order, precision, and beauty. The secret is revealed in the following images.

Image 113: Grape seed water salt (400x)

——— THE SILENT LANGUAGE OF LIFE ———

7. Swirled Plant Salt in Water (Yam Salt)

The *Dioscorea batatas* salt (yam, crushed, with salt added) available in organic shops is swirled with water in a hand swirler. It should be tested in which way a plant root changes the salt water.

Image 114 shows a drop image of swirled yam salt water.

Image 115 shows in magnification strong "animated" salt circles around a delicate square in the center. In this and the following images we see an extraordinarily powerful animation of the salt-water structures by the plant root and the swirling.

Image 116: Small clumps, always growing out of the corners of the salt crystal, spread rhythmically. Two "seed fields" meet each other at a black line. Here we see that materiality seized by living forces forms boundaries. In the drying drop it can be observed microscopically how the flowing fields grow toward each other, slow down and finally stop in front of each other. In watching such processes, one is deeply touched; one experiences—in the soul—a moral force in that one living growing thing perceives and respects another. In this salt process it is "natural law" and quite "normal."

Image 117: Similar structures are shown here, in which, however, what follows is already foreshadowed.

Images 118 and 119: Now we gain a deep insight into what is actually happening. We see the wonderfully geometrical-exact form of a salt crystal that has become empty (image 118).

It shines in a delicate blue into the black surrounding circle. At its corners we discover four small triangular structures—like little fir trees—which grow into four large "trees"—clearly separated from each other—in the surrounding space. (Compare also image 113.) It remains a great mystery *how* this happens, because the quadrilateral form is—as always in such cases—unharmed.

Image 114: Yam salt water (40x)

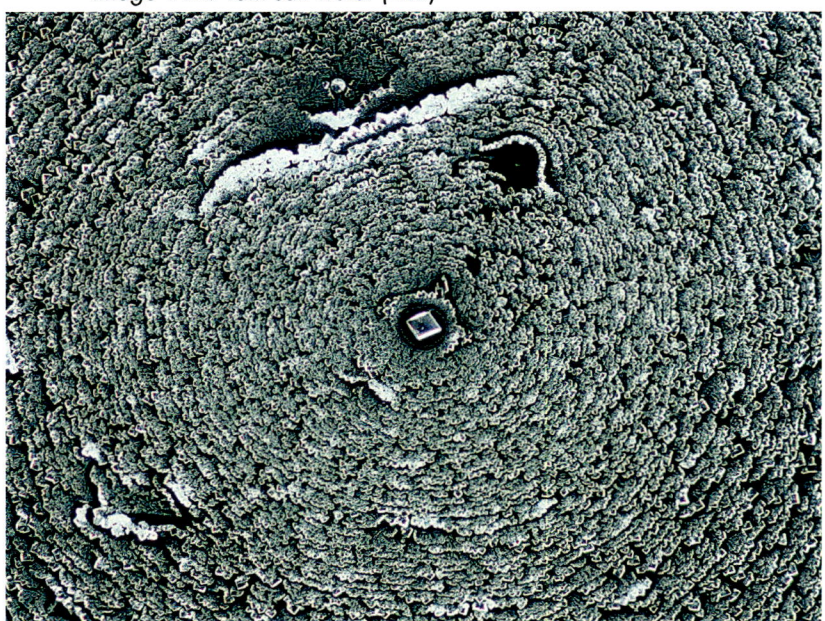

Image 115: Yam salt water (100x)

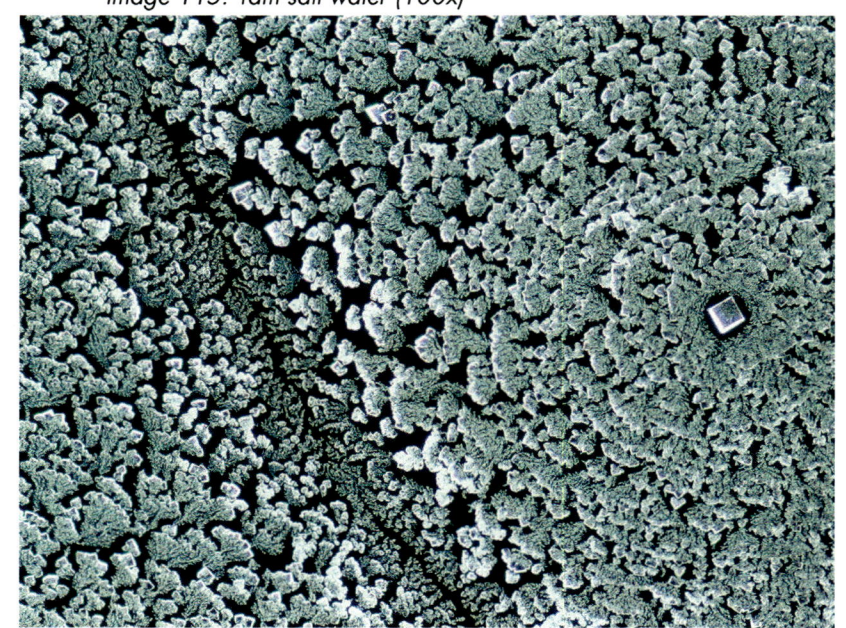

Image 116: Yam salt water (200x)

Image 117: Yam salt water (400x)

Salt between Substance and Form

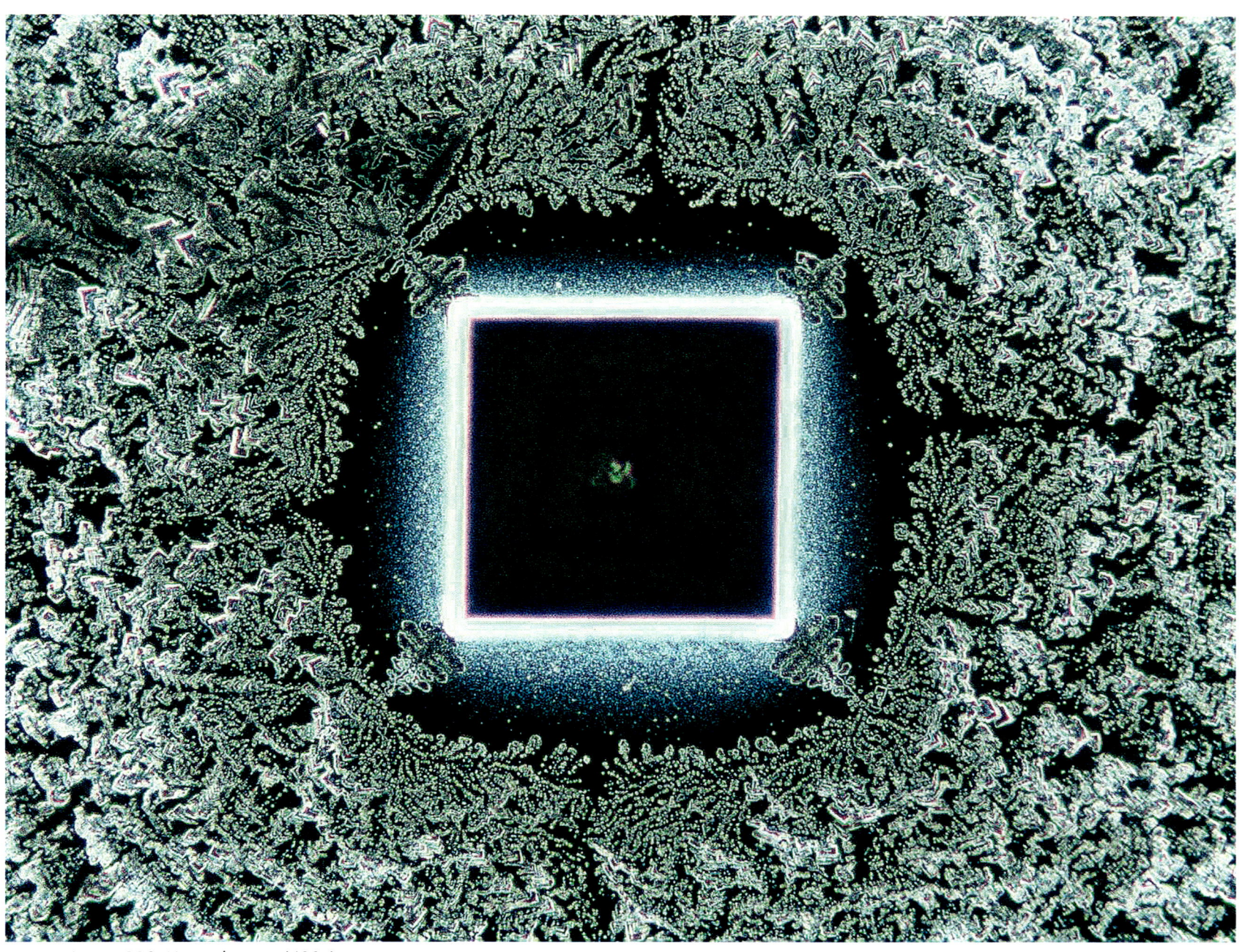

Image 118: Yam salt water (400x)

Image 119: Yam salt water (200x)

In this image we see a scientifically exact representation of a fact. Its reproducibility in the same way and also in variations is given in many other images. This can become a "sense-image" for us, which I can let arise in my soul before my inner eye. This activity of the soul gives me the realization that living, etheric forces have a sucking effect on the formation of salt.

The closed, white square border poses a riddle. Is it only the salt substance that has been deposited as the outermost layer of the step formation? If so, it would also have dissolved into the surroundings via the black diagonals. But what substance is used to "write" this quadrilateral form? It is not the water, probably not the salt substance either. The following image could give us a clue.

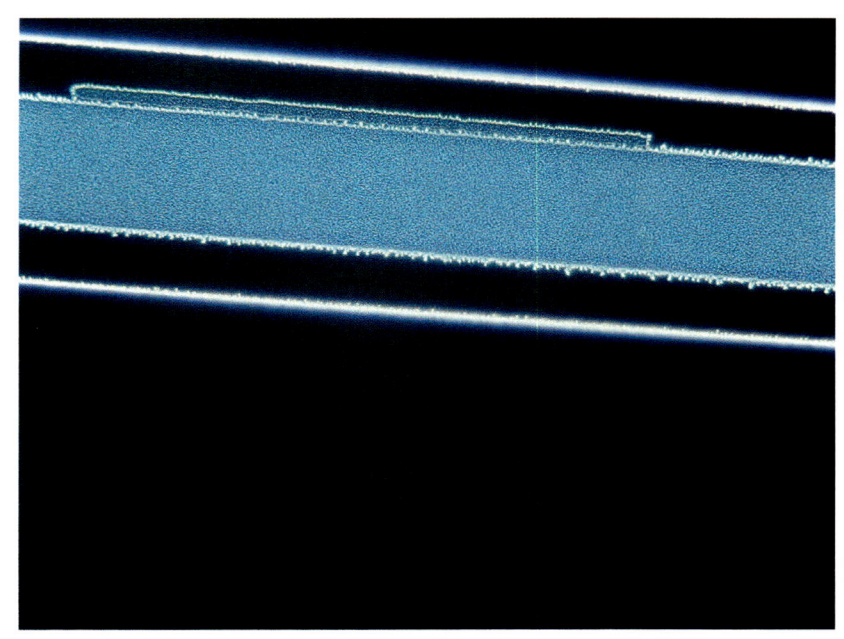

Image 120: Phenomenon (400x)

Image 121: Yam salt water (400x)

Image 120 shows a section of a "phenomenon" (as it is called here) that suddenly appeared on the slide as one of a total of three in the course of the years of research. An elongated, precisely shaped figuration had settled without reference to the drops. The *whole* "phenomenon" and the two others are documented elsewhere (see p. 190).

Why does this detail appear here? The two boundary lines above and below the light blue band are comparable in their expression to the lines that border the deflated salt crystal. The question remains, what kind of "materiality" underlies these exact lines?

Image 121 shows that triangular salt crystals also form in yam salt water. They often appear in the animate water environment and also in seawater. (The triangular salt crystal will be dealt with below).

8. Salt and "Mineral Water"

As we will see in the next chapters, water is not only stimulated by plant parts but also by minerals, precious stones, and metals (see chapter III) to form plant-like shapes. The burning question then arose as to whether the water of these mineral preparations could also "animate" the salt.

A tourmaline lay in the water for four months (the images that appeared later [chapter III] appeared after more than nine months).

Image 122 shows its drop pattern with regular structures spreading linearly from a center in a dense texture. This water is dripped on and a second drop of physiological NaCl solution is added as above.

Images 123 and 124 show the altered salt behavior. So etheric power is also present in mineral water?

The process is similar to that of the yam root. Here, too, suction forces are active that empty the salt crystal.

This can be seen more clearly in the magnification (image 124). However, in this case four triangular salt fields remain behind in the square.

Image 125 enchants with its rhythmic alternation of growth and salt deposition.

What also appears in other tourmaline salt drop images are "suction vortices"—that's what they are called here.

Image 126 shows such a "suction vortex." If we compare it with the well-known salt circles (e.g., in "Salt and sun" etc.), we notice that the prongs of the salt circles now point inward. Gradual dilution occurs toward the center, which no longer contains any salt. This gives us an image of what happens in every real vortex: Substance transformation in the sucking center.

Here, however, there was no vortexing at all. So, we should assume that living forces are also at work in the tourmaline water, which exert a sucking influence on the salt action?

In the tourmaline (image 122) itself we see growth structures spreading linearly from a center to the periphery. This linear, one-dimensional development of force could give us an image for the light ether, as we will also see it effective in the stem image of the plant in the water. Through the addition of salt, the linear action changes into a tuft-like, rhythmic arrangement. Compressions occur as an effect of the earth substance: salt (image 125).

Salt between Substance and Form

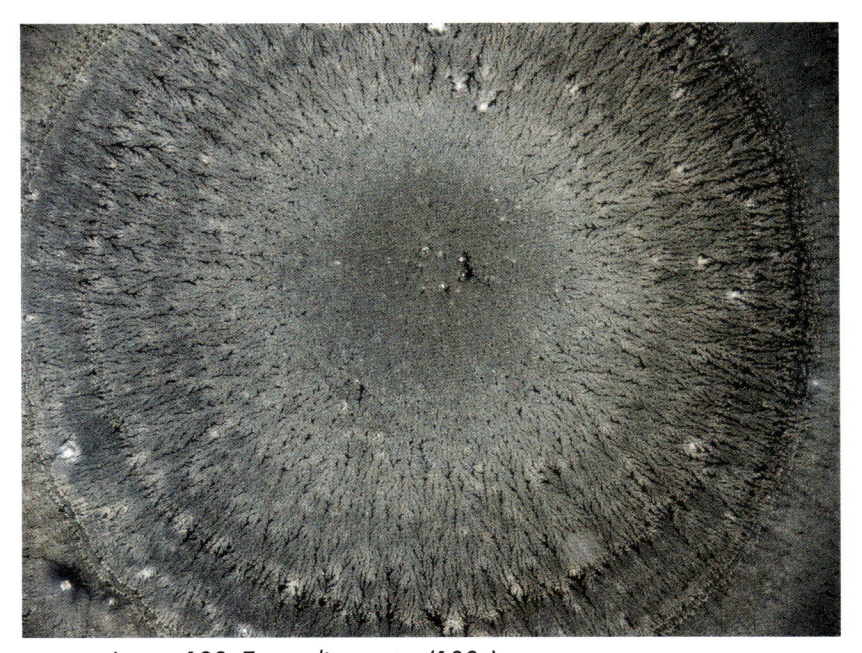

Image 122: Tourmaline water (100x)

Image 123: Tourmaline water salt (200x)

Image 124: Tourmaline water salt (400x)

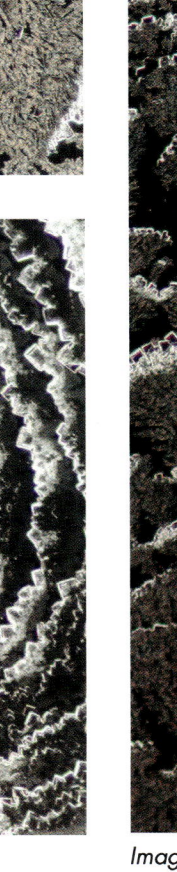

Image 126: Tourmaline water salt, suction vortex (200x)

Image 125: Tourmaline water salt (200x)

9. Salt Ash Water

At the end of this chapter we see in some images what happens when these three mineral substances come "into conversation" with each other. The suggestion for these experiments comes from Rudolf Steiner (*Lectures and Courses on Christian-Religious Work,* vol. 2).[22]

We see two different experiments with the same experimental arrangement (images 127 to 138):

1. Lime blossom ash salt experiment (images 127 to 132).

Drops of water are dropped onto a microscope slide. Water is taken from the supernatant (the ash has settled on the bottom of the test tube) of a test tube containing ash (lime blossom ash, many months in water) and added to the drops of water dripped on. A third drop from a 0.9 percent NaCl solution is also added. So we have three drops in one—i.e., a mixture of three mineral substances.

Image 127 shows the 0.9 percent NaCl solution, dripped on.

Image 128 shows the lime blossom ash mixture, dripped on.

We are familiar with the image of a drop of water.

What happens immediately after mixing is revealed after drying in images 129 to 132. Nothing can be seen of the actual ash character. The salt is visible, of course, but it has been strongly rhythmicized and has settled into spine-like formations. Fern-like branches grow out of these.

We see growth processes in a purely mineral environment. Is it the ash resulting from a fire process that causes these living structures?

Image 127: NaCl drops, 0.9 percent solution (40x)

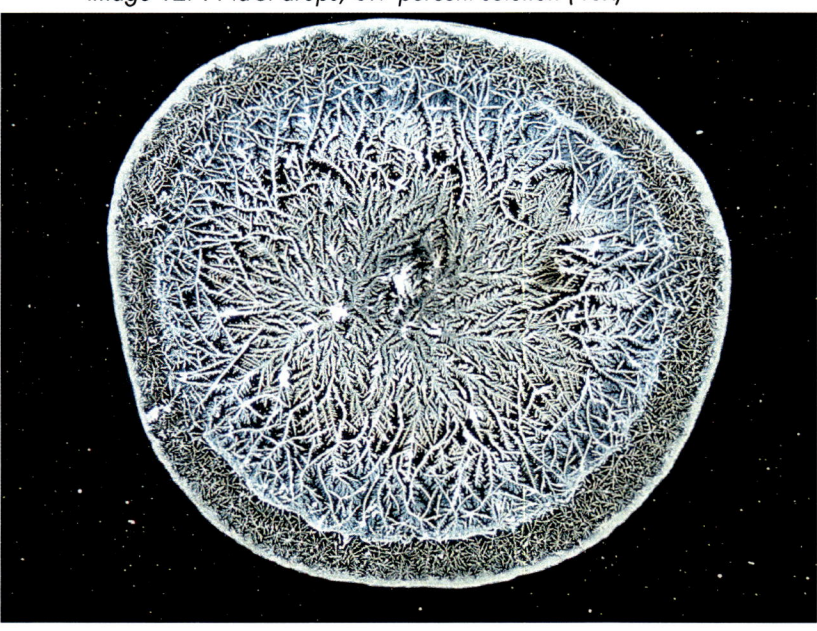

Image 128: Lime blossom ash (40x)

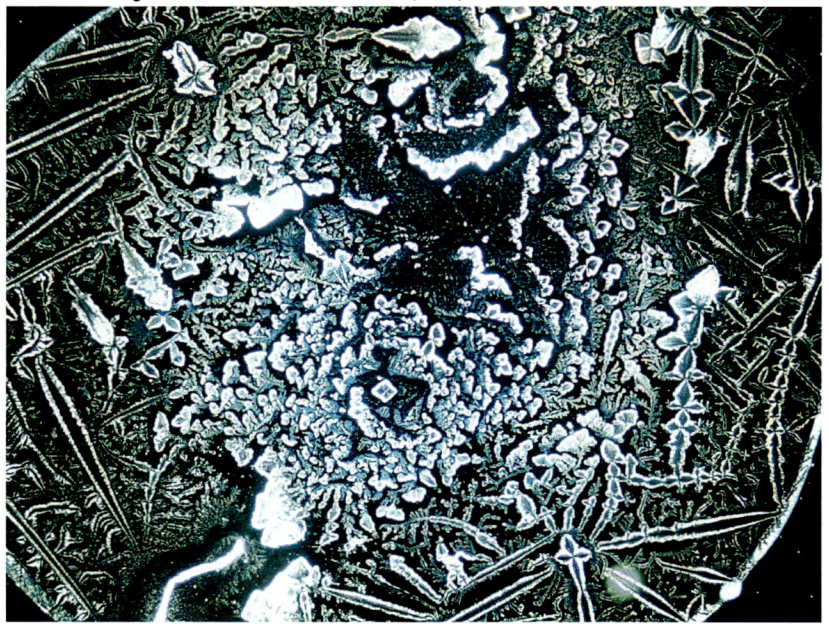

Image 129: Salt ash water (100x)

Salt between Substance and Form

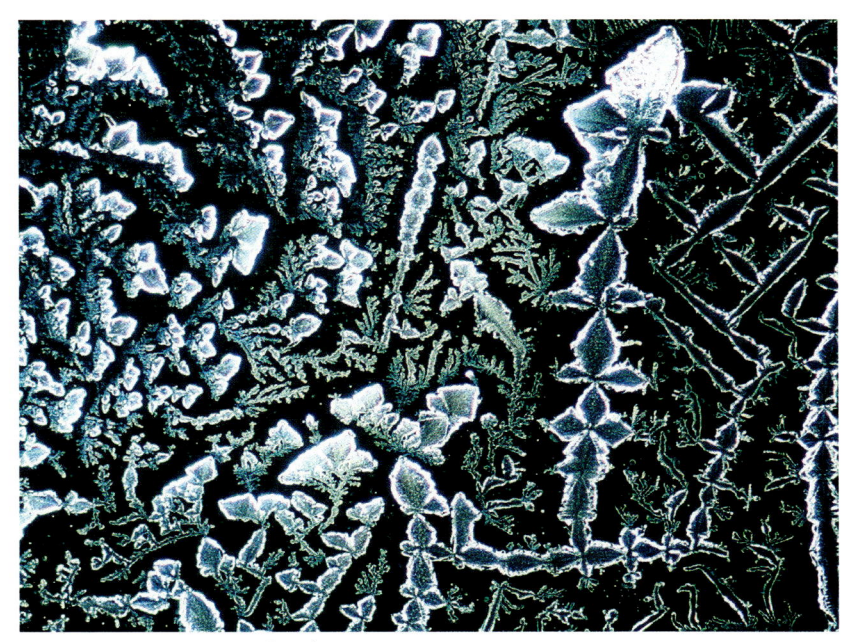

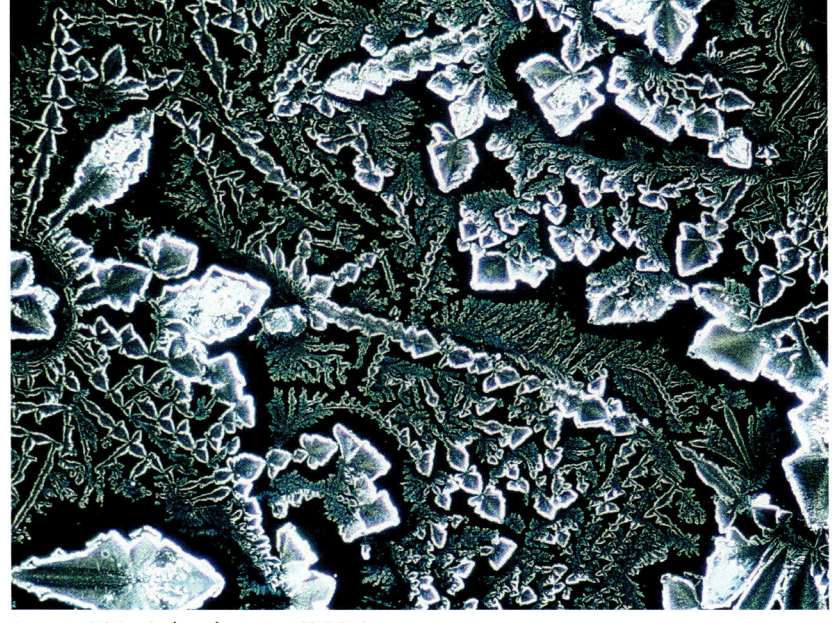

Image 130: Salt ash water (200x)

Image 131: Salt ash water (200x)

Image 132: Salt ash water (200x)

— THE SILENT LANGUAGE OF LIFE —

2. Rosewood-ash-salt experiment (digital camera, images 133 to 138). We see the same experimental arrangement with rosewood ash.

Image 133: Rosewood ash.

Image 134 shows a similar event as the images with lime blossom ash. The rhythmic salt action is clearly formed into spinal column-like shapes. In between there are hints of growth structures.

Images 135 to 138: Here, the same action, but the 0.9-percent salt solution has been diluted 1:10. We now see the plant-like structures more clearly, but also salt crystals, partly as hexagons (image 137).

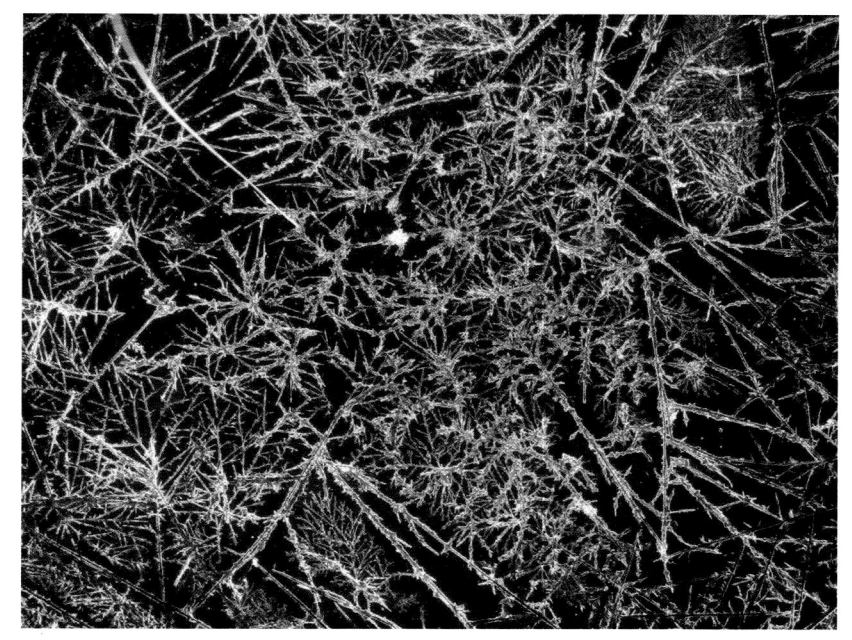

Image 133: Rosewood ash (200x)

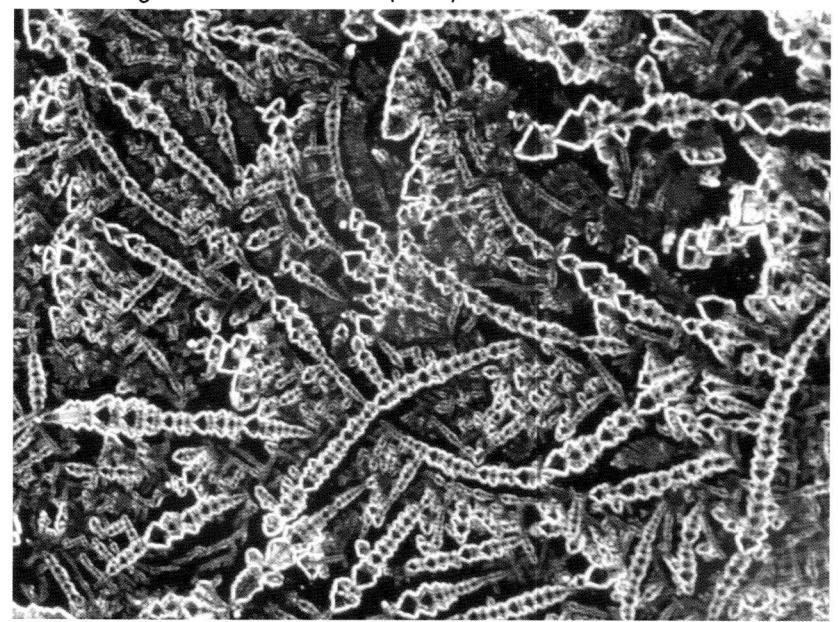

Image 134: Salt ash water (200x)

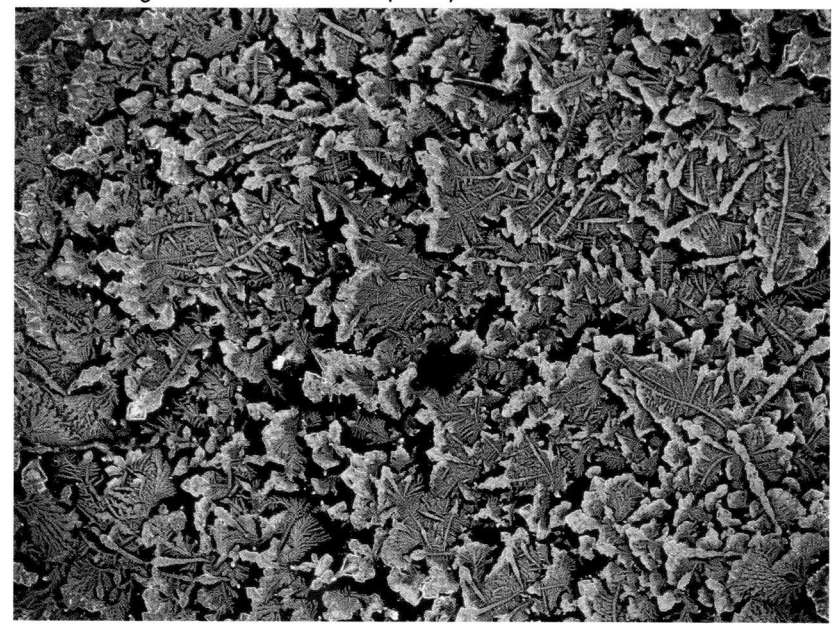

Image 135: Salt(1:10) ash water (200x)

Salt between Substance and Form

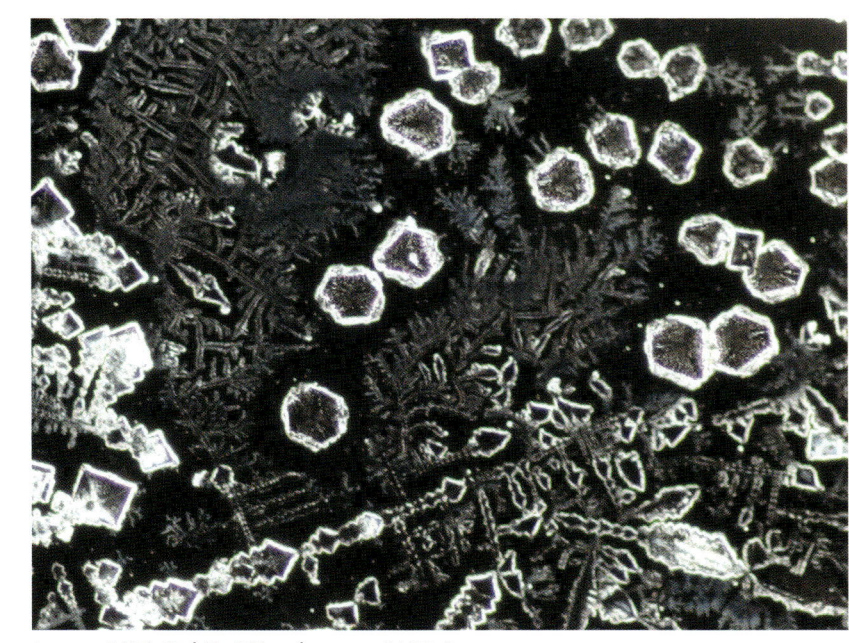

Image 136: Salt(1:10) ash water (400x)

Image 137: Salt(1:10) ash water (400x)

Image 138: Salt(1:10) ash water (200x)

Image 139: Salt crystal (400x)

10. Reflections on Chapter II

Can we build a bridge from these salt experiments with their fascinating possibility of shape transformation to the salt process in man? Can we perhaps also read from the individual crystal forms that appear under different conditions an inclination toward more earthly or more cosmic form shaping? This could lead to a deeper understanding of the salt process in humans.

First of all, it could be seen how salt swirling and dilution led to a very harmonious-rhythmic peripheralization and to a filigree refinement of materiality—up to black "trails" (images 79 to 85).

Heat effects transformed the crystallizations into nearly dematerialized "spheres," while the salt was deposited in other drop areas, forming new centers by irradiating needles (images 87 to 91). Dilution, movement, and heat transform the salt. These are three preconditions for the fluid organism of man.

Let us look again at individual salt crystals (images 72 and 77 as well as image 139): An idea, a spatial idea of the highest symmetry is mysteriously filled with substance, seen here in the plane. The square shows the salt crystal in its "earthly design."

The cube is a hexahedron with six faces, eight corners, twelve edges, three perpendiculars to the center of the six faces (the three directions of space), four diagonals, and 24 right angles. Every salt cube is formed according to these numerical ratios. This perfect, mirror-symmetrical form can be built up in the mind's eye, as a *mental spatial shape* in the imagination. The cube is one of the five Platonic solids.

You cannot encounter the salt substance in this way. You have to touch it, taste it, physically grasp it. In this way we become aware of *substance* and *form* in different ways—mentally as form, physically as substance. The special feature of salt is that it dissolves in water. In the process, its "physical form" is lost. Does its spiritual form power also get lost? That this is probably not the case could be seen in the images.

When salt is diluted, swirled, and "revitalized" by plant and mineral water, it seems to assert its spiritual forces of form in the liquid in contrast to the forces of form that act from the water of origin. These rather handle the materiality in the most diverse ways. In the playfully designed growth processes of the plant-salt encounters, it becomes apparent in a new way how the dissolved salt expresses its formative forces in the living. One could say that it allows itself to be *enlivened*, but does not lose its spiritual formative power in the salt process.

Here is another example of the salt-plant water (digital camera). In images 140 to 142, we see rose hip preparation water mixed with salt 0.9%, drop by drop. We get the impression that "the salt wants to become a plant." It wants to pass over into living processes.

Vitalization of the substances happens in the human being through his etheric body. In this respect, Rudolf Steiner says:

> It [the etheric body] is what calls the inorganic substances into living existence, brings them up out of lifelessness in order to thread them on the thread of life.[23]

This refers to all inorganic substances—here we are only concerned with the particularity of salt. This "threading" occurs and shows itself as an extraordinarily rhythmic, partly linear process!

Salt between Substance and Form

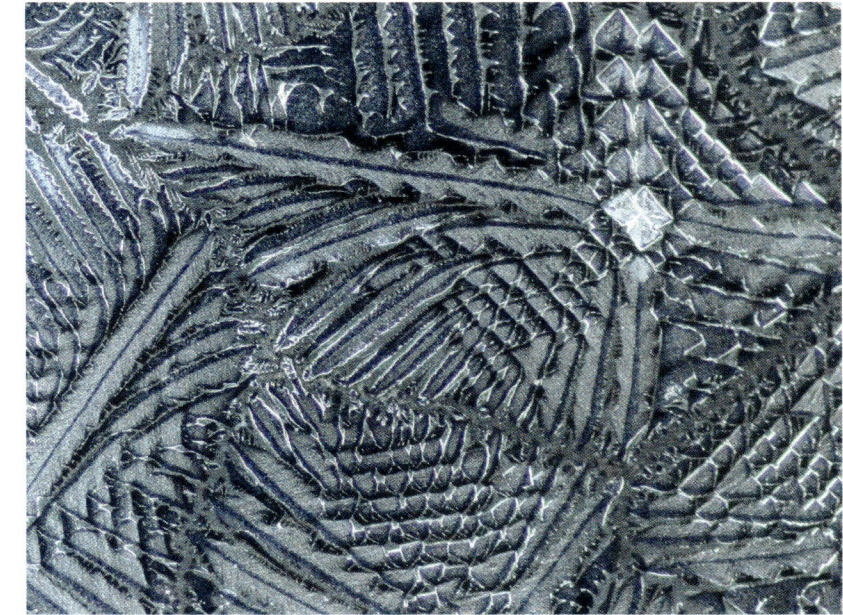

Image 140: Rosehip water salt (200x)

Image 141: Rosehip water salt (200x)

Image 142: Rosehip water salt (200x)

— THE SILENT LANGUAGE OF LIFE —

Image 143: Salt crystal, triangular, with 3-star in the center (200x)

Image 144: Salt crystal, triangular (200x)

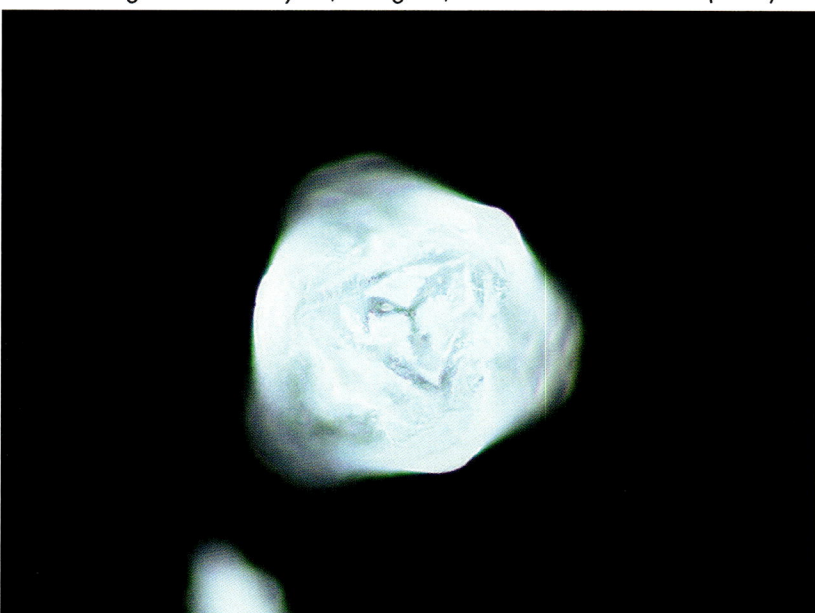

Image 145: Salt crystal, triangular, with 3-star in the center (400x)

Image 146: Salt crystal, triangular, with 3-star in the center (400x)

What also happens are crystal variations. The square with its right angles is abandoned and triangular and hexagonal salt crystals appear (images 143 to 146).

Certain circumstances of an enlivened milieu (plant preparations, tourmaline preparations, but also seawater) transform the cube shape into a triangular spatial shape. Images 143 and 144 show a salt solution to which a trace of blood serum diluted 1:10 has been added. Here, too, we see the salt in *action*: black lines radiate from the three corners into the circumference. It can be assumed that here, too, suction effects are expressed, which are on their way to empty the salt crystal. Some of these triangular figures look like small "spherical formations."

The triangular salt crystals are puzzling and led to the following considerations: The structure of water molecules is based on the formation of tetrahedrons (Greek: tetrahedron). Four triangles of equal area (trigones) form this platonic solid.

In water, these tetrahedra arrange themselves in hexagonal structures of the molecular structure. An example of this is given by snowflakes as platelets (hexagons), as stars with six rays, and as hexagonal prisms in the cirrus clouds. Cosmic-etheric peripheral forces reveal crystallizing water based on hexagons, six-pointed stars, and prisms. It is also the structure of tourmalines, which, like quartz, belong to the silicates. The latter make up the largest part of our earth's crust.

Image 147: Honeycombs

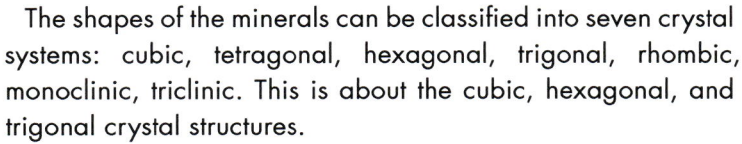

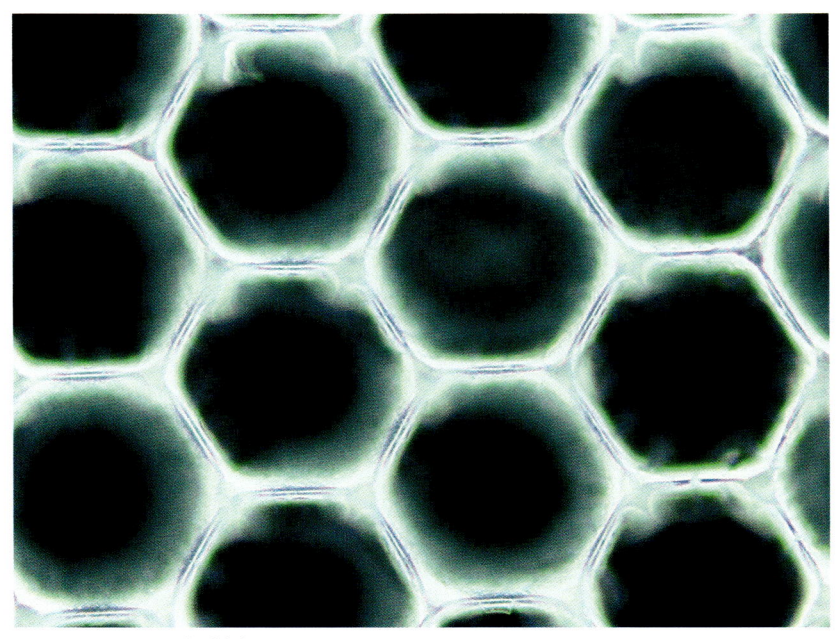

Image 148: Air bubbles in serum (400x)

The shapes of the minerals can be classified into seven crystal systems: cubic, tetragonal, hexagonal, trigonal, rhombic, monoclinic, triclinic. This is about the cubic, hexagonal, and trigonal crystal structures.

The silicates (SiO4) crystallize as tetrahedra and can join together to form all kinds of hexagonal shapes.

The water-clear quartz (SiO_2) crystallizes trigonally (deep quartz) *and* hexagonally (high quartz) into prismatic forms. The Greeks called rock crystal *krystallos*—ice! Until the seventeenth century, people were still firmly convinced that it must be fossilized ice![24]

Tourmaline—as a silicate—also crystallizes trigonally and forms prismatic shapes. In the tourmaline slices, a three-star appears in the center. This also appears in the triangular salt crystal (images 143, 145, and 146)!

A relationship between silicates in their strict crystal lattices (hexagonal and trigonal) and water in its tetrahedral-hexagonal molecular structure is recognizable, as it appears in ice needles and snowflakes.

And the salt? It dissolves in water and can crystallize out of it, usually as a cube. Under special circumstances, however, it can also appear as a triangular crystal. The "circumstances" have to do with a "living milieu" in which the earthly quadrangular form is transformed into the more cosmic triangular form.

Thus the Quadrity and the Trinity reveal themselves in the salt of the earth.

And the sixth-ness? Can we get closer to its cosmic structural formation? In alchemical Rosicrucian works, the signs for fire and water are two triangles:

Fire: △   Water: ▽

Together they form the hexagram, an occult sign: ✡

The fourth element, fire, interpenetrates with the second element, water. In water, the life or ether principle can be seen, it carries the oxygen. From fire, the all-pervading warmth, the human being could develop his blood warmth and maintain it in the face of all external temperatures—as the basis for his "I." Rudolf Steiner speaks about this in the Esoteric Lesson of November 29, 1907.[25] In Record B we read that man, as a microcosm, could attain knowledge of himself and his connection with the macrocosm if he meditated on the hexagram.

The outer connection of the six corners of the hexagram creates a hexagon, as we know it from the honeycomb. The larvae of the bees develop from the eggs in these special "power houses" (image 147).

Image 148 shows small air bubbles in a dried human serum drop!

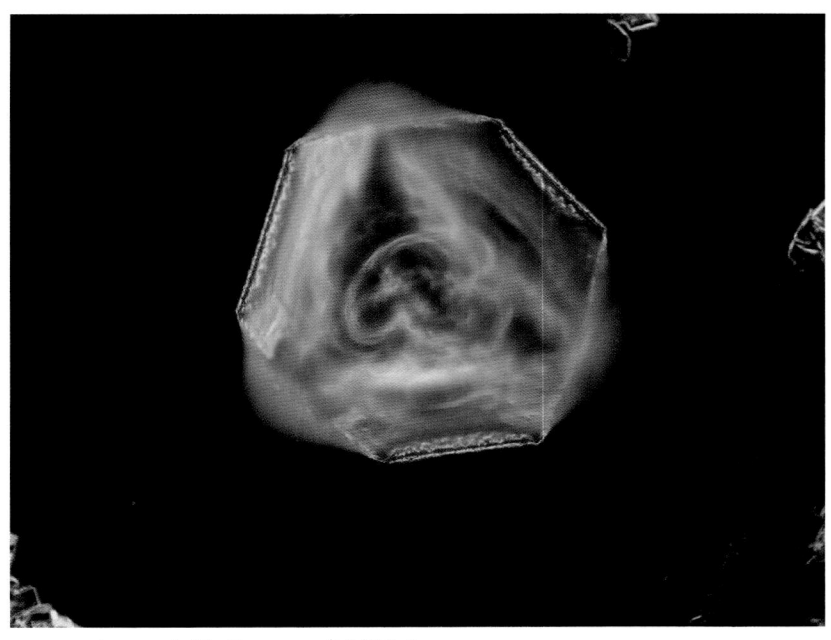

Image 149: Liquor, salt (400x)

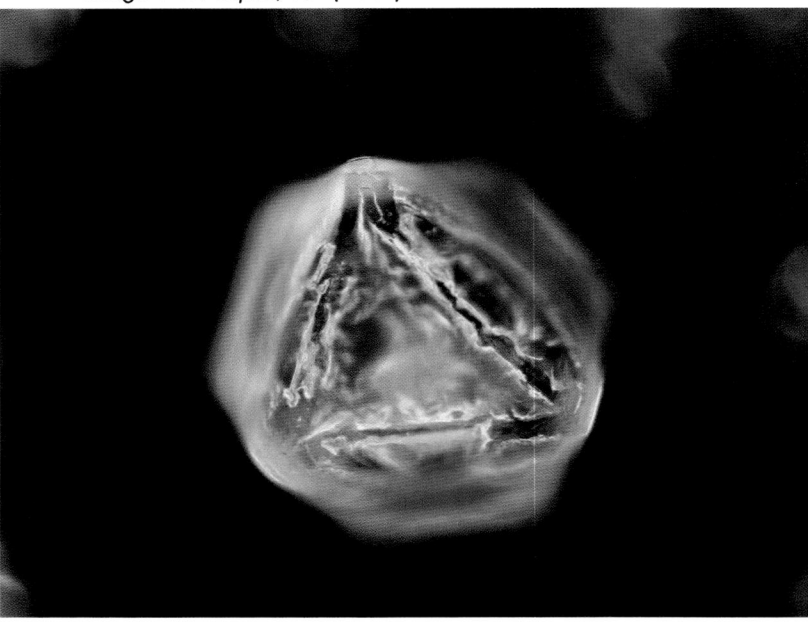

Image 150: Liquor, salt (400x)

Image 151: Liquor, salt (400x)

The hexagon seems to be based on a transformation of two triangles running in opposite directions. This is shown in the following three images (149 to 151, digital camera). These inscribed structures can be seen by changing the depth of field in the microscope. The two triangles—symbols for fire and water—seem to merge playfully. In this way, they make the hexagon appear. In this case, it was human spinal fluid (CSF) (diluted 1:10), to which NaCl 0.9% (diluted 1:10) was added drop on drop.

Thus we see that the cube (or square), which is an earthly form, can very well change into the cosmic form of the hexagon (image 137). This is shown by two more examples out of many.

Image 152: Bean starter water with salt added (NaCl 0.9%), drop on drop.

Six triangles are delimited from each other here. If we focus the view on the center, we see a tilted cube, as always in such cases.

Images 153 and 154: Water of red and black berries of the woolly snowball (*Viburnum lantana*) mixed with salt (NaCl 0.9%), drop on drop.

The hexagon is based on the six-pointed star, as can be seen (image 154, digital camera).

Image 155 now shows the salt also in six-pointed star shape. It is a plant-applied water with salt added (NaCl 0.9%), drop on drop.

Finally, the salt also appears as a pentagon: Image 156: Crataegus (hawthorn berry) salt, 0.9%.

Thus, on the one hand, the salt shows a great willingness to enter into living processes. On the other hand, in the highly astonishing crystal variations, a kinship can be seen with the structure of water itself and the structure of silicates, which unfold strong form-giving forces in nature and man.

Rudolf Steiner leads us into the deeper understanding of the salt process:

> The water dissolves the salt. In every salt process, that is, whenever salt settles somewhere—and we can certainly use the general name salt here for everything that settles in this way—that is, whenever salt settles, it means that the salt also gives off a spiritual-etheric content to the environment.[22]

In purely physical terms, light and heat are released during salt crystallization. These are two etheric entities! Rudolf Steiner continues at this point:

> So the salt that is dissolved in the liquid, in the water, is, as one knows through the imagination, wisdom-bearing. The dissolved salt is wisdom-bearing. As the salt coagulates, as the salt settles, the real wisdom evaporates, so to speak, into the environment, and the salt is

Salt between Substance and Form

Image 152: Salt crystal, hexagonal (400x)

Image 153: Salt crystal, hexagonal (400x)

Image 154: Salt crystal, hexagonal (200x)

Image 155: Salt crystal, six-pointed star (400x)

Image 156: Salt crystal, pentagonal (200x)

without wisdom. You must think of all this as being linked more to the *process* than to the substance, for this is a process that goes on in the most eminent sense in one's own human organism. And when you think, when you develop thoughts, you are only filled with thoughts because salt is deposited in you. The denser the development of thought, the more salt is deposited.[22] (emphasis I. J.-N.)

We struggle to understand such living processes between substance and thought development. Spirituality becomes free through the process of salt crystallization and dissolution in the human being! The I with its spiritual formative power handles the mineral in the liquid through heat.... Could it be that in this physiological salt process a transformation from the quadrangle into the more cosmic hexagonal form plays a role, as the images show us physically?

Without salt and the salt process, man would have no consciousness. How often have a few grains of salt on the tongue restored consciousness to a person who has fainted from dehydration. This happens through the dissolving power of salt, which awakens the fluid organism to new life. In the plant-salt encounters we also saw how the spiritual formative powers of the plant call the earthly substance into living form.

Elsewhere we find the remarks Rudolf Steiner made with the physician Dr. Ita Wegman (1876–1943) in their joint book *Extending Practical Medicine: Fundamental Principles Based on the Science of the Spirit*.

> It is of the greatest importance to know that ordinary human powers of thought are refined powers of configuration and growth. A spiritual principle reveals itself in the configuration and growth of the human organism. And as life progresses this principle emerges as the spiritual power of thought.[26]

There may now be a few more aspects to the significance of salt for the human being. This is a very abbreviated presentation—as an incentive to explore the background of the pictorial appearances.

In the Gospels and in Paul's letters there are a total of four passages that mention salt: "You are the salt of the earth," says Matthew (Matt. 5:13). Immediately afterward, we read: "You are the light of the world" (Matt. 5:14).[27]

The salt process in man becomes the basis of his light of consciousness. The way is through life, as John describes it in his prologue of the Logos: "In him was life, and the life was the light of men" (John 1:4).[27] It is the true light in its spiritual potency.

Three other salt passages in the New Testament illuminate in a wonderful way the salt process in man as the basis of his soul-spiritual activity.

Mark 9:49: "For everyone will be salted with fire. Good is the salt; but if the salt loses its saltiness, wherewith will you restore it? Have salt in yourselves, and live in peace with one another."[27]

Luke 14:34-35: "Salt is good. But if it loses its saltiness, whereby could it be restored? Neither for the land nor for the dung heap is it to be used; it is thrown away. He who has ears to hear, let him hear!"[27]

Col. 4:6: "Let your speaking always be done with devotion, seasoned with salt, so that you may know how to answer each individual person."[27]

The shape of the cube also has great significance in the Apocalypse of John. Here it is not the salt but the diamond, which can also crystallize in the shape of the cube among other forms.

The city, the new Jerusalem, consists of four high walls with twelve gates. The city was measured with a golden measuring rod. Its length and width and height are equal to each other. "As a square the city lies there." 144 cubits is the measure of the walls. "This is the measure of man, and at the same time the measure of the angel."[27]

We also look at the seventh seal as sketched by Rudolf Steiner in his course on the Apocalypse of John.[28] In the lecture of September 16, 1907, in Stuttgart, he describes this seal in more detail. The seventh seal proceeds from a glass vessel in the form of a cube. "This space, into which the divine word of creation is spoken, is represented by the occultist in the form of a cube as clear as water."[29]

The dimensions of the three perpendicular directions of space are countered by three dimensions. "These counter-rays represent at the same time the primordial germs of the highest members of the human being. The physical body, crystallized out of space, is the lowest. The spiritual, the highest, is the opposite; it is represented by the counter-dimensions."[29]

The further forms of the seventh seal emanate from this cube.

Salt between Substance and Form

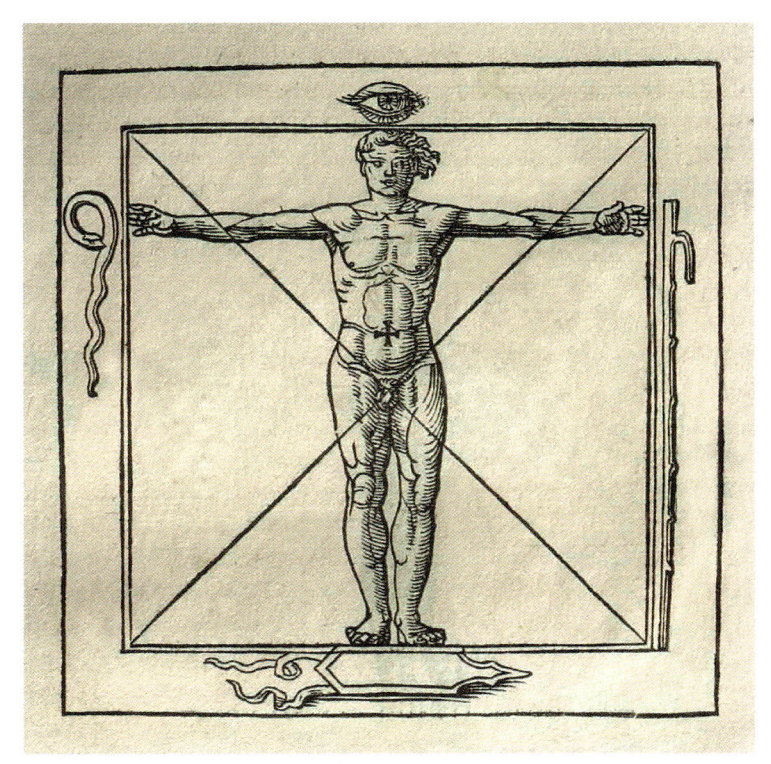

Image 157: "I think the speech."

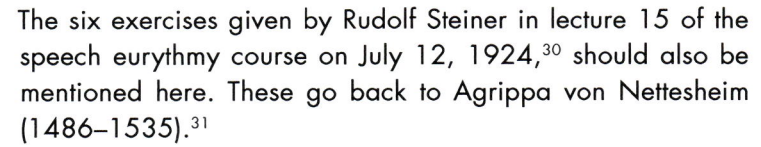

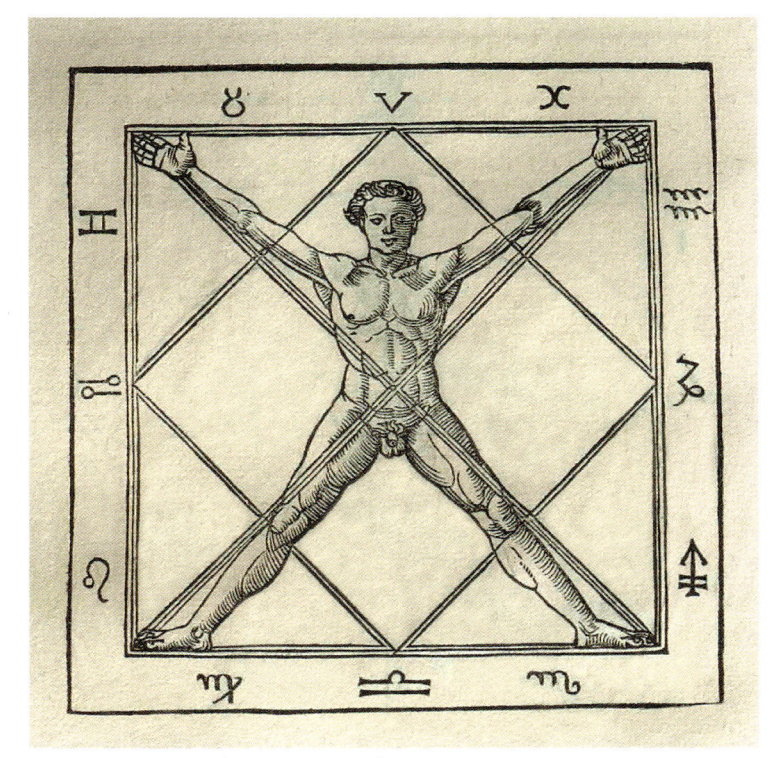

Image 158: "I seek my origin in the spirit."

The six exercises given by Rudolf Steiner in lecture 15 of the speech eurythmy course on July 12, 1924,[30] should also be mentioned here. These go back to Agrippa von Nettesheim (1486–1535).[31]

In the first and fourth position of these exercises (order according to Rudolf Steiner, images 157 and 158) we can see a connection to the salt images.

First exercise: "I think the speech"—the human being stands in the salt crystal, limited by the four sides. His height and the span of his arms are ideally equal to each other. You can measure this on yourself (image 157).

Fourth exercise: "I seek my origin in the spirit"—the person places himself in the power diagonals of the salt crystal. He grows beyond himself (image 158).

The process of salt formation was also of special significance for the medieval Rosicrucian. He experienced the pure forming power of the earthly substance in his soul like a prayer. He saw the processes of putrefaction and decomposition that were at work in the soul being purified and ordered by his thoughts directed toward the spiritual. "The process of overcoming the forces leading to decay through spirituality, that is the microcosmic formation of salt."[32]

In the course of this chapter, which began with a salt crystal, the salt experiments led us through dilution, swirling, heat, and a living milieu to the salt process in the human being. This in turn perhaps gives us an idea of the extraordinary importance of this earthly substance for the soul-spiritual activity of the human being.

Image 159: Onyx water (400x)

Chapter III
Minerals, Gemstones, Metals in Water

This chapter alone may show the viewer the variety of drop appearances from the realm of minerals, precious stones, and metals (images 160 to 195). The question here was: What do the non-soluble earth substances reveal in the water?

Only individual references to some of the images are given. In this way, the eye can wander through the diverse designs and perhaps discover connections between a well-known mineral or gemstone and its drop image. Minerals do not dissolve in water like salt—but after many months they cause the medium to create unique forms.

The images of the preparations are called "mineral water" (e.g., onyx water). We look at forms, some of which can also be seen in the plant chapter.

Thus, it becomes clear that the entire mineral world once crystallized out of a living, spirit-infused, fire, air, and liquid elemental creation. After many months, a faint memory of these living primordial states of past earth times may be awakened in the water.

Image 161: The agate druse shows a delimited interior. The druse is filled with water. Such interior spaces show their image in the drop. In other drop images, this interior space shows animate structures.

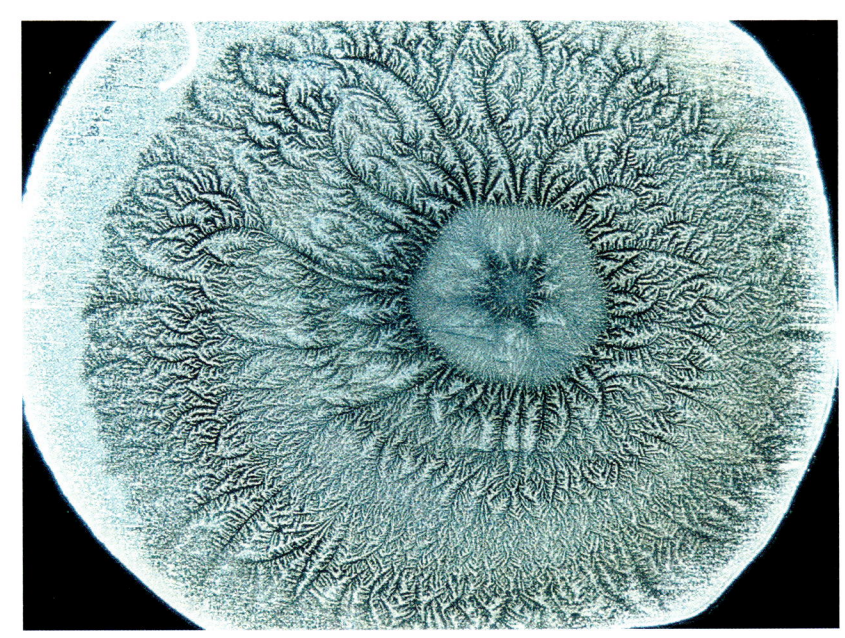

Image 160: Green onyx water (100x)

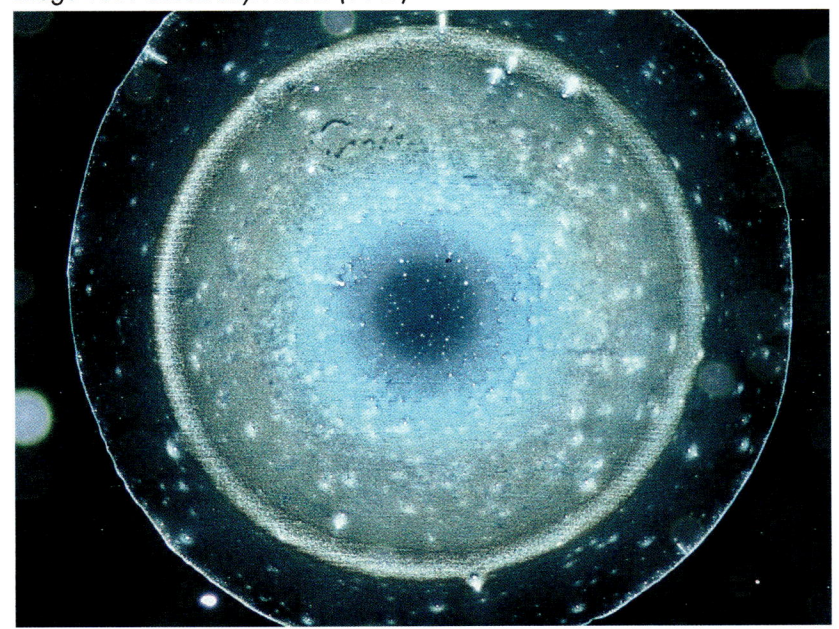

Image 161: Agate druse in water (water-filled) (100x)

73

THE SILENT LANGUAGE OF LIFE

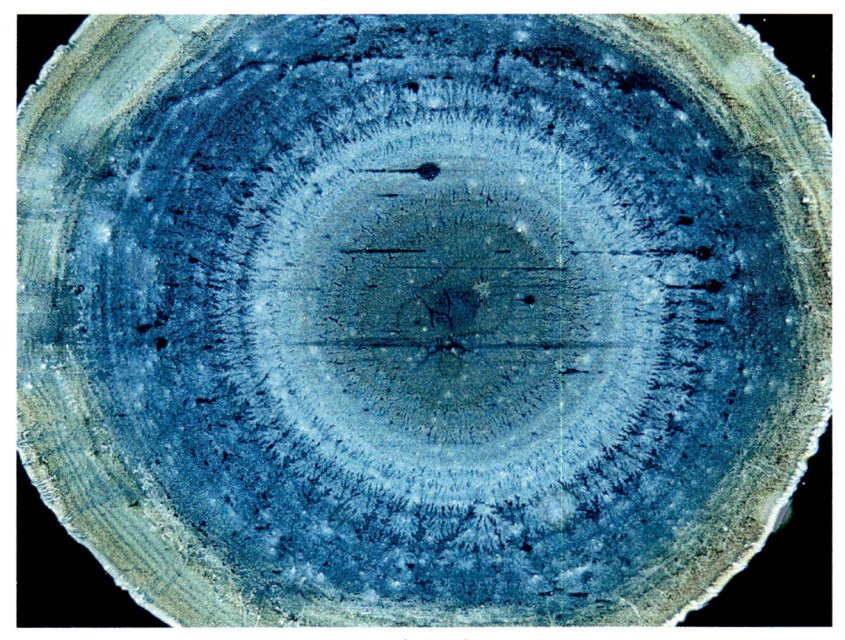

Image 162: Hematite water (100x)

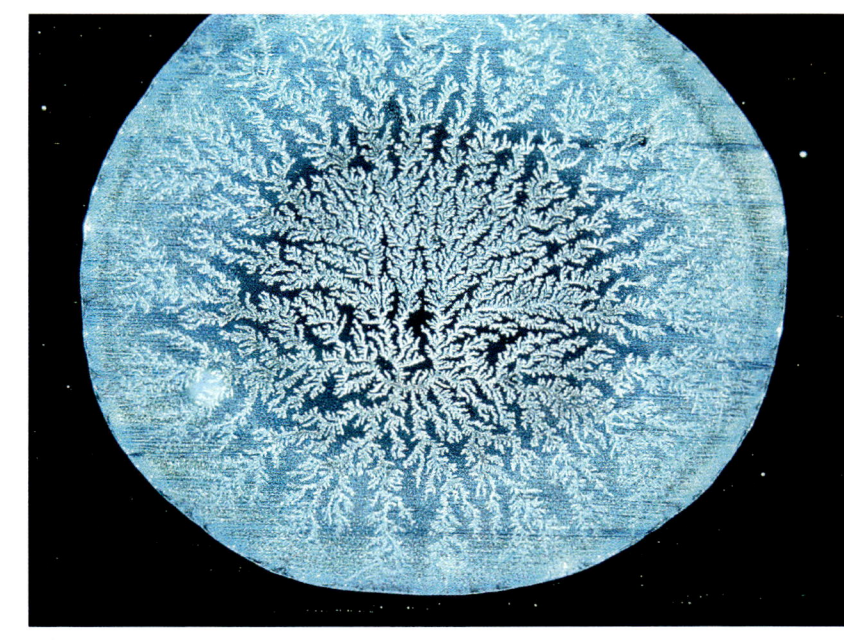

Image 163: Lava water (100x)

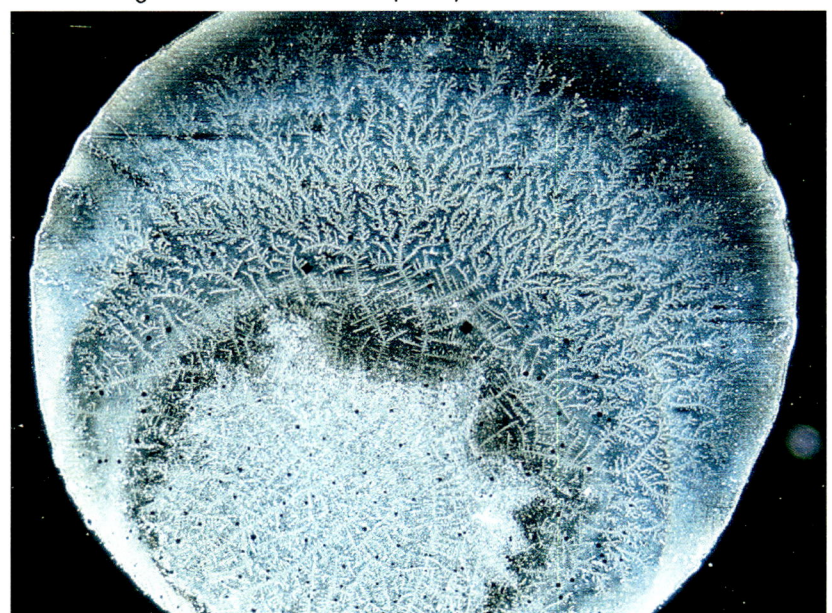

Image 164: Uranium pitch water (100x)

Image 165: Sapphire water (100x)

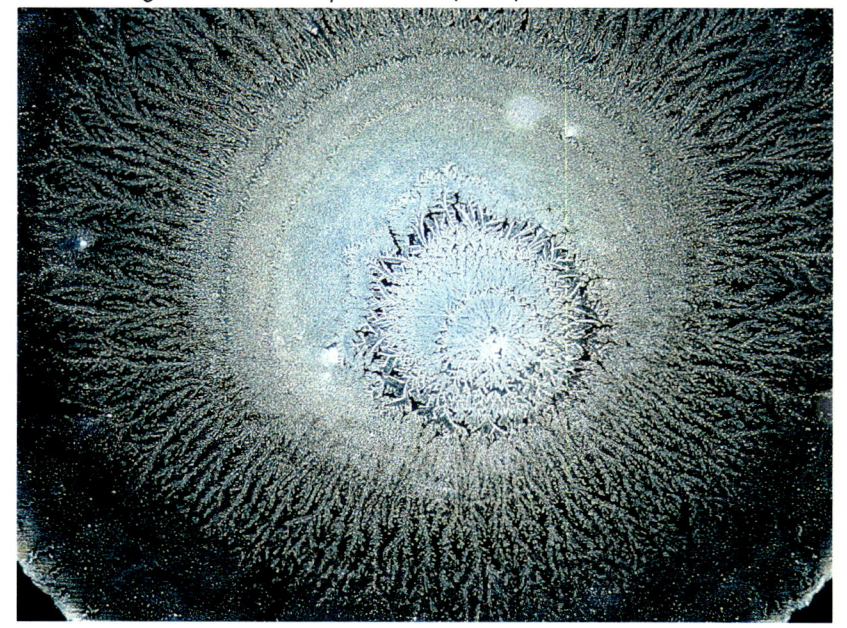

Image 166: Fluorite crystal water (100x)

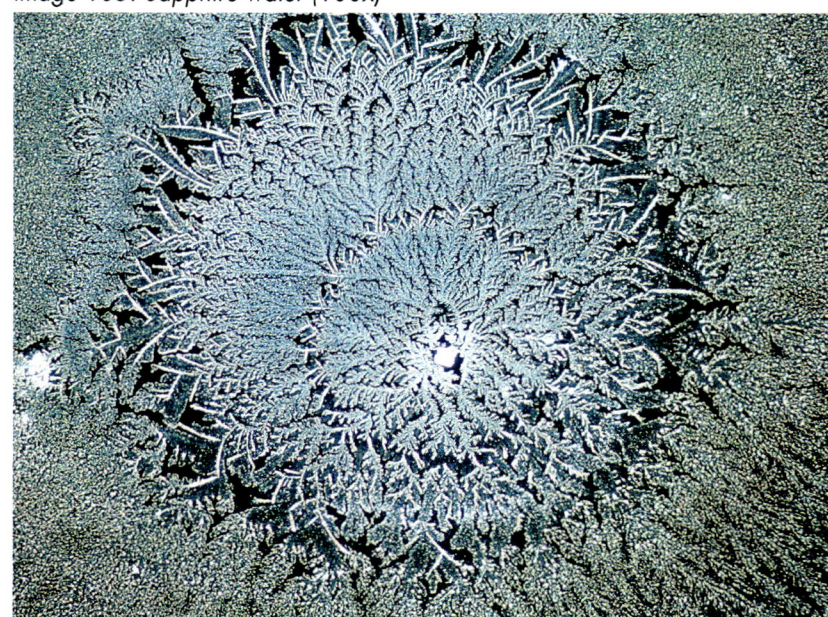

Image 167: Fluorite crystal water (400x)

74

Minerals, Gemstones, Metals in Water

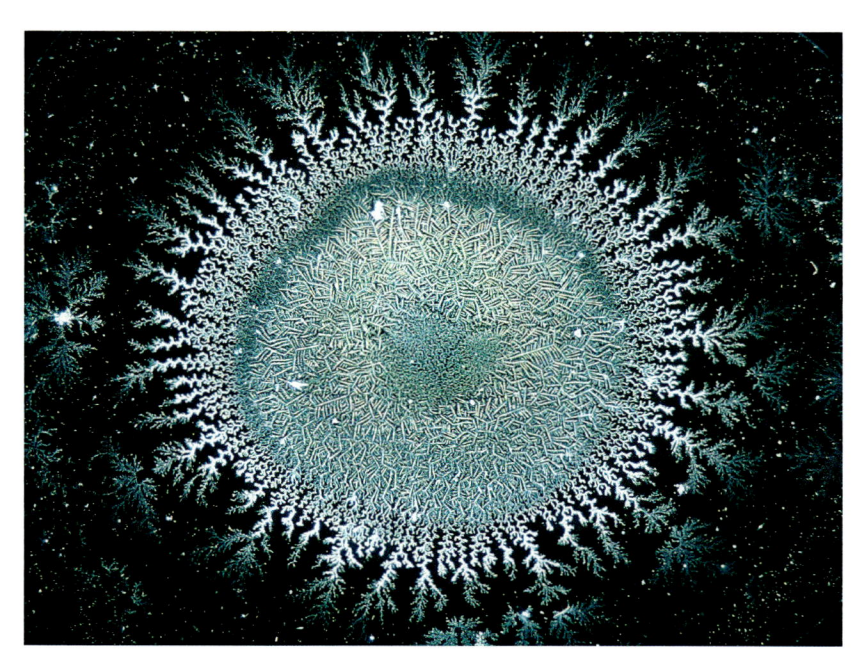

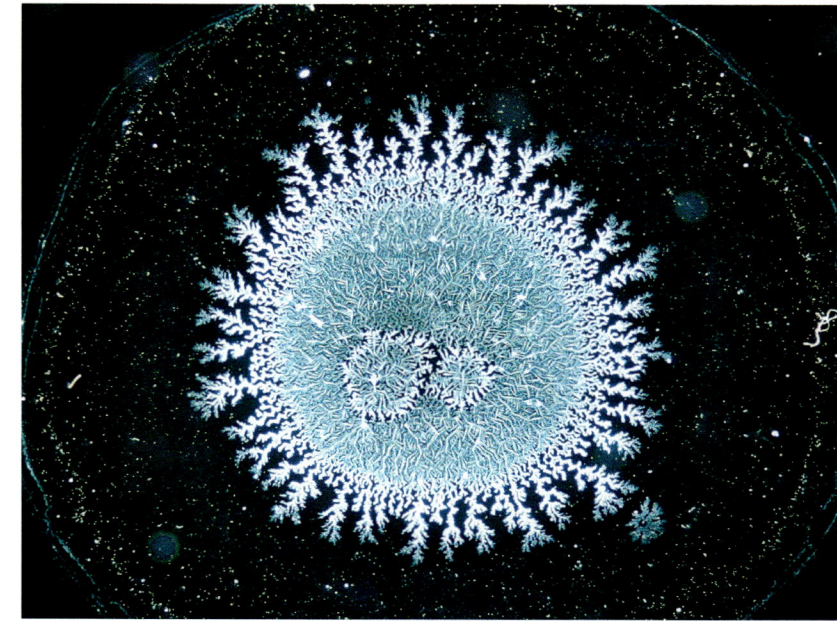

Image 168: Onyx water (100x)

Image 169: Onyx water (100x)

Image 170: Onyx water (100x)

75

———————— THE SILENT LANGUAGE OF LIFE ————————

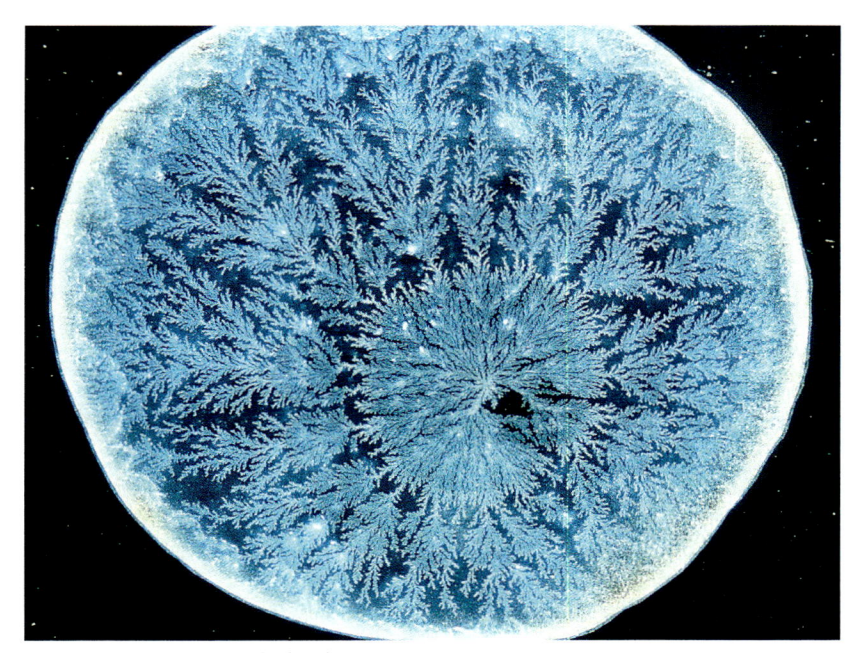

Image 171: Chalcedony water (100x)

Image 172: Chalcedony water (100x)

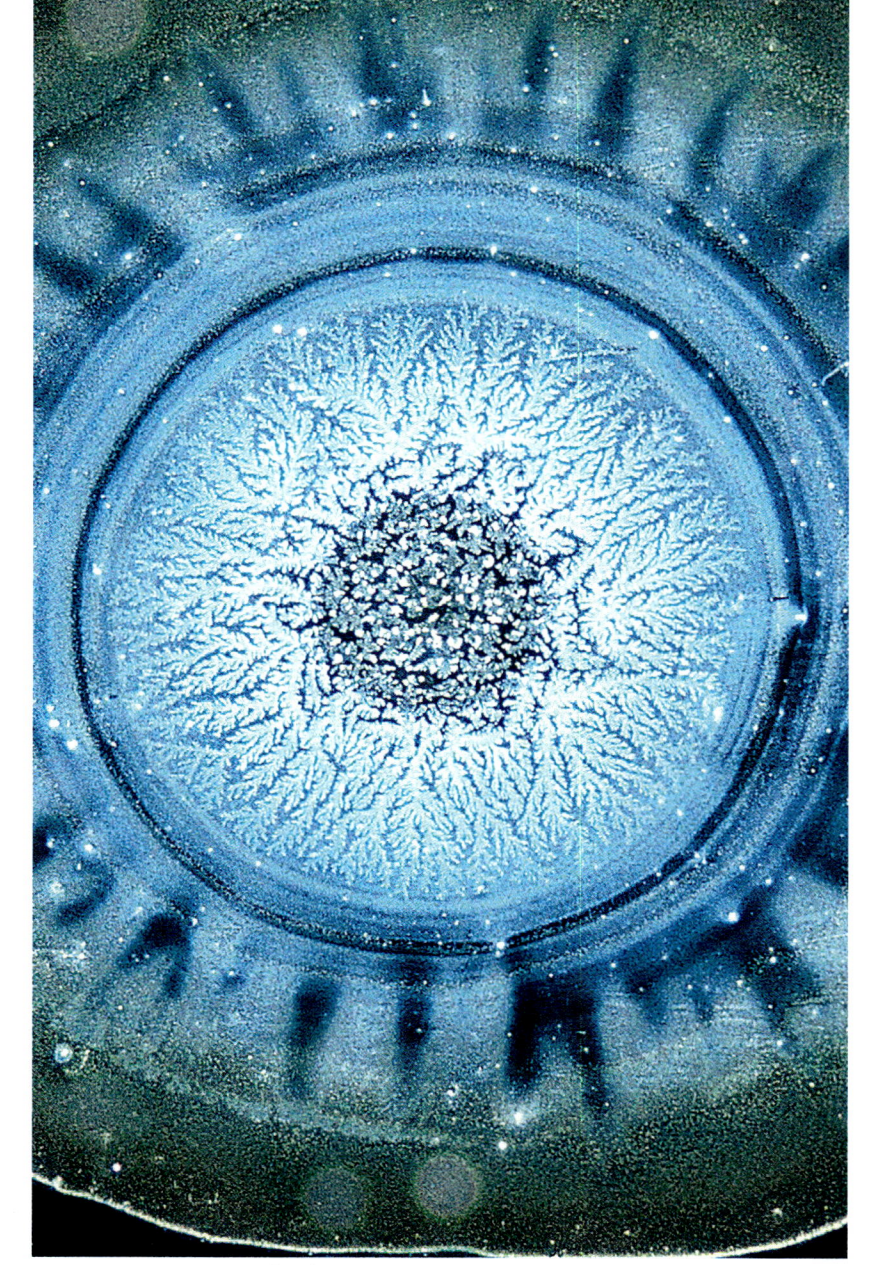

Image 173: Obsidian water (100x)

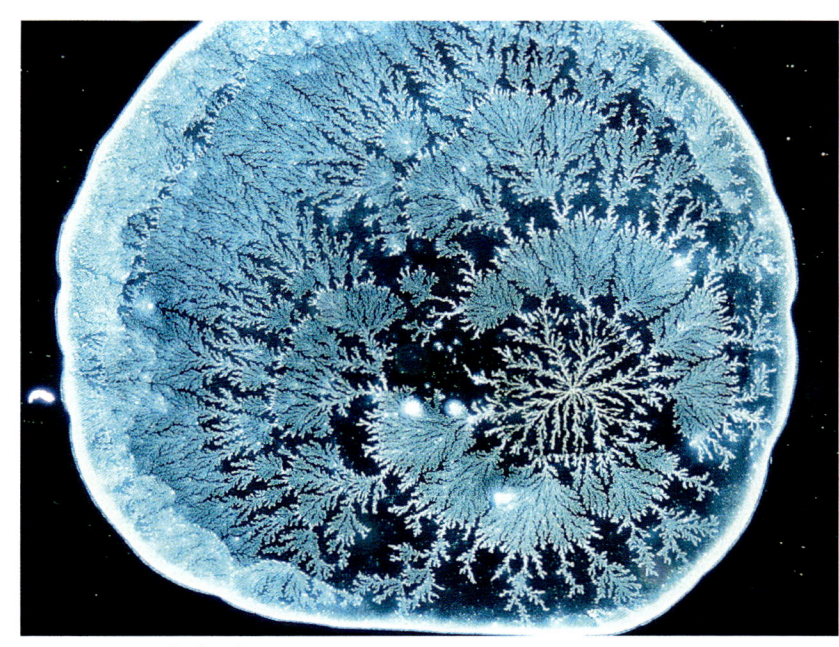

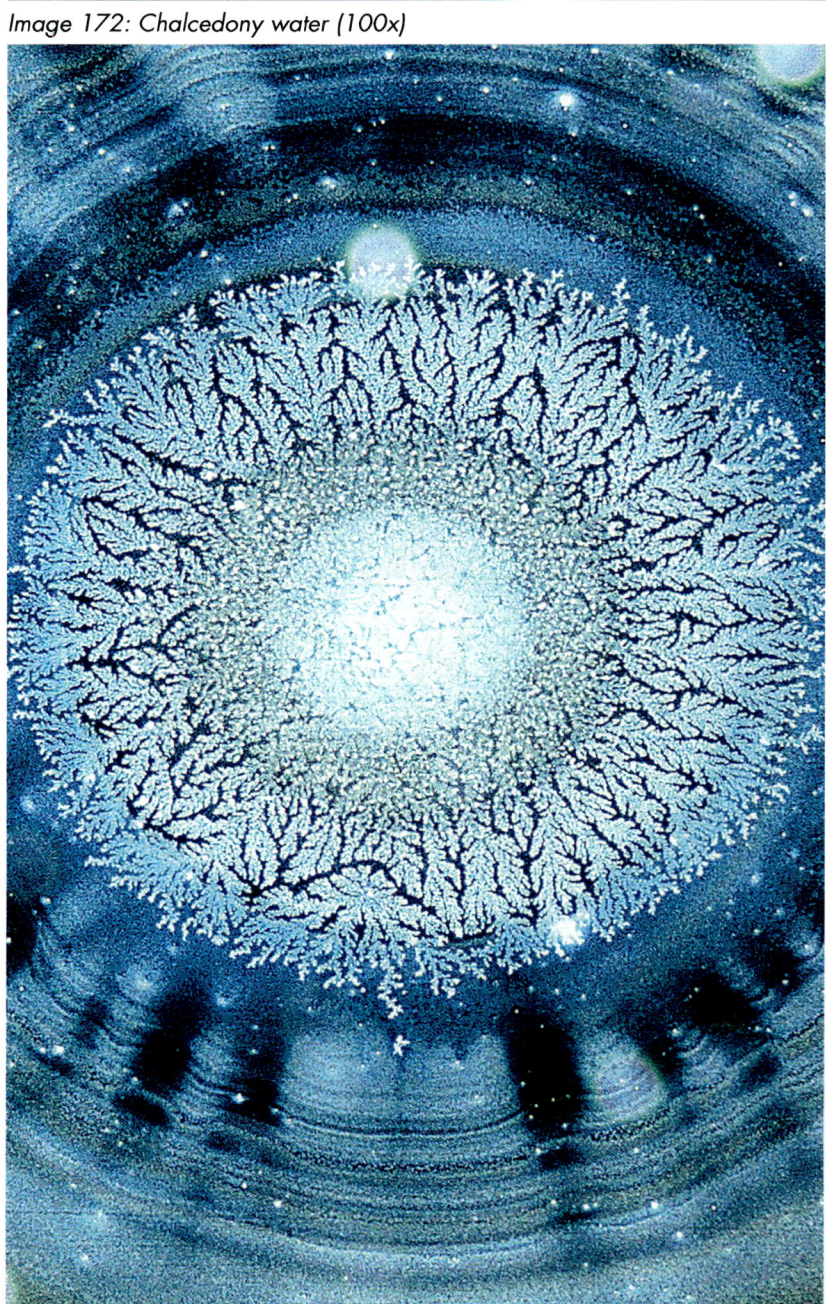

Image 174: Obsidian water (100x)

Minerals, Gemstones, Metals in Water

The special thing about the tourmaline (images 175 to 182) is that all the variously colored crystals show flower shapes: red tourmaline (rubellite), blue and green tourmaline as well as the black schorl, as can also be found in a book by Dr. Friedrich Benesch (1907–1991), *The Tourmaline: A Monograph*.[33]

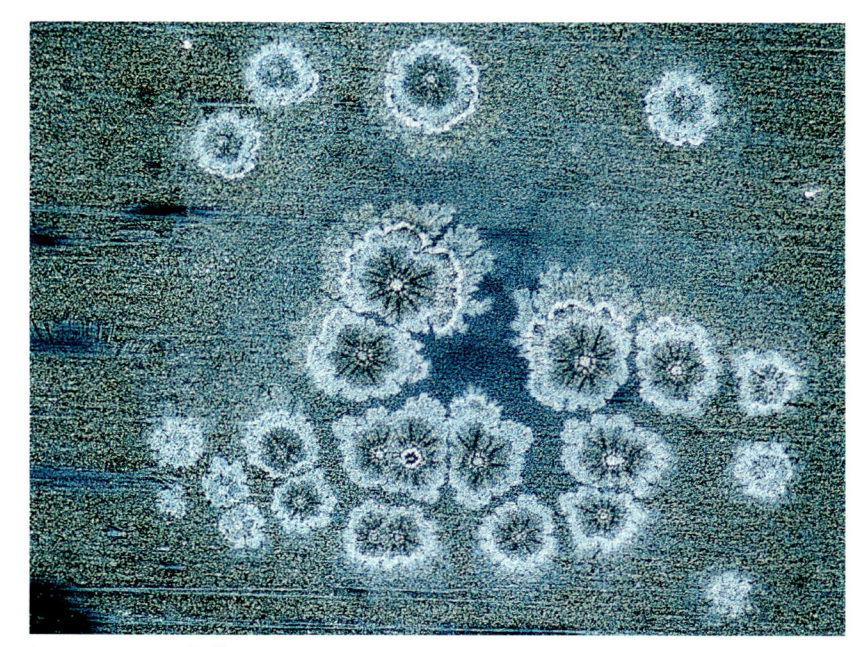

Image 175: Rubellite water (200x)

Image 176: Rubellite water (400x)

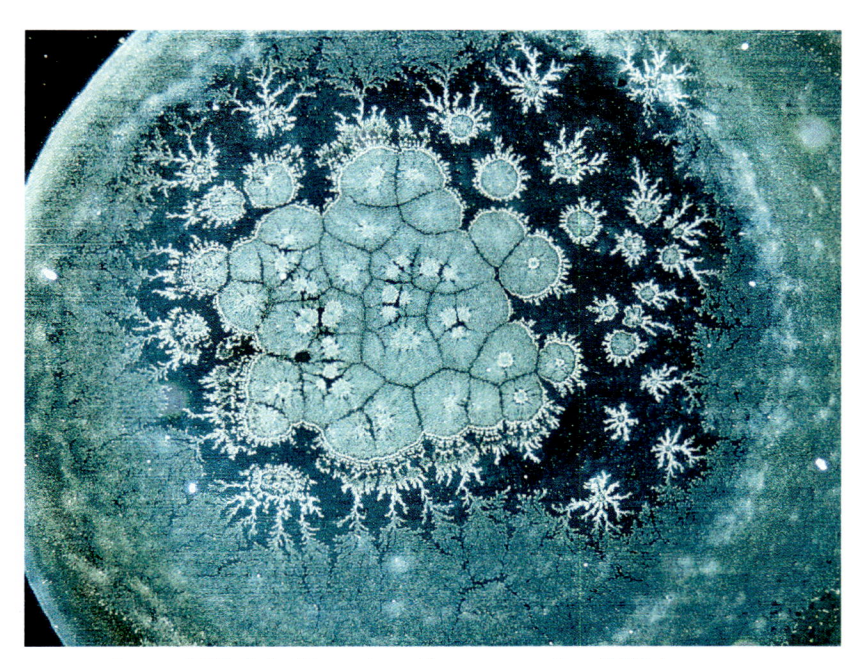

Image 177: Rubellite water, other preparation (100x)

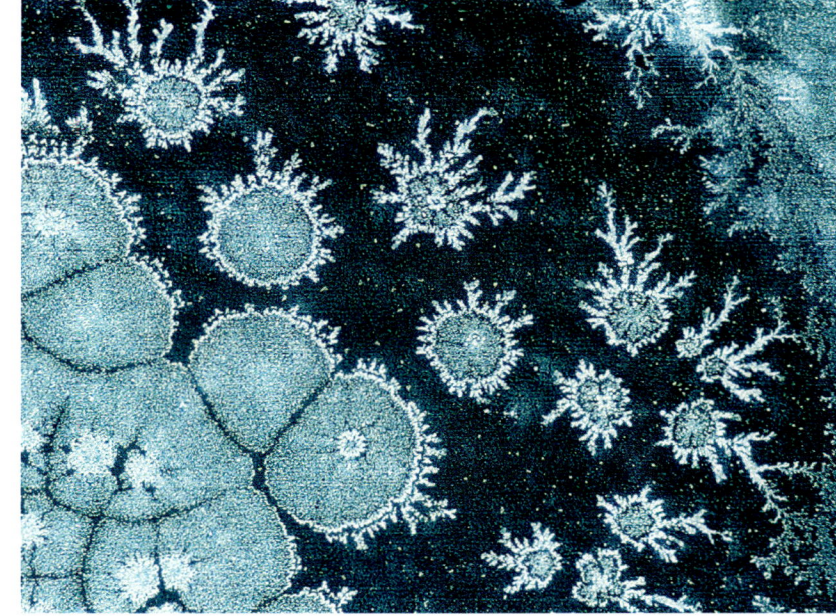

Image 178: Rubellite water, other preparation (200x)

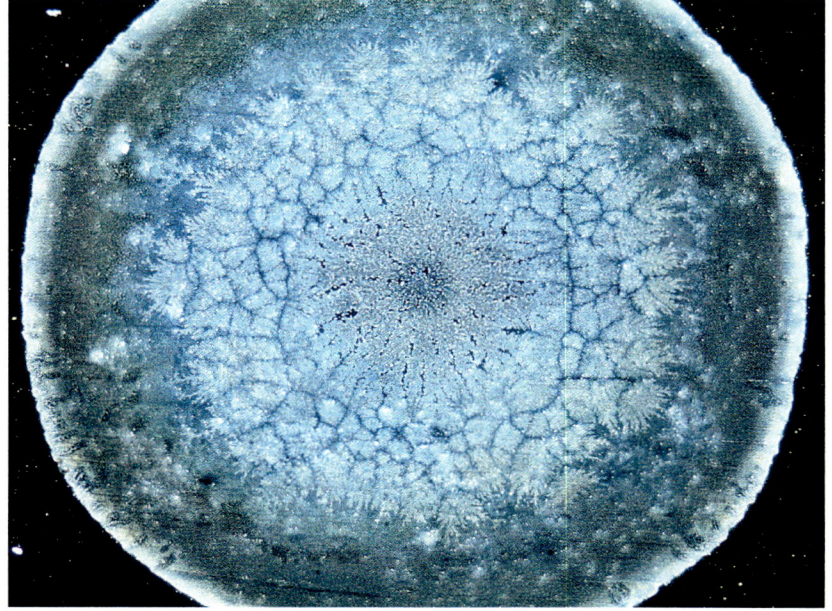

Image 179: Blue tourmaline water (100x)

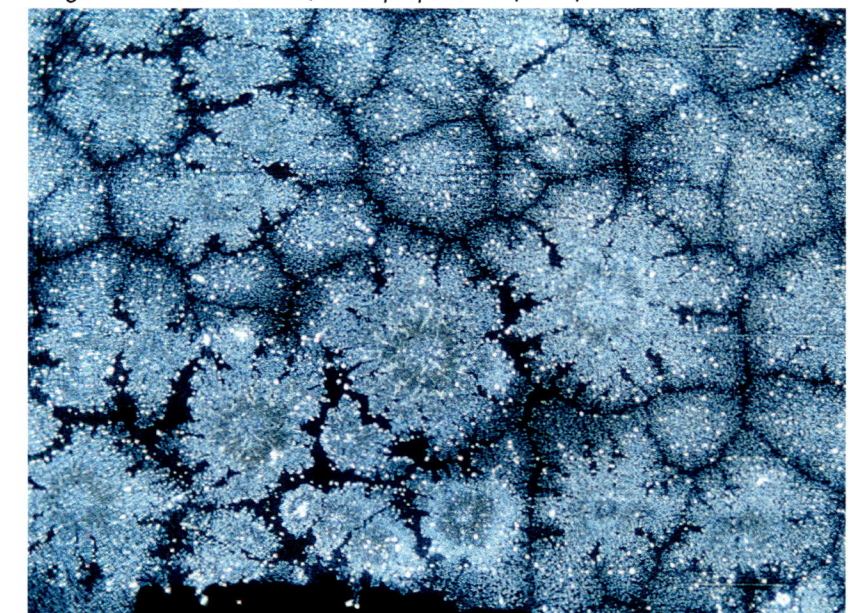

Image 180: Blue tourmaline water (400x)

Image 181: Schörl water (200x)

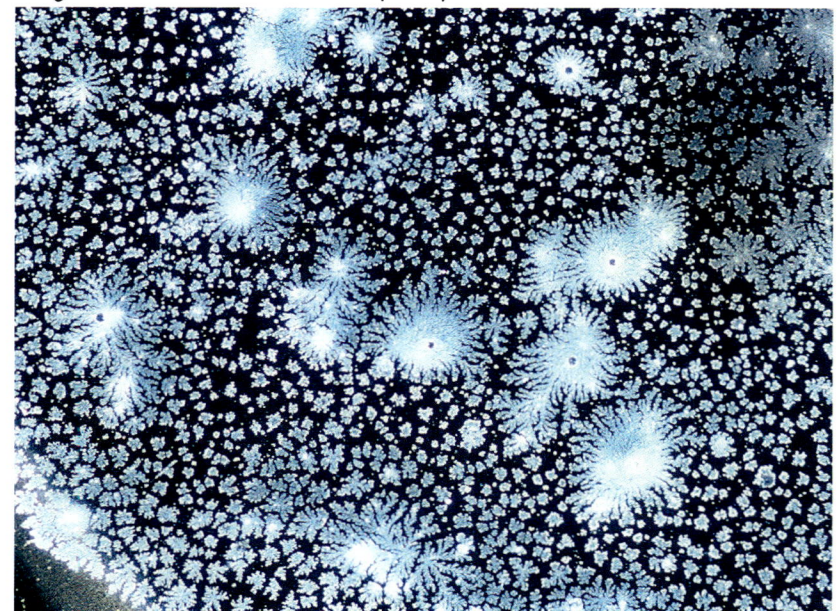

Image 182: Green tourmaline disc above the water (200x) (see p. 26)

Minerals, Gemstones, Metals in Water

Concerning the metals it should be noted: Image 185 shows the preparation water in which a small meteorite spent many months.

Image 186 shows a boiling infusion. The small meteorite was doused with boiling water and, after cooling, the water was dripped on. Vegetative growth spreading from a center characterizes the image.

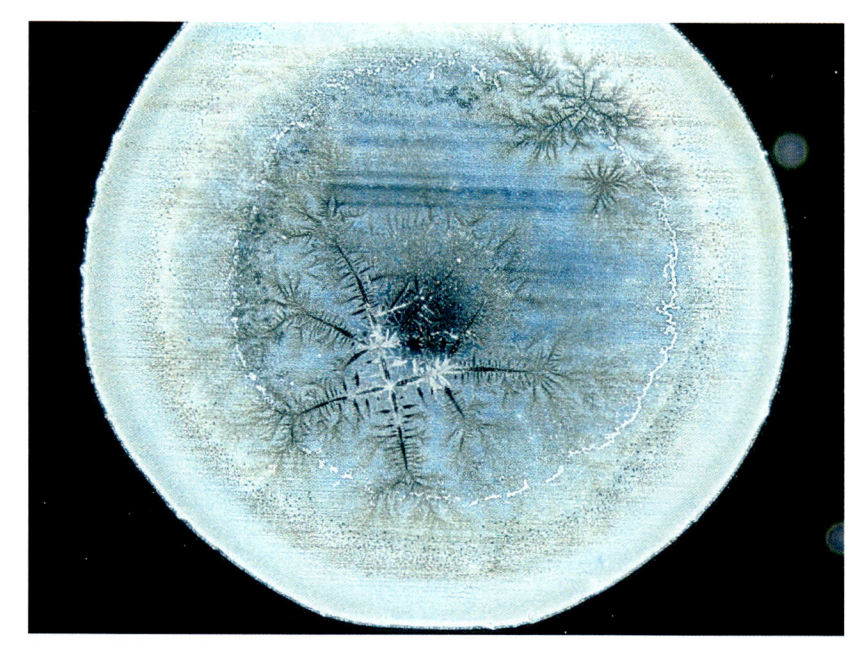

Image 183: Lead water (100x)

Image 184: Tin water (400x)

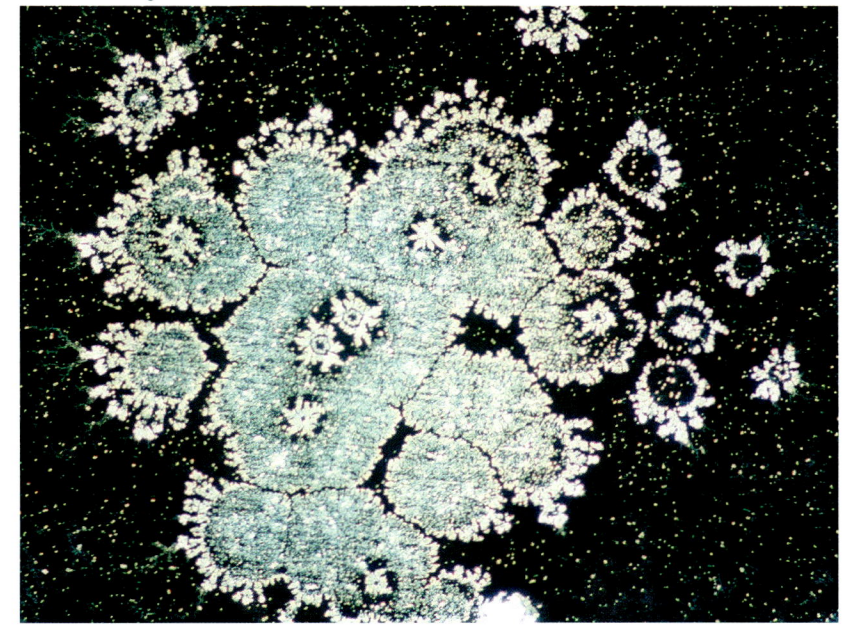

Image 185: Meteorite water (400x)

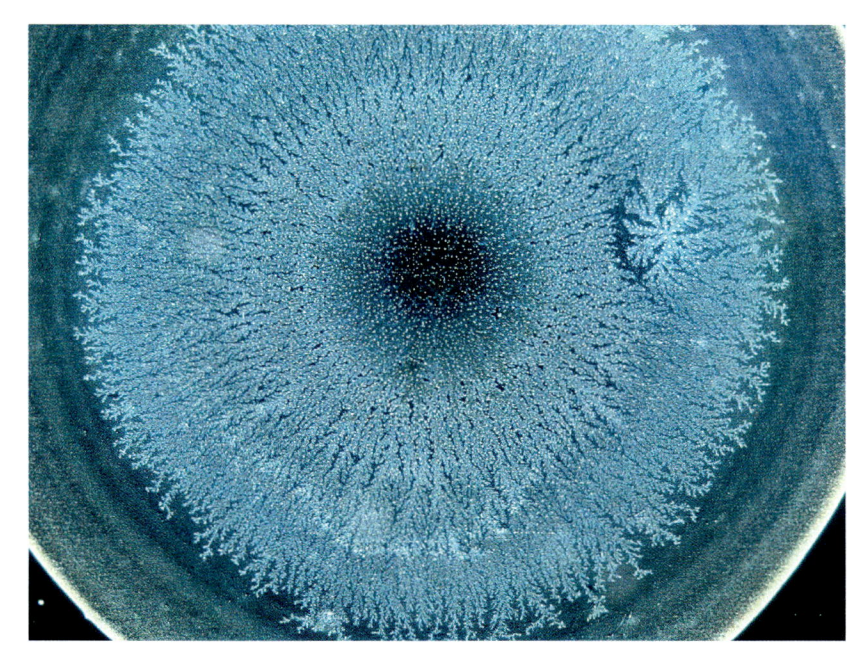

Image 186: Meteorite boiling infusion (100x)

THE SILENT LANGUAGE OF LIFE

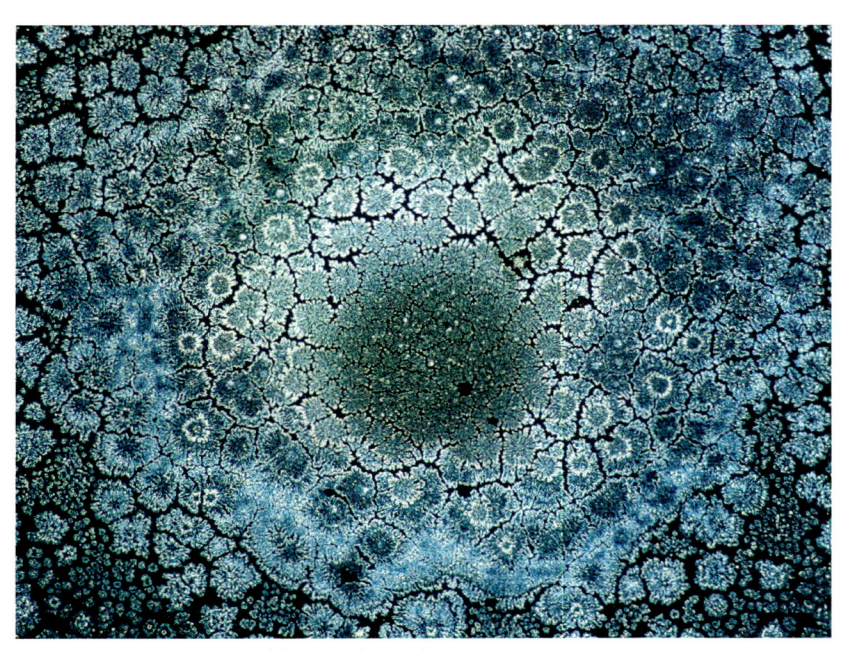

Image 187: Gold water (200x)

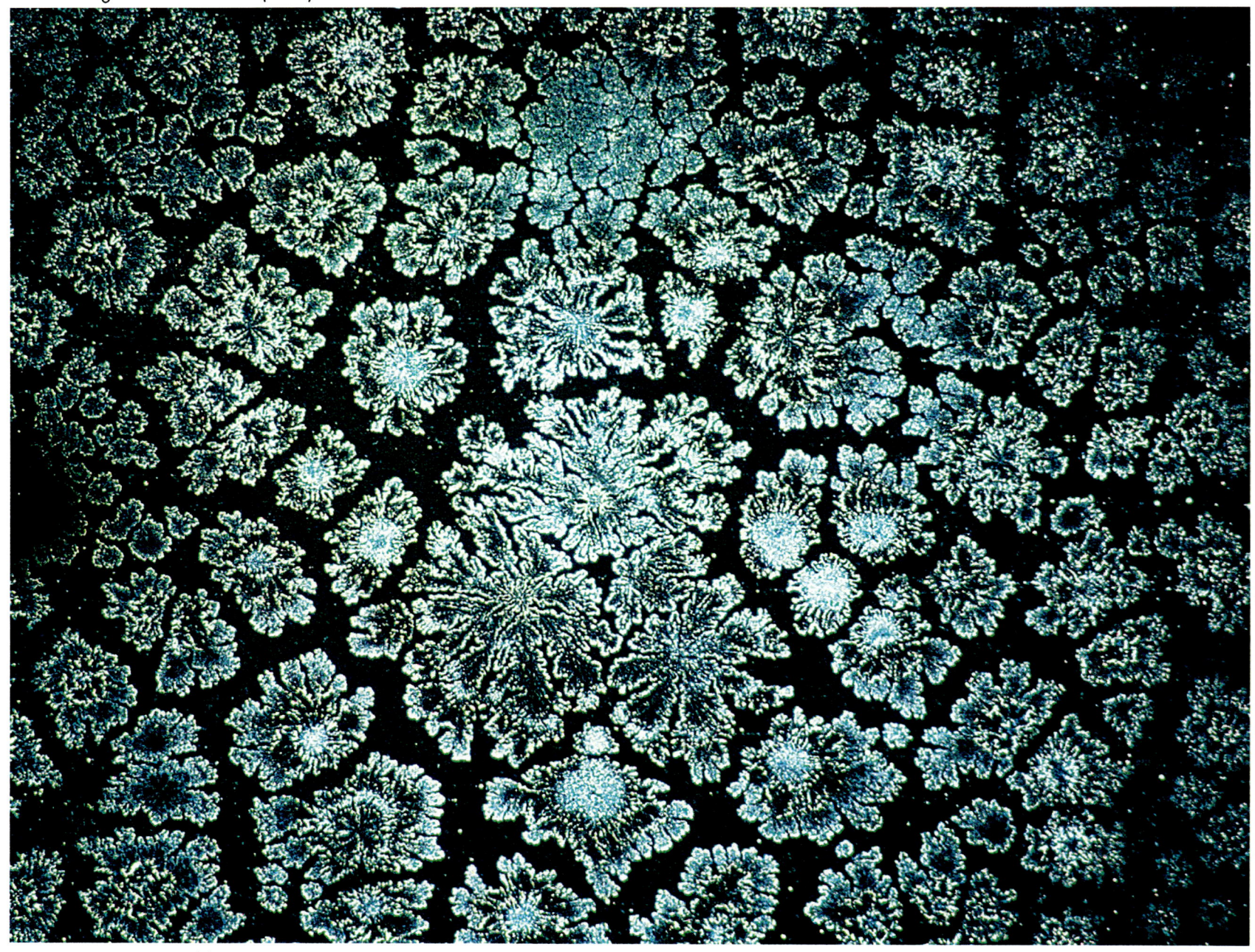

Image 188: Gold water (400x)

80

Minerals, Gemstones, Metals in Water

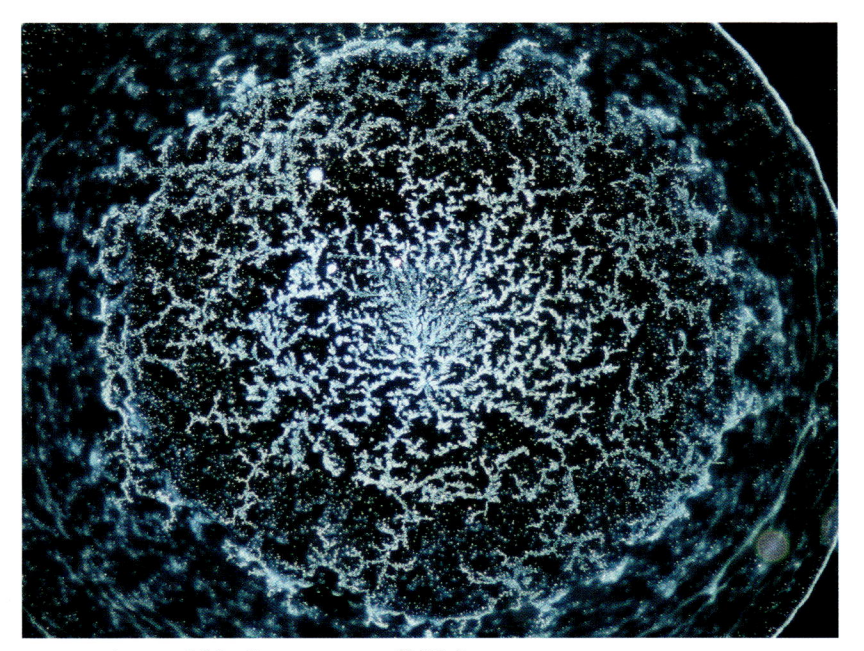

Image 189: Copper water (100x)

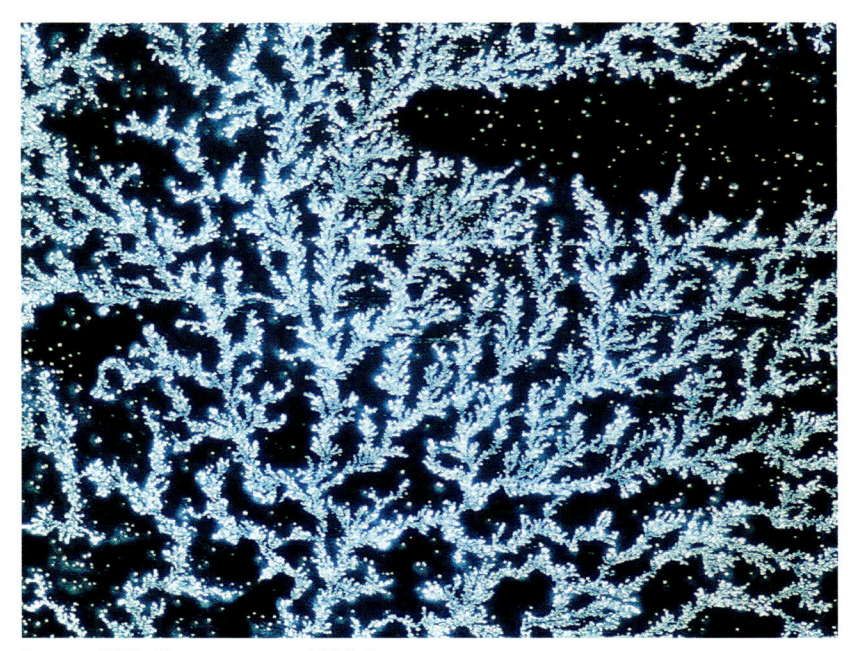

Image 190: Copper water (400x)

Image 191: A thick, shiny mercury sphere lay in water for many months.

Image 192 shows a water deposit in which the mercury lay fragmented into many small globules for months. The whole drop surface is covered with small ring formations (digital camera).

Interestingly, two quite opposite structural formations appear in the silver water (images 193, 194, 196, and 197).

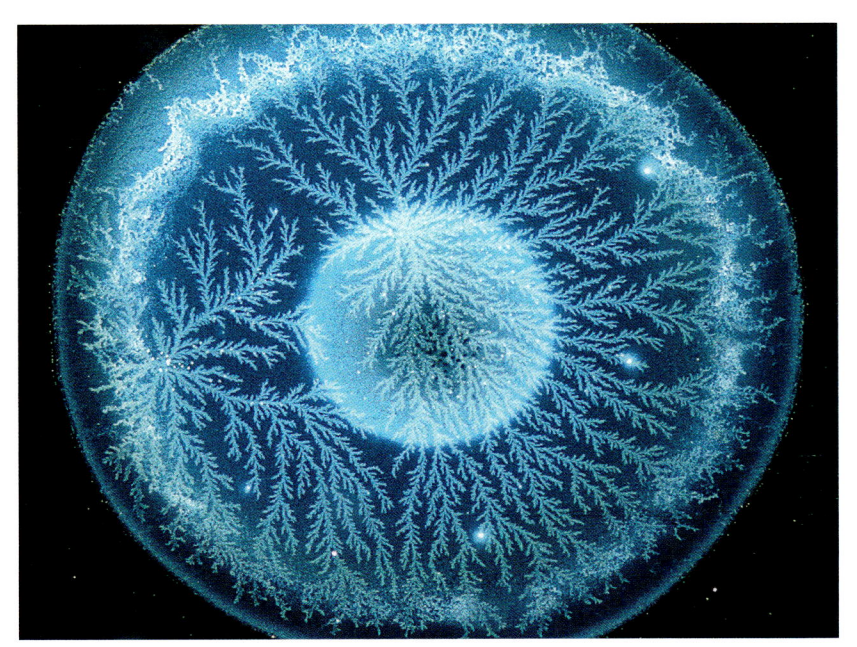

Image 191: Mercury water (100x)

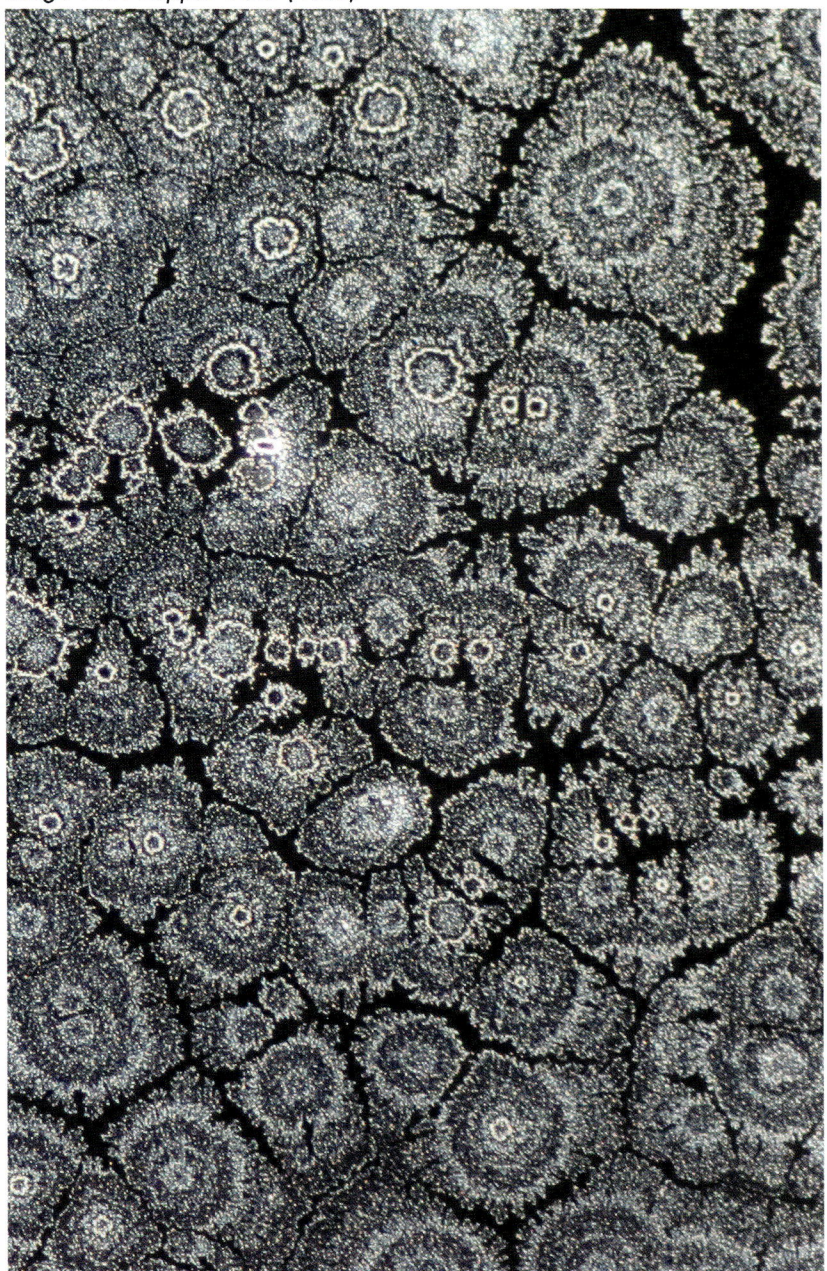

Image 192: Mercury water, other preparation (400x)

— THE SILENT LANGUAGE OF LIFE —

Image 193: Silver water (40x)

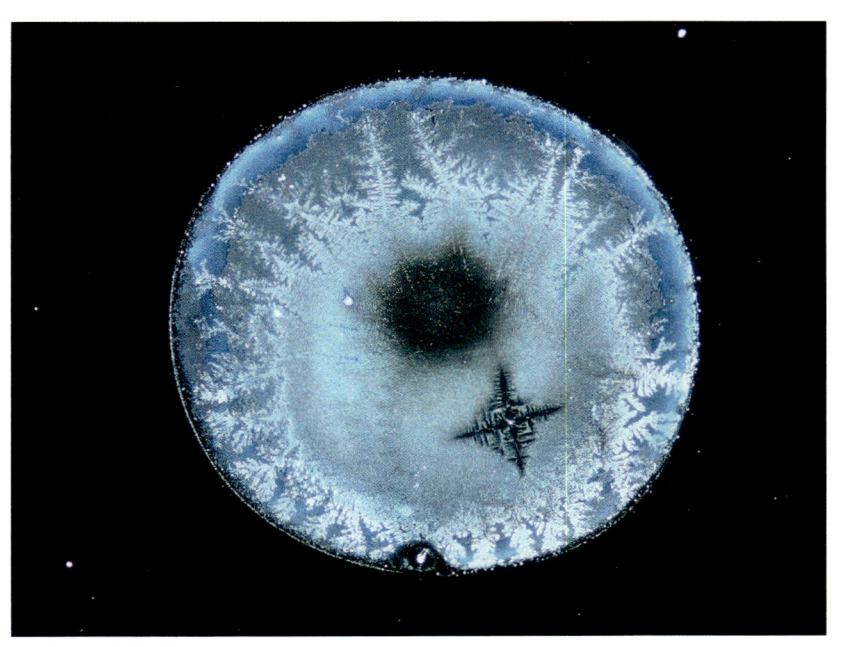

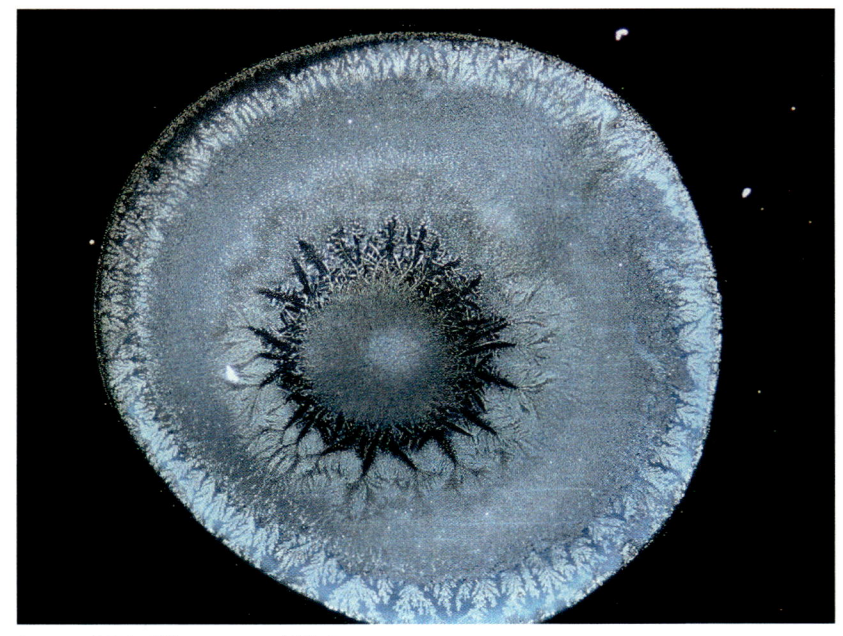

Image 194: Silver water (40x)

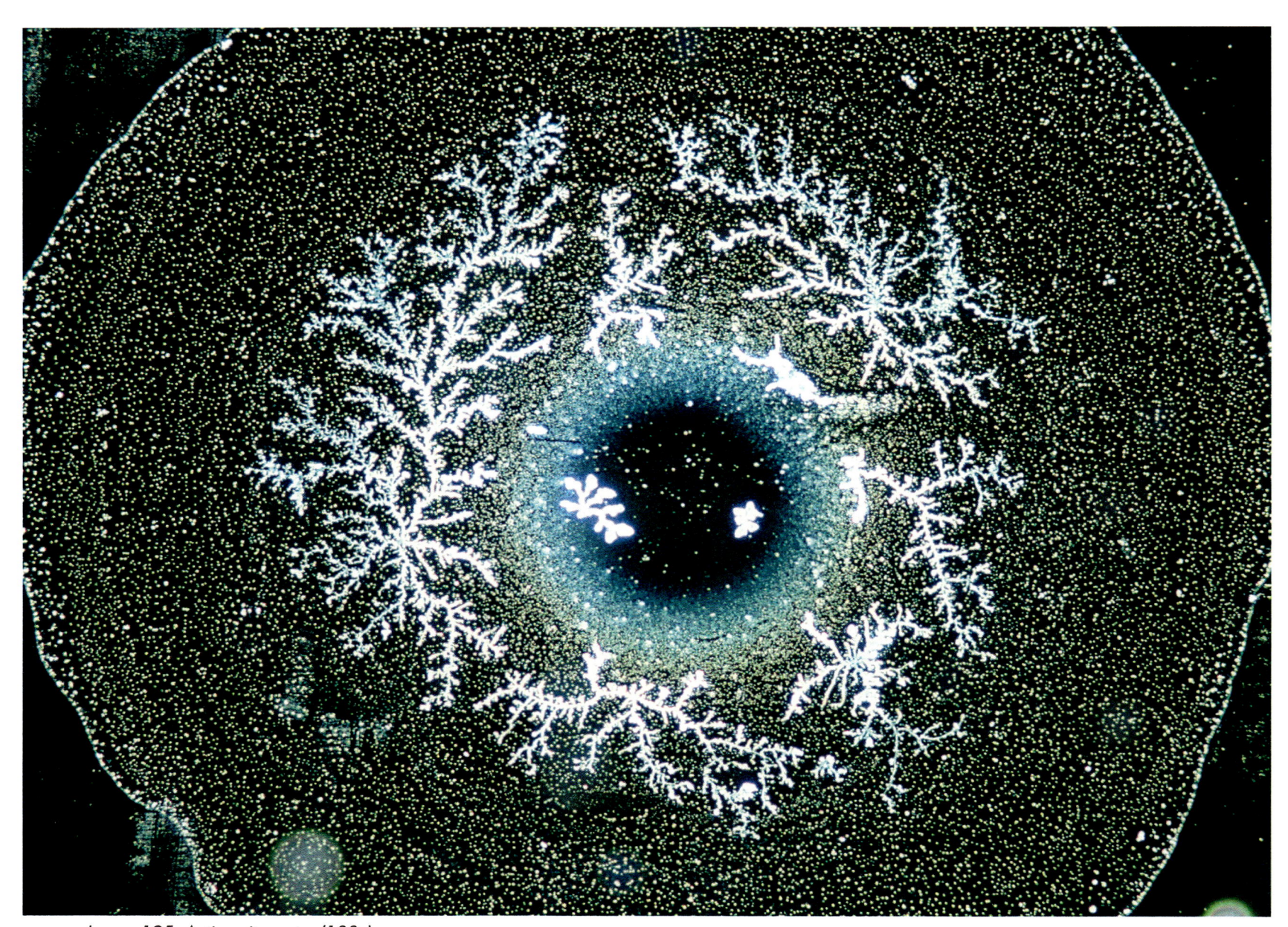

Image 195: Antimonite water (100x)

Minerals, Gemstones, Metals in Water

Image 196: Silver water (400x)

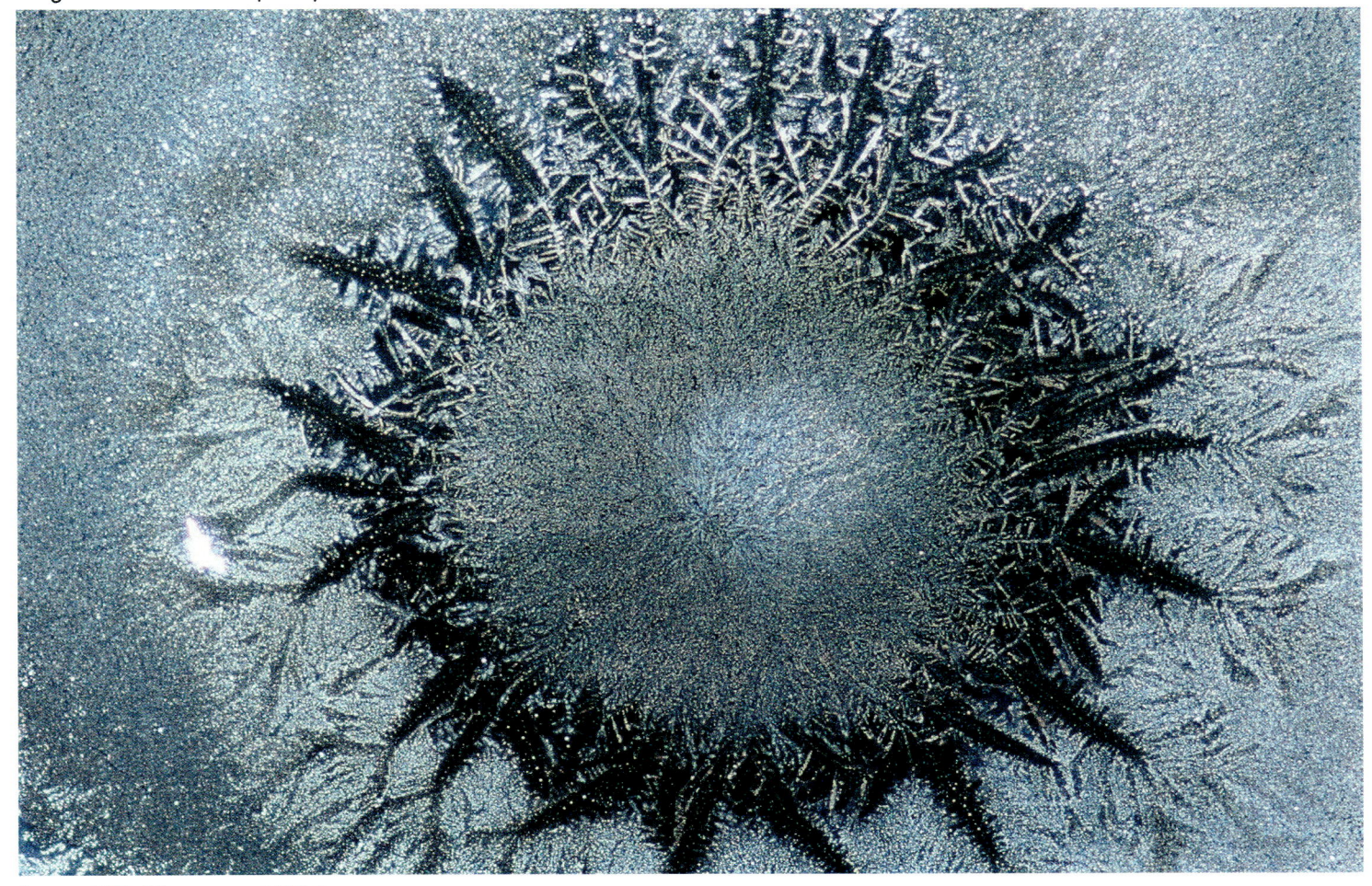

Image 197: Silver water (400x)

Image 198: Lavender branch, flowering (400x)

Chapter IV
Plants, Fruits, Seeds, Barks, and Ashes in Water

At the beginning of chapter II, on salt, was the desire for "solid ground" in the exploration of "flooding images." This solid ground had to be left in the salt *process*, some of which came to view in the images.

What was gained through this? The spiritual form of the cube, where the square is on the surface—abandoned by the material. Added to this are the salt transformations in triangular and hexagonal form in the living milieu. These shape variations were brought into a comparative relationship with the molecular structure of water and silicates. Thus, on this molecular level, a relationship between water, silicates, and salt was revealed. The interplay between material and form forces was further demonstrated when salt went into solution in the water of the plant and mineral preparations. In this way, the boundary of physical calculability was transcended into living form in the surrounding area, which could become an experience. Salt transformation and revitalization thus formed a bridge to the salt process in the human being.

What we see in the previous as well as in the following drop images is a small insight into the connection between two worlds. These are normally separated from each other for everyday perception—the sensory world on the one hand, and the formative or idea world on the other. Goethe had the ability to see both worlds together. For him, the sensory and suprasensory unite through the training of perception in natural phenomena to form a higher kind of seeing *(viewing power of judgment)*.

Rudolf Steiner writes in his book *Goethe's World View*:

> In Goethe's view, the natural scientist should not only be attentive to how things appear, but how they would appear if everything that works in them as *ideal driving forces* were also really to appear outwardly. Only when the bodily and spiritual organism of the human being is confronted with the appearances do they reveal their inner selves.[34] (emphasis by I. J.-N.)

The water seems to reveal in the drop at least a part of this interior, of these ideal driving forces of the supersensible. We are also made aware that the interior of phenomena cannot be experienced solely through our bodily organization (sense organs). J. W. v. Goethe writes at the end of his poem "Schillers Reliquien" (Schiller's relics):

> What more can man gain in life,
> Than that God-nature reveals itself to him:
> How it turns the solid into spirit,
> How it firmly preserves that which is spirit-generated.[35]

In this chapter IV, it is the plant that leads us on the search for the lasting, superior, the "ideal driving forces." Can we find something of this in the drop images?

As a living being, endowed with a fluid life or etheric body, it unfolds between root and seed, between earth and heavenly space.

Let us look at the elements in which this happens: Root in the dark, moist earth; shoot, stem, and leaf development in the watery sap flowing through air to light; the blossom in light and warmth; the fruit with ripening seed in the completed inner becoming of the plant in warmth.

The plant withers, the seed falls out of the living context and sets off in thousands of variations on its way into the dark earth from which it once came.

The etheric forces immerse themselves in this elemental activity, which can only be hinted at briefly: Life ether primarily in the earth element of the root; light ether working primarily in the shoot and stem toward the sun; chemical or sound ether primarily in the unfolding of the leaves; warmth ether working in the whole, penetrating everything, especially in the ripening of the fruits and seeds.

What has been separated here in a few words actually works in unison in a living interaction, but in a differentiated way. This differentiation should in no way disturb the unbiased observation. On the other hand, in the imprints of individual plant parts in the water, much of this interplay of etheric forces can be experienced.

Now individual organs of the plant are put into the water. Their "conversation" with this medium is shown in the following images.

1. Plant Stem and Plant Root, Flowering Plant, Larva, Butterfly, Bee

Plant Stalk and Root

In the *hypocotyl* (Greek: *hypo*=under, *kotyle*=cup, bowl) of the seed plants a great decision has taken place: The root strives downward into the earth, the shoot upward toward the sun. If the plant is cut at this point, we have the root on the one hand and the stem with the leaves on the other. These two parts of the plant are placed separately in the water. It is a small oak tree in this experiment.

The root is completely immersed in the water and the container with a flower pot is put over it, because the root lives in the dark soil. Next to it, the oak stem with its leaves is placed upright in the water and light. After seven days, the water is dripped on.

Image 199: Oak stalk (200x)

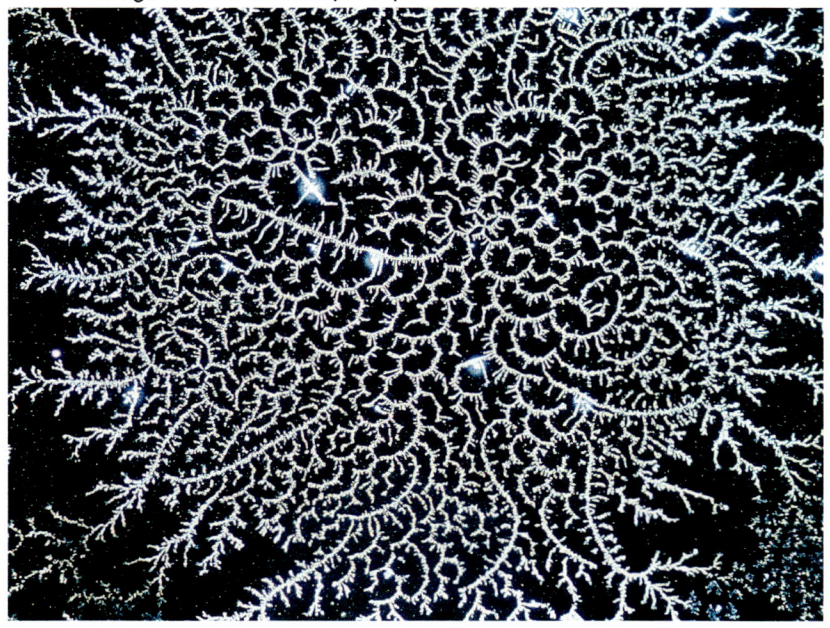

Image 200: Oak root (200x)

Image 201: Mistletoe (100x)

The first two images (199 and 200) show us a polar design.

Image 199: Oak stem in the water causes a design that grows linearly from a center to the periphery. The impression is "radiance."

Image 200: Oak root in the water shows a crystalline hexagonal structure, linked to rounded branches. The overall impression is: nothing is growing here, here it is rounding, even in the hexagonal shape itself. It is the typical root structure, as we will also see in the germinating seed (see below). The first thing the seed does is to stretch its little root into the earth.

The inspiration for this fundamental experiment came from a lecture in which Rudolf Steiner talks about stem and root growth of the plant.

> The striving upward, the striving downward. And if today we were really already so far advanced in the physical investigation of nature that we would apply the methods of investigation sometimes used for less essential things to something like the stem growth of the plant upward, the root growth of the plant downward, we would find the connections in the universe that, in turn, by coming into relation with the human being, actually first make this totality — human being and world, macrocosm and microcosm — comprehensible.

And further:

> And in the upward striving linear direction we must see the presence of the solar forces on earth. In the root striving toward roundness we must see the presence of the lunar forces on the earth.[36]

Plants, Fruits, Seeds, Barks, and Ashes in Water

Image 202: Ginkgo branch (100x)

Image 203: Garden sage stalk (200x)

Image 204: Ginkgo leaves, fresh (200x)

Image 205: Ginkgo leaves, wilted (200x)

The resulting image resulting in its clear statement was a surprise. *Before* the experiment, there was no idea of what would appear. Through Rudolf Steiner's explanations, a more vivid understanding of these radiating, growing effects of the sun and the more rounding, crystalline, shaping effects of the moon on the plant awoke in the viewing.

In the drop image, these polar processes in the surface are illustrated in a meaningful way. They were confirmed in further experiments, so that one can speak here of stem and root structure.

The following images (201 to 203) now show the peripherally striving processes of different branches.

Image 201: Mistletoe branch in water.

Image 202: Ginkgo branch in water.

All plant stems (or branches) in the water reveal this "stem or branch type" (as it is called here, referring to this research alone), even if in variations. The individual plants cannot be distinguished from each other in this respect.

Image 203: Garden sage stems in water.

Now the question arose again and again: What gives rise to the composition of the image? Is it the plant *substance* that plays with the water via molecular interaction? ... or is there something else to the composition of the image? The cellulose of the stem does not dissolve in water!

So fresh and withered ginkgo leaves with their long stems were placed in the water. After a few days, different shapes appeared.

Image 204: Fresh ginkgo leaves placed in water.

Image 205: Withered ginkgo leaves placed in water. The wilted leaves were bright yellow and of a firm consistency, but no longer showed any vitality in the water drop. The "substance" of the leaf stalks does not seem to play a role.

In comparing these two images, we can notice how growing and withering lead the soul into two quite different realms of sensation.

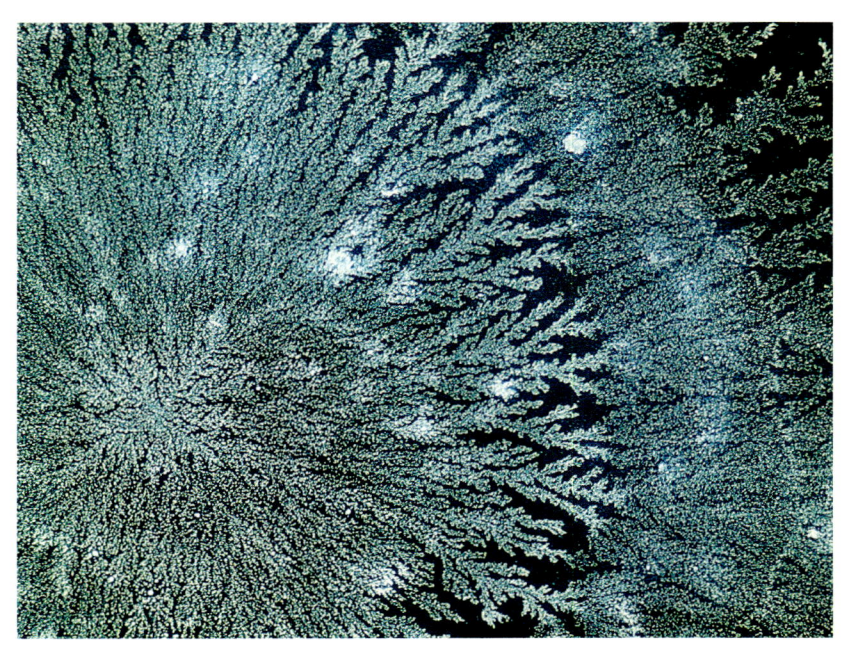

Image 206: Lavender stem, upright (200x)

Image 207: Lavender stem, inverted (200x)

Once again we ask about the origin of the creative impulse. We take two lavender branches before flowering and put the first one upright, the second one upside down in water for a fortnight.

Image 206: Lavender stem, upright in water.
Image 207: Lavender stem, upside down in water.

The comparison of the two images takes us beyond a purely molecular event in the water, when the stem's juice stream can flow upward (image 206).

Thus we can see that it is not the materiality of the plant stem that shapes the water, but its living sap flow, which strives upward against gravity and pulls the plant out of the earth, as a result of the activity of etheric forces.

Flowering Plant, Larva, Butterfly, Bee

Now there was a big surprise when a lavender branch in the water started to bloom after more than 14 days and transformed the drop image. This led to further experiments and considerations.

Image 208 and image 209 show the beginning of flowering and the change of the drop center.

Image 208: Lavender stem (100x)

Image 209: Lavender branch begins to bloom (200x)

Plants, Fruits, Seeds, Barks, and Ashes in Water

Image 210: Lavender branch, flowering (200x)

Image 211: Lavender branch, flowering (400x)

Images 210 to 213: The lavender branch is now in full bloom and reshapes the linear structure from the center into extraordinarily rhythmic, arching lines that form small hollows. In addition, lemniscates appear in many cases. We have not encountered these designs before.

Let us first ask about the background. What has happened?

In the sunlight and warmth, the inflorescence of the lavender branch has grown considerably. Colors and fragrances appear, attracting numerous insects: Bees, bumblebees, butterflies…. The flowering plant meets the animal world.

Can a relationship, a connection to the animal world appear in the drop image with its rhythmic-lemniscatic design? Is it the animal world itself that gives a new impulse to the plant being?

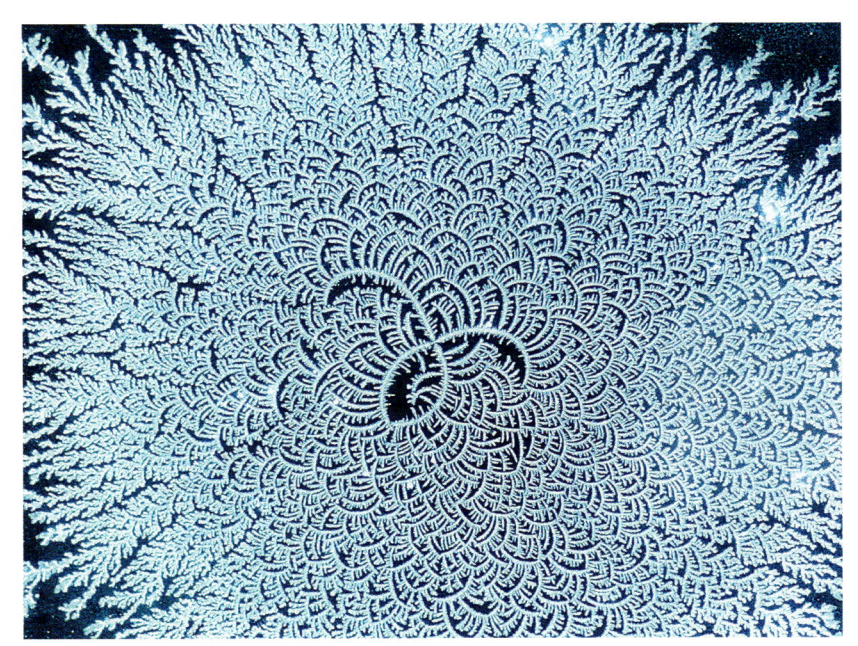

Image 212: Lavender branch, flowering (200x)

Image 213: Lavender branch, flowering (400x)

89

— THE SILENT LANGUAGE OF LIFE —

Image 214: Gall formations, oak (week 7) (100x)

Image 215: Gall formations, beech (week 7) (100x)

Image 216: Lavender branch, flowering (200x)

The two following images (images 214 and 215) show how an animal, the gall wasp, changes the actually linear spreading plant type from the center. It lays its eggs in the leaf blades of the plant and stimulates them to form a sphere. The egg and larva can develop inside this sphere. On the outside, the gall formations appear on the leaves. These are placed in water for three weeks.

Image 214: Gall formations of the oak in the water.
Image 215: Gall formations of the beech in the water.

In both images, these mobile, rhythmic, arching figures appear again, which change into linear-growing structures only in the periphery. Did the larva, which is developing inside, play a part in the design here?

In image 216 we look again at a drop center of flowering lavender to have a comparison to the following one. It happened one day that a small green caterpillar crossed a slide, depositing its liquid excrement on the glass. This quickly dried and was looked at in the microscope. The caterpillar was placed in the green.

Images 217 and 218 show caterpillar excrement!

Now these moving, rounding rhythmic formations can also be seen in the excrement of the larva, as in flowering lavender and in gall formations. Do the animal and the flowering plant give the same structural impulses to the water?

Let's go beyond the plant again into the surrounding area—to the butterflies and bees.

The butterfly has developed from a caterpillar. This caterpillar eats the green leaf and has metabolized the sunlight via the chemical process of photosynthesis, internalized it. The caterpillar is an animal—i.e., a being with the ability to move and with instincts, desires, sensation, and soulfulness. In addition to the physical body and etheric body (life body), animals and humans have an astral body, which can also be called the star body or soul body. Star and planetary forces come to internalization through it and form all organs as the basis of sensation and feeling.

When the plant begins to blossom, it touches the astral realm, whose formative forces penetrate the etheric sea of the earth. The forces of growth recede and instead multiform flower forms, scents and colors appear. The plant encounters the world of insects.

In the rhythmic, arching drop images of flowering lavender and caterpillar, we can recognize a structure, right up to the lemniscate, that no longer only grows in a peripherally striving manner, but creates small interior spaces. The formation of inner space is the prerequisite for the ability to feel and for the soul. In these structures, the sound ether could come to expression in its two-dimensional unfolding of forces on the surface.

The caterpillar now provided the stimulus to watch its pupation in the zoo. Since the hatching of the butterflies could also be observed next to the caterpillar display case, further questions arose about an experiment:

Plants, Fruits, Seeds, Barks, and Ashes in Water

Image 217: Caterpillar excrement (200x)

Image 218: Caterpillar excrement (400x)

When the larva has eaten enough, it retreats and anchors itself to a plant stem. The pupa forms inside. The outer caterpillar skin tears open and is shed as an old shell. The silkworms (e.g., *Bombyx mori*) spin a thread of "light" and wrap themselves completely in it in strong rhythmic movements. The caterpillar dies in the *cocoon*, completely wrapped in "spun light." It dissolves in this rather hard casing. Tranquility has set in. Soon, the outer surface of the cocoon already shows the future organs of the butterfly: wings, antennae, legs.... Finally, it slips out of its shell and at first hangs limp and watery in the heaviness. The wings unfold as a few drops of liquid fall to the ground: It is called "butterfly blood."

Here, these drops were collected in a filter bag at the zoo after a banana butterfly hatched, dried, and redissolved in distilled water from a small petri dish. (Animal and human substances are only added to distilled water. The mineral substances of the Volvic water are superfluous here, because the substrates bring their own substances.) The following images appear after dripping.

Images 219 and 220: "Butterfly blood," dripped on. What a difference to caterpillar excrement! There is nothing left of the caterpillar formation. Delicate, elongated lines with fine branches spread out on the drop surface, along with a small "eye" here and there, just like the wing markings of the banana butterfly! In comparison to caterpillar excrements, we experience a calm, straightforward, sparse design. But the butterfly is also an animal—with astrality?

What context can help us understand the completely new designs? The butterfly, "born of light," lives entirely in colors and forms in the light. The voracity of the caterpillar is alien to it; it hardly needs food. Its fluttering in light and air is entirely borne by the environment. The drop image of the "butterfly's blood" shows no centering. The delicately branched, black lines appear to float. They are without materiality (hence black)

Image 219: Butterfly blood (200x)

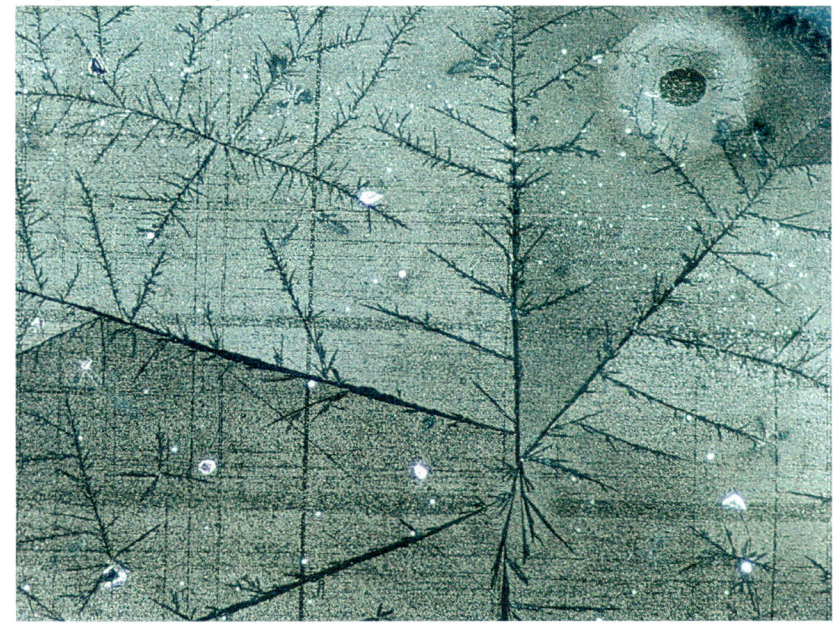

Image 220: Butterfly blood (400x)

— THE SILENT LANGUAGE OF LIFE —

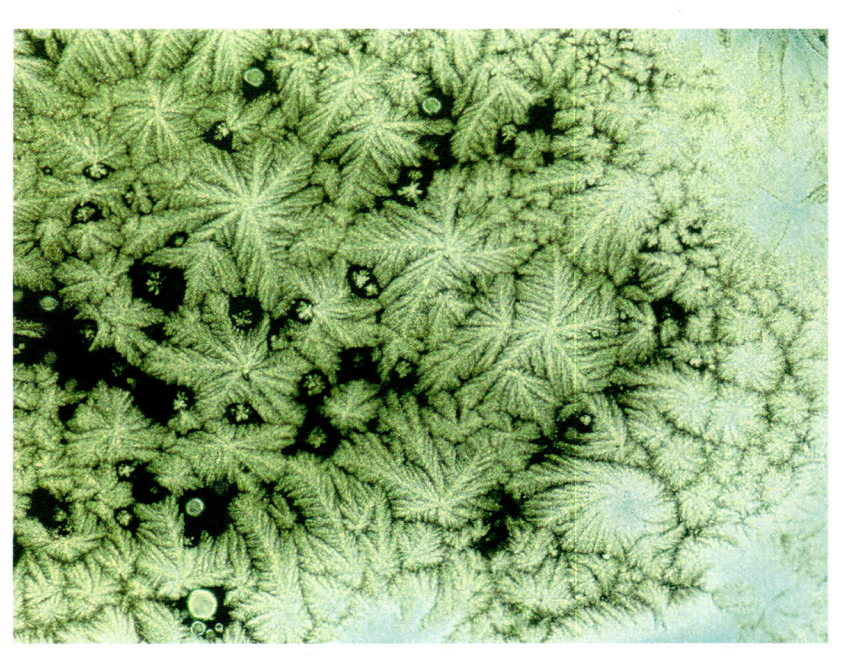

Image 221: Bee pollen (200x) (Ph)

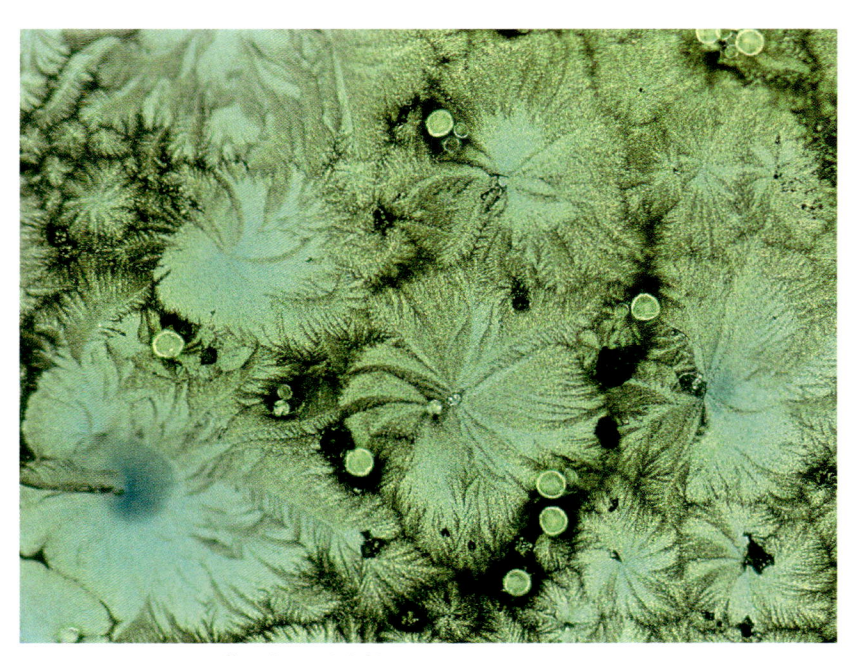

Image 222: Bee pollen (200x) (Ph)

and perhaps give us an image of the *light ether* in its linear, one-dimensional development of forces?

Rudolf Steiner speaks of the butterfly breathing through tubes that go into its interior.

> In the butterfly, the outer air with its light content spreads throughout the entire inner body...the butterfly is not only the animal of light outwardly, but it spreads the light carried by the air everywhere in its whole body, so that it is also light inwardly.[37]

This could be illustrated in the drop image. How differently does the caterpillar deal with the light in the leaves? It eats it, metabolizes it, and can spin a silken thread from it (e.g., *Bombyx mori*, mulberry silk moth).

Let's see what the bees tell us in the drop image.

One of the many tasks that have to be fulfilled in the beehive, with its constant temperature of 37°C, in a division of labor, is the collection of pollen. It is easy to observe how the entire body of the bee inside the flower is powdered with pollen and how the bee briskly cleans and strokes the dust adhering to its hairs with its proboscis so that it can be glued to the hind legs to form pollen. This pollen can be collected in a bee-friendly manner.

The pollen is placed in a test tube with distilled water. Fermentation and a strong sulphuric odor occur. After two to three weeks, the preparation water from two different preparations is dripped on.

The images 221 to 224 show us many small, centered "star and flower forms." These are extraordinarily animated. They are small swirls when the water has matured. These images were not taken with the dark field setting in the microscope (DF), but with the phase contrast setting (Ph).

If we take a closer look at the bee-being that lives in the warmth, these images touch us deeply. Above all, we are impressed by these micro-movements radiating from many small centers. Materiality is actually not to be seen! The phase contrast setting (Ph) gives an impressive image of this.

Image 225 shows the dark field setting (DF). Hardly any materiality is embedded in the dark vortices! Do thermal forces also lead to dematerialization here? (See chapter I, Bovist/Heat Drying.)

The question arises: Is perhaps every *pollen* able to bring such small "power wheels" to form as in the pollen of bees?

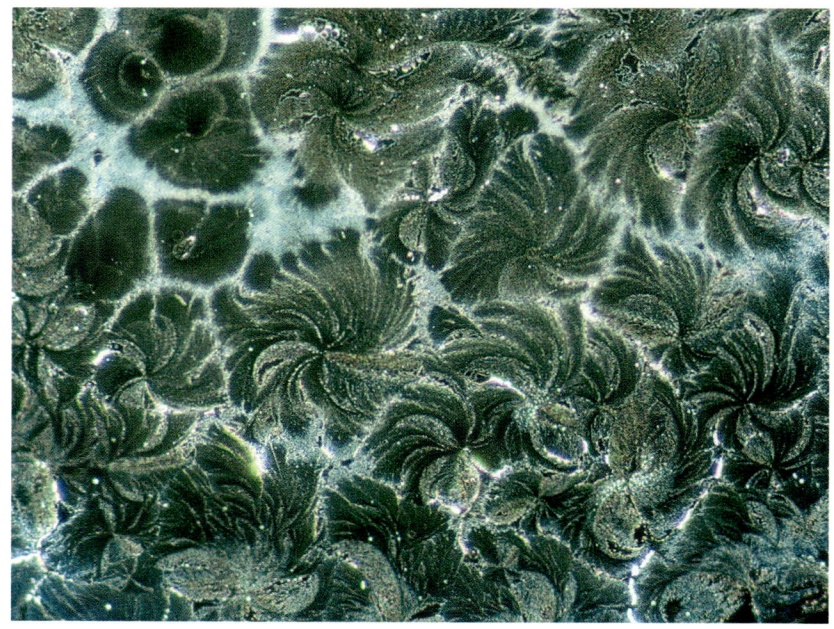

Image 225: Bee pollen (200x) (DF)

Plants, Fruits, Seeds, Barks, and Ashes in Water

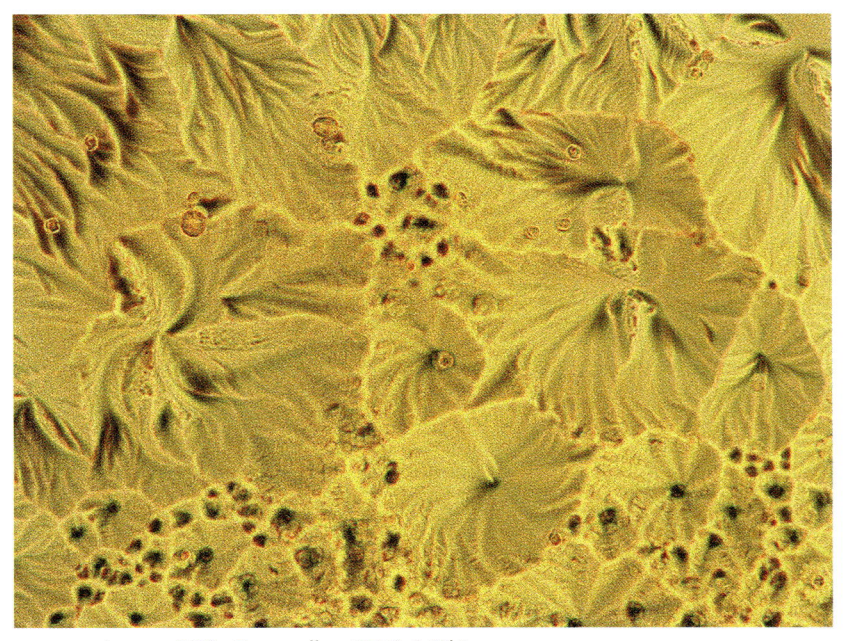

Image 223: Bee pollen (200x) (Ph)

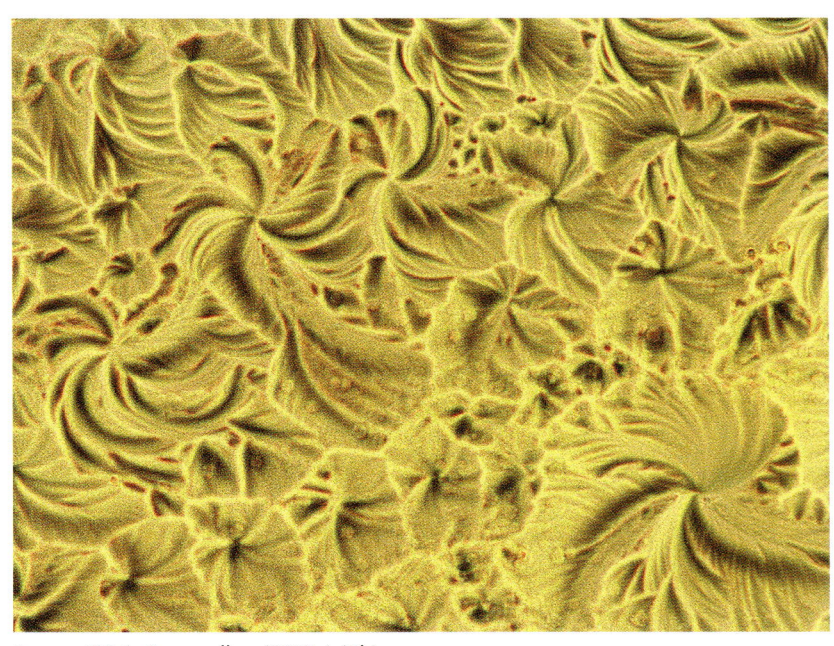

Image 224: Bee pollen (200x) (Ph)

Images 226 and 227 show a sunflower pollen preparation (digital camera). We see something quite different: linear, angled, playfully ordered black lines with rhythmic branching. This is how we must clearly distinguish the pollen from the pollen of the bees. So the bee has interwoven the pollen with its powers by sticking it together with nectar to form bee pollen and attaching it to its hind legs.

Could light forces (flower dust) and warmth forces (pollen) underlie the different imagery? What have we experienced here?

It is what happens in an intermediate realm. Animals and plants come into intimate contact—it is the realm where corresponding elemental beings reside.

Rudolf Steiner speaks of it being significant for the progress of evolution when "...a certain contact of the animal kingdom with the plant kingdom takes place...," and further:

> When the ox eats the grass, however, there is also a contact of the animal kingdom with the plant kingdom; but this is, so to speak, a prosaic, regular one, which lies entirely within the normal progress of evolution. On quite a different page of world evolution is that contact which takes place between the bee and the flower, because the bee and the flower are much further

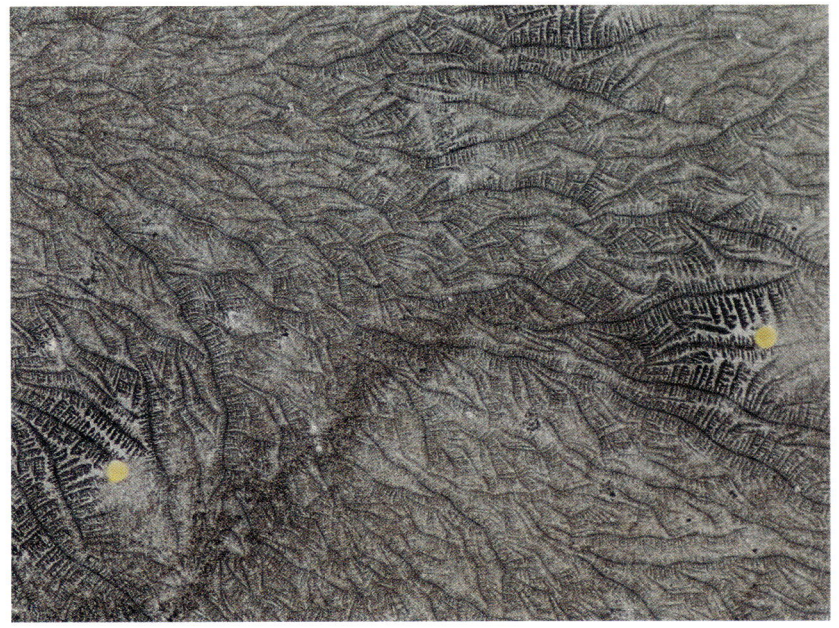

Image 226: Pollen (200x)

Image 227: Pollen (400x)

93

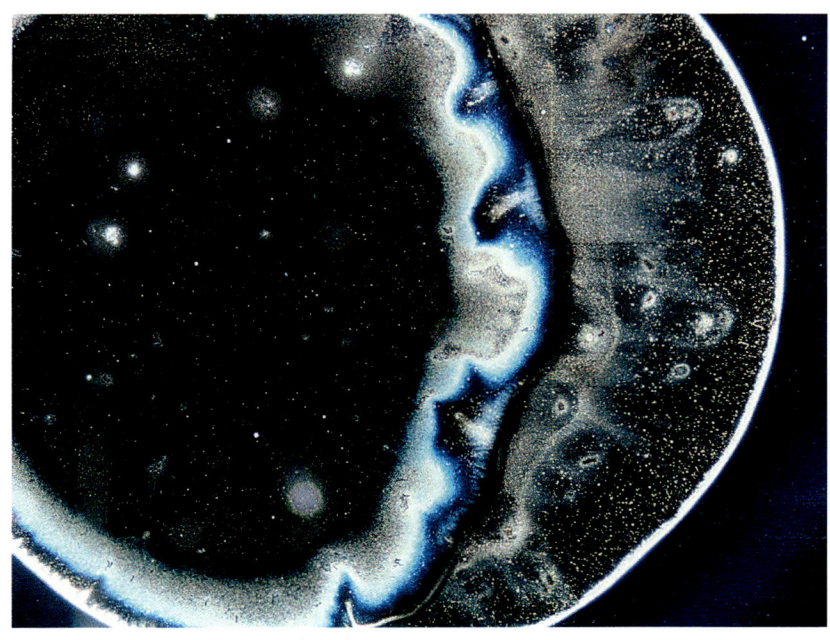

Image 228: Lavender blossoms (100x)

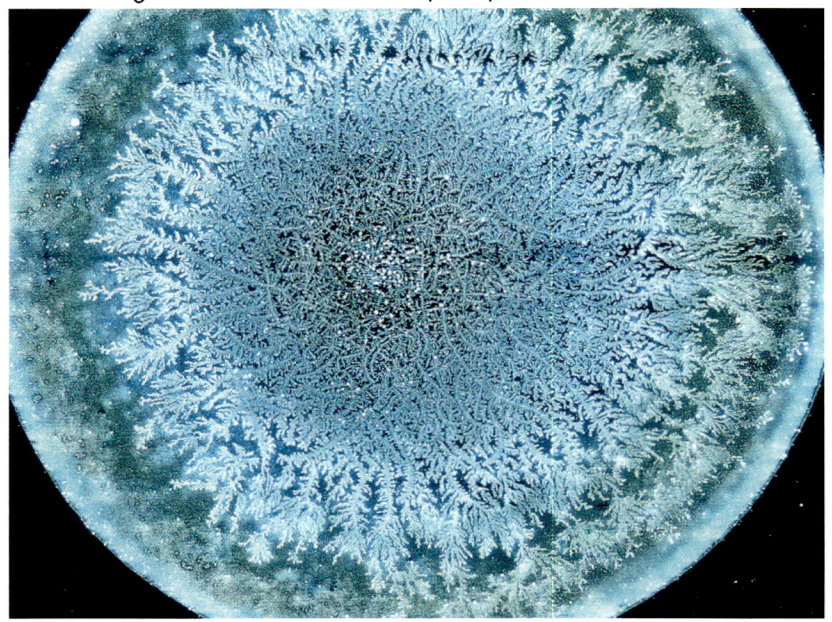

Image 229: Christ thorn blossoms (100x)

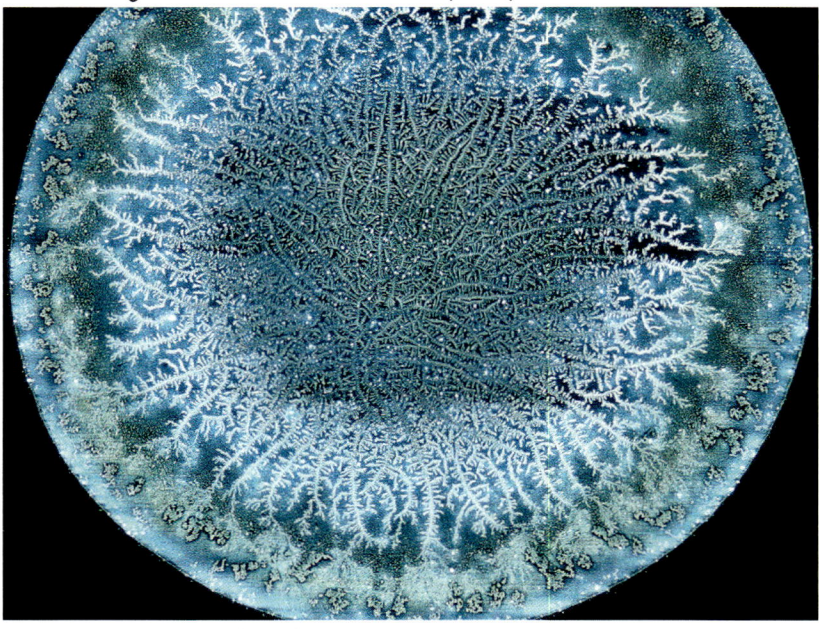

Image 230: Christ thorn blossoms (100x)

apart in organization and subsequently come together again, and because in the contact of bee and flower—though only for the occultist—*a quite marvelous power* is unfolded.[38] (emphasis I. J.-N.)

Thus we are given a small glimpse into the realm between blossom and bee in the drop image. The first impression is of powerful life radiating from many centers and limiting itself. Compared to "Butterfly Blood" with its delicate, floating lines that seem to come from nowhere and want to go nowhere, here in the bee-blossom encounter we are instantly centered and awakened to join in the magical whirling and spinning.

Then we saw how the flowering lavender branch calls the water to formations similar to the excrement of a caterpillar. Again, mysterious connections between flowering plant and animal are revealed, as well as in the images of the gall formations in the water. The plant-animal encounter is expressed in a variety of ways in the drop image design.

To "complete" the lavender plant, we also see a drop image of the lavender flowers themselves placed alone in the water (image 228). In the drop image there is no growing or rhythmic structure at all! What could be the reason for this?

If we follow Rudolf Steiner's explanations given to the doctors in Stuttgart on October 28, 1922, we do not find this surprising. He describes that the plant undergoes a process of devitalization in its development from root to flower.

> We have the strongest vital force in the root. And we have a gradual devitalization process from the bottom upward; and if we approach the petals, namely those petals *which contain strong essential oils*, then we also have the strongest devitalization process in such plants.[39] (emphasis I. J.-N.)

Thus, we cannot expect any living structures in the drop image of the lavender flowers. The fragrant blue flowers are of extraordinarily fine materiality, and the pure growth processes have reached an end in the flower. Lavender essence and lavender oil give us the aromatic end product of this devitalization process, so to speak.

Flowers of plants that do not contain essential oils show quite delicate structures in the water (images 229 and 230).

Plants, Fruits, Seeds, Barks, and Ashes in Water

2. Berries in Water, the Dark Core in the Center

Let us now follow the plants after they have flowered and see what the water can tell us here. From the abundance of fruits and seeds we will see a small selection and step by step in different experiments we will be able to experience the meaning of this drop formation.

Seeds ripen in the small berries of shrubs. A hawthorn bush delights us in spring with its splendor of blossoms. When autumn approaches, we see the bush covered with red berries. These are placed in water. Beginning around the third week, we look at their drop images.

In images 231 and 232 we see hawthorn berries (*Crataegus fructus*) in the water. A free structure, completely detached from its surroundings, has clustered in the center of the drop as a materiality. It appears as a "light structure" in the darkfield microscope. It is clearly distinguished from its surroundings by a delicate border. The edge of the drop shows a yellowish ocher.

If we hold the slide with the berry drops against the light, we see a dark core in the center of each drop; against a black background, the drop centers appear bright! This is an artifact of the darkfield microscope, which renders what is physically dark as brightness (see "Material and Method," page 19).

Image 233 shows Crataegus fructus in the brightfield setting.

If one disregards the optical conditions of the darkfield microscope, the often pronounced "luminosity" of the center can lead to erroneous interpretations. However, these interpretations were untenable, and only the uncovering of this "light glare" opened the way to a better understanding of what is clustered there in the center of the drop in the fruits and seeds: an individually shaped materiality interwoven by special forces, as will be seen in many center magnifications.

It is a very impressive, calm, and inward appearance, marked by the red berries of the hawthorn bush in the center of the water drop. We have not seen such a clear center before.

It turned out that all the specific berry variations chose the drop center for their design. It is also clear in this drop design that the carefully formed core area is in harmony with the drop *edge*. It is just as characteristically formed as the center itself. The free middle zone of the drop is remarkable. Nothing happens in this zone, but only at the center and at the perimeter—i.e., the core and the edge of the drop are visible.

Over the years, it was a joy to recognize again and again the preparations, because the berries and seeds of the plants showed their own unique formations. Through their fruits, one could now distinguish the various plants. This was not possible with the plant stems.

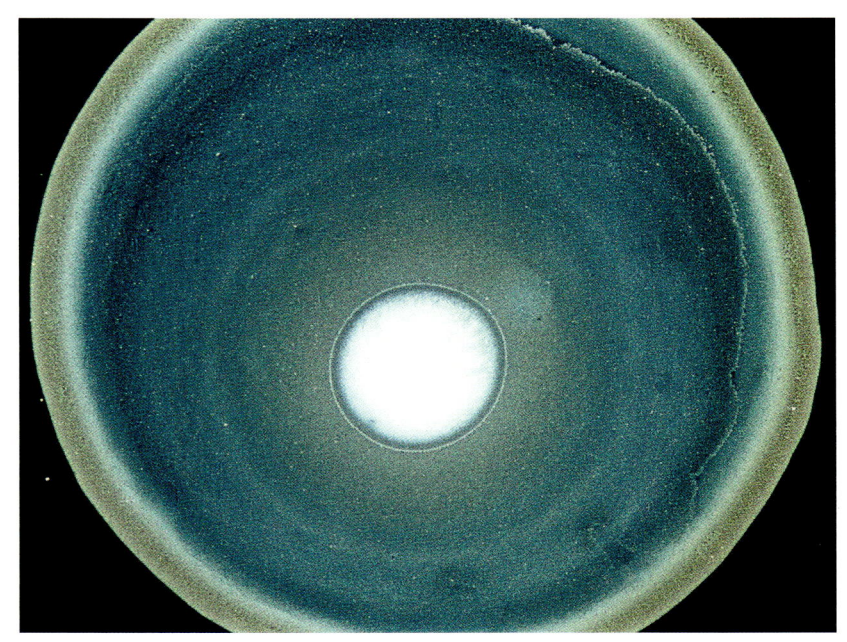

Image 231: Crataegus fructus (100x)

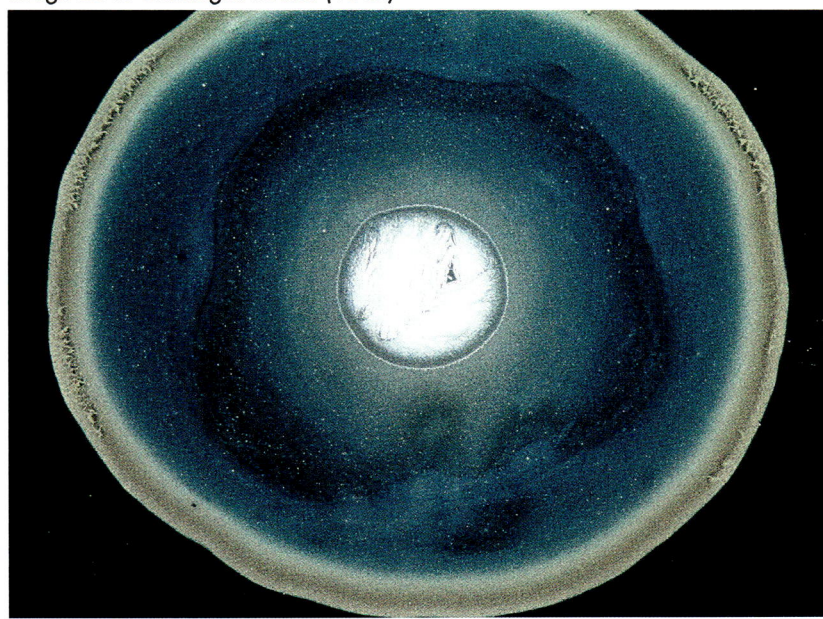

Image 232: Crataegus fructus (100x)

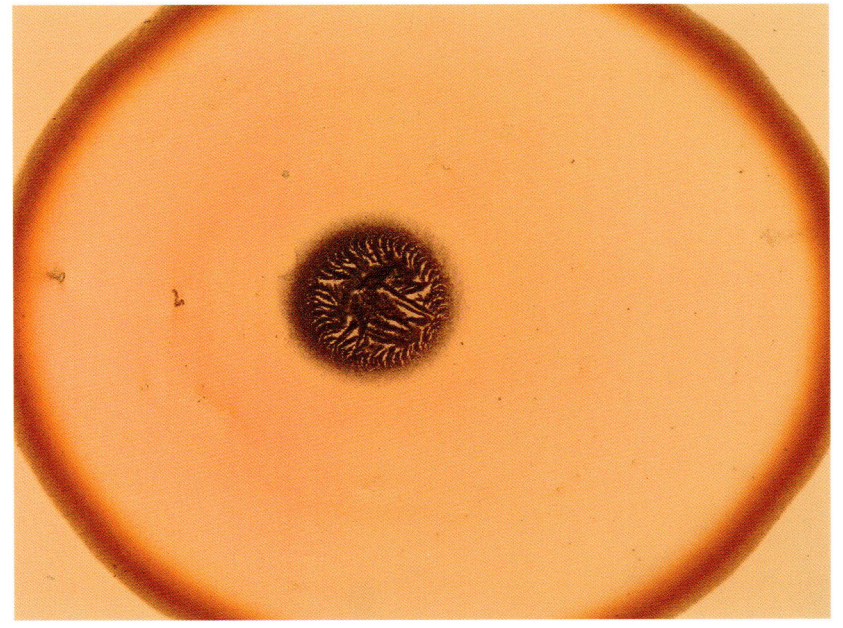

Image 233: Crataegus fructus (brightfield setting) (100x)

THE SILENT LANGUAGE OF LIFE

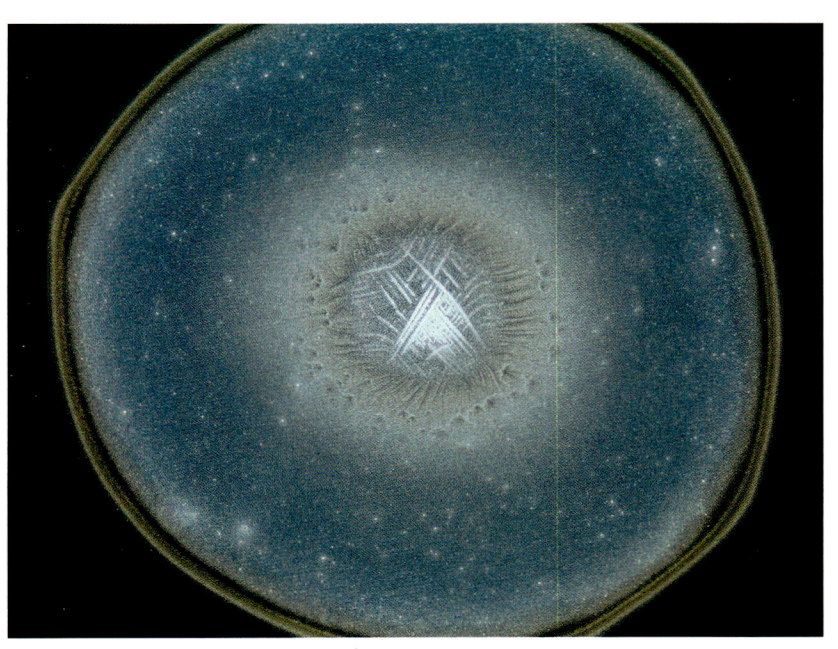

Image 234: Belladonna fructus (100x)

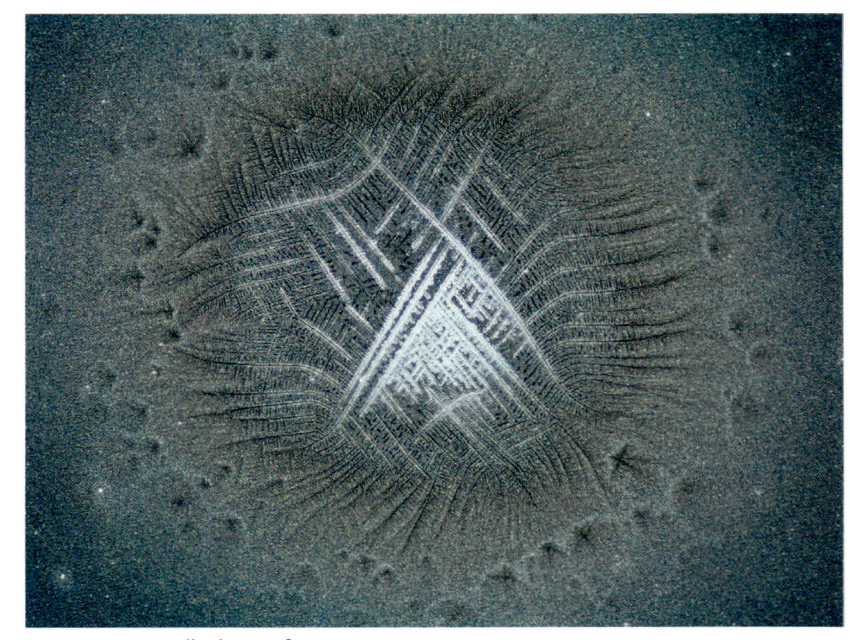

Image 235: Belladonna fructus (400x)

Images 234 to 238 show belladonna berries (Belladonna fructus) in water—the center magnifications allow us to see a delicate crystalline pattern all of its own (images 235, 237, and 238).

Images 239 to 242 also show Belladonna fructus, but in a different way.

When autumn was already drawing to a close, Belladonna berries were once again harvested from the bush and placed in water. Images 239 to 242 show us the late harvest.

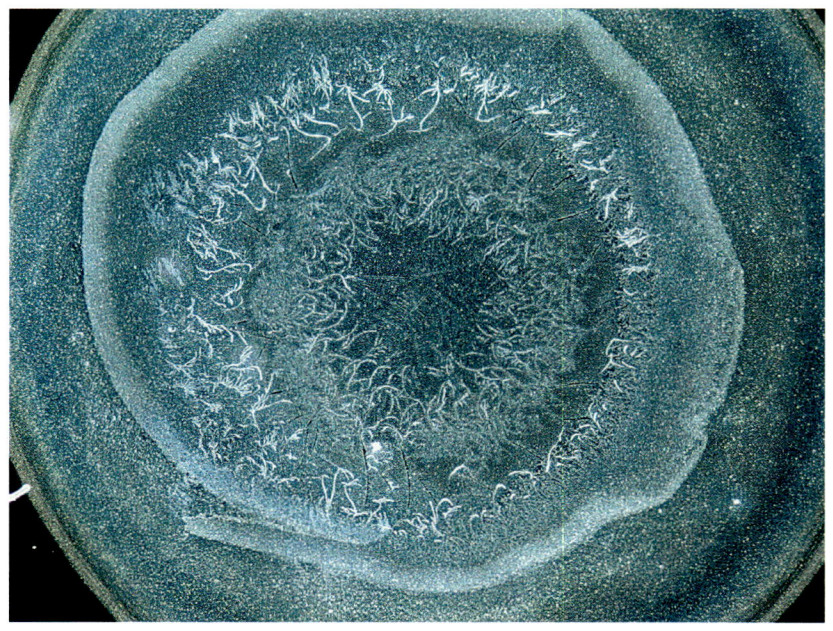

Image 239: Belladonna fructus, late harvest (100x)

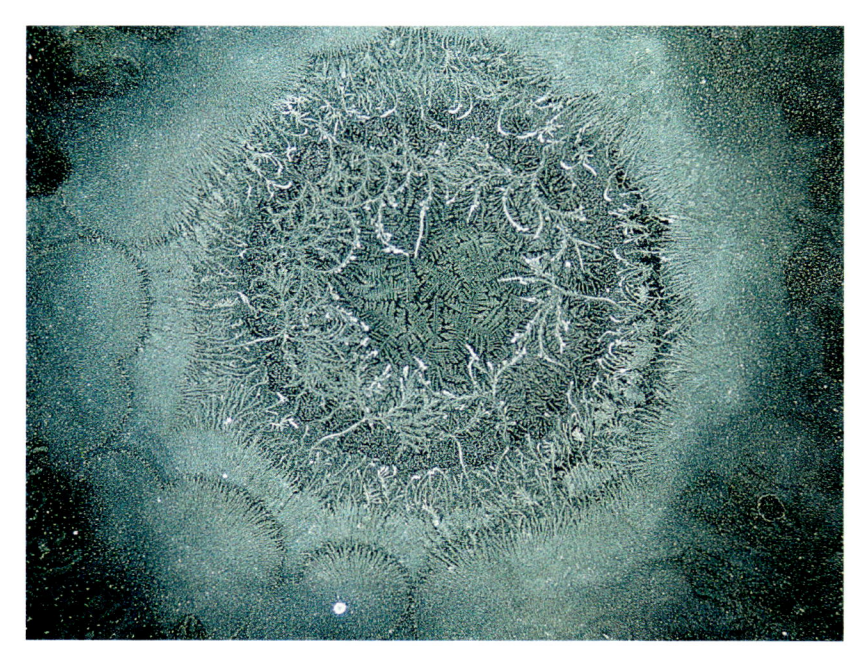

Image 240: Belladonna fructus, late harvest (200x)

Plants, Fruits, Seeds, Barks, and Ashes in Water

Image 236: Belladonna fructus (100x)

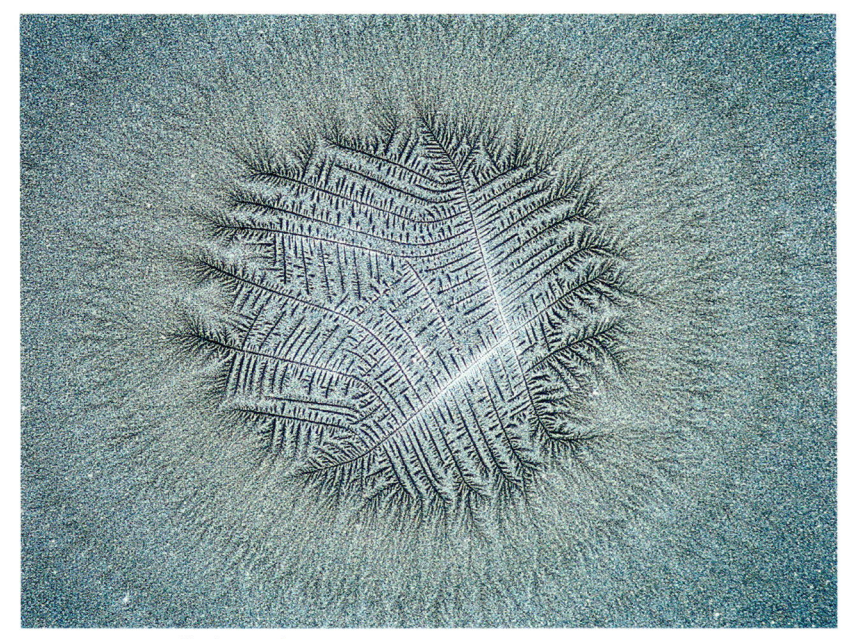

Image 237: Belladonna fructus (400x)

The advanced ripening of the berries on the bush changes the water. The images are now enlivened with loosened, playful drawings. This was also to be discovered in many other berry fruits.

The early harvest: mostly crystalline events in the center. The late harvest: imaginative vegetative structures leaving the center area. In the period in between, a more homogeneous core has very clearly gathered in the center, sometimes with a delicate border. We will return to its real secret later.

Image 238: Belladonna fructus (400x)

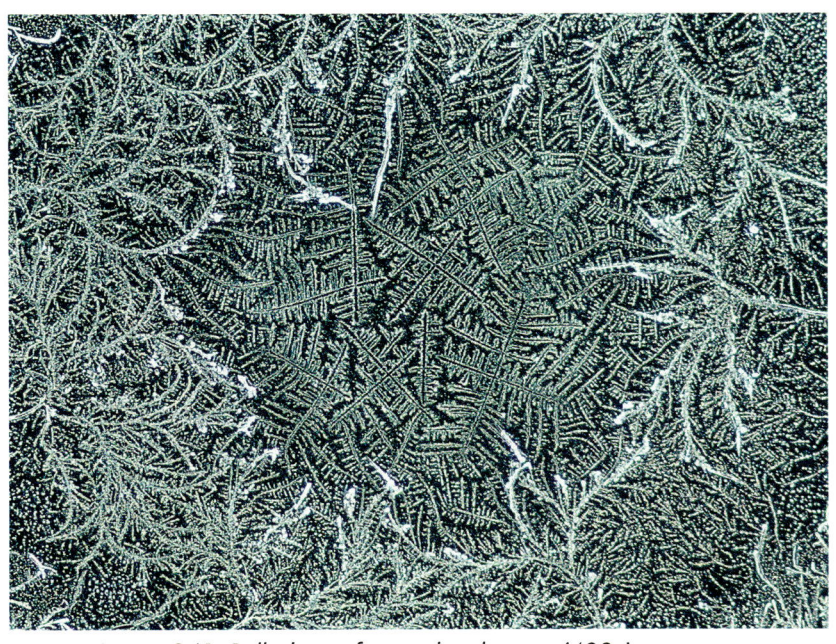

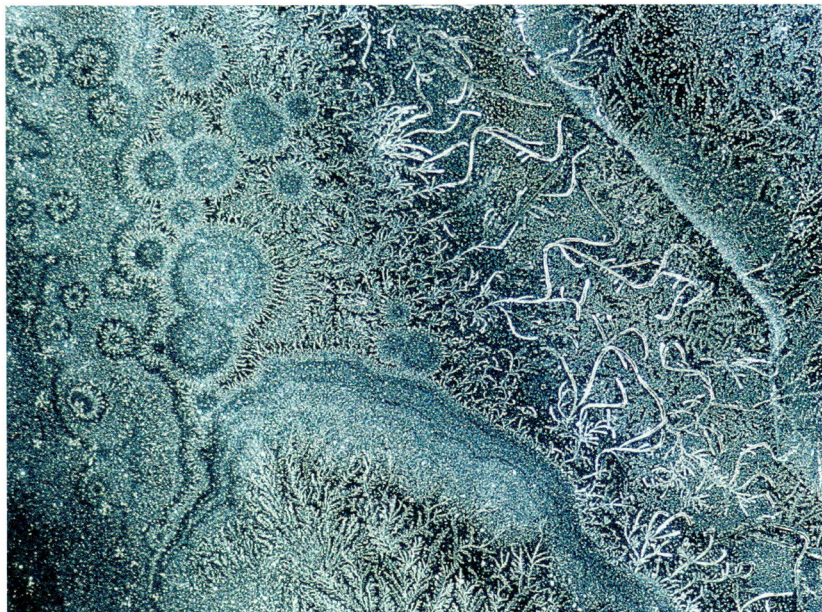

Image 241: Belladonna fructus, late harvest (400x)

Image 242: Belladonna fructus, late harvest (400x)

THE SILENT LANGUAGE OF LIFE

Image 243: Juniper berries with twig parts in water (100x)

Image 244: Juniper berries with twig parts in water (100x)

Image 245: Juniper berries with twig parts in water (100x)

Plants, Fruits, Seeds, Barks, and Ashes in Water

Images 243 and 244 (digital camera): Here we see juniper berries with chopped twig parts in water, dripped on after three weeks. The brightly appearing center is lattice-like and crystalline, surrounded by a ring of tufts of radiating needles.

Image 245 shows a drip of the preparation water another three weeks later. You can now see how the central area expands. The lattice structures "fade" (cf. image 244), as a certain dematerialization has occurred. The tufts of needles radiating in from the ring of the drop center field, which appears as a pointed blue, are now more pronounced. Substance transformation from crystalline to vegetative? "Maturation" also in the water of origin?

Images 246 and 247: Here, this preparation water was kept in a syringe (without objects) and dripped on again after ten weeks. The water drops now show a delicately animated inner space into which the playful designs grow from the outside inward to branch out finely in the center of the drop.

Thus, even without juniper berries and twigs, there still appears to be an active-vegetative creative force at work in the preparation water! The center cluster is no longer detectable.

Image 248: Cotonica berries in water—does the design change under the influence of heat?

Image 249: Cotonica berries, heat drying on a hot stone. There is hardly any change, the center formation remains.

The fruits are joined here by potato bovist (*Scleroderma aurantium*), a poisonous fungus that ripens its spores inside to preserve the species. It was placed in water and floated on the surface in the jar for ten days.

Image 246: Juniper berries with twig parts in the water, ten weeks in the syringe (100x)

Image 247: Juniper berries with twig parts in the water, ten weeks in the syringe (100x)

Image 248: Cotonica berries (100x)

Image 249: Cotonica berries, heat drying (100x)

THE SILENT LANGUAGE OF LIFE

Image 250: Bovist (100x)

Image 251: Bovist (100x)

Images 250 to 254 show the Bovist entrusting its playful, crystalline inner life to water. A clearly delineated drop center appears. We already encountered it in chapter I with the heat drying.

Image 252: Bovist (200x)

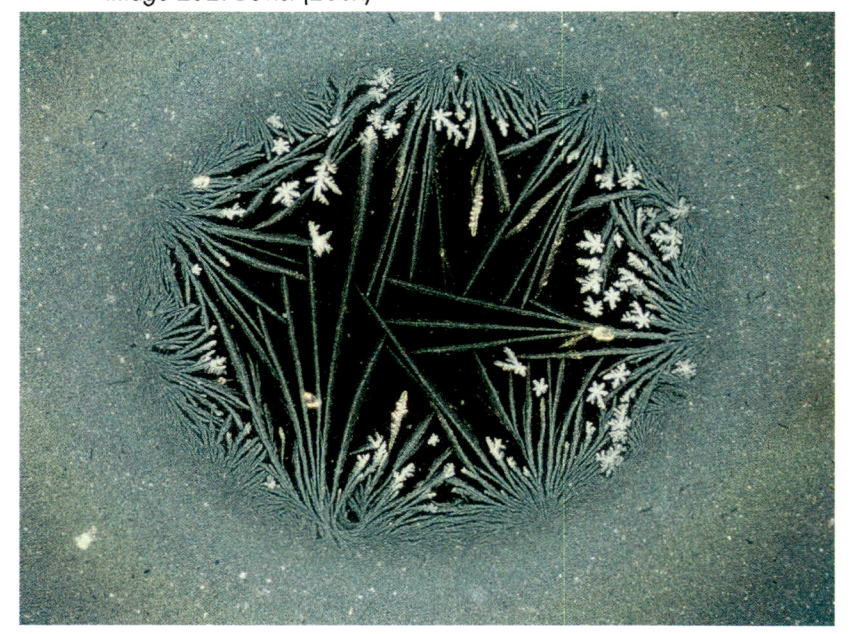

Image 253: Bovist (200x)

Image 254: Bovist (400x)

Plants, Fruits, Seeds, Barks, and Ashes in Water

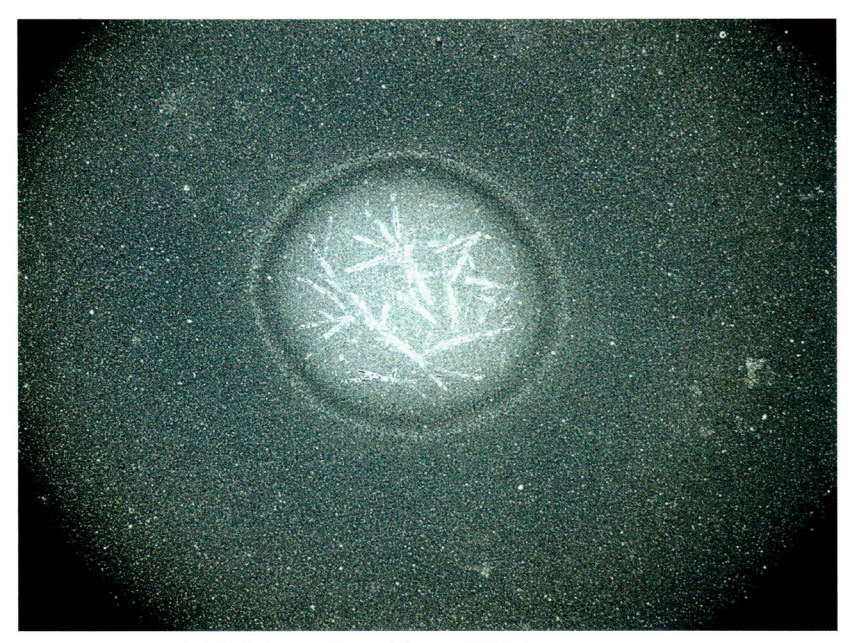

Image 255: Solomon's seal fruct. (100x)

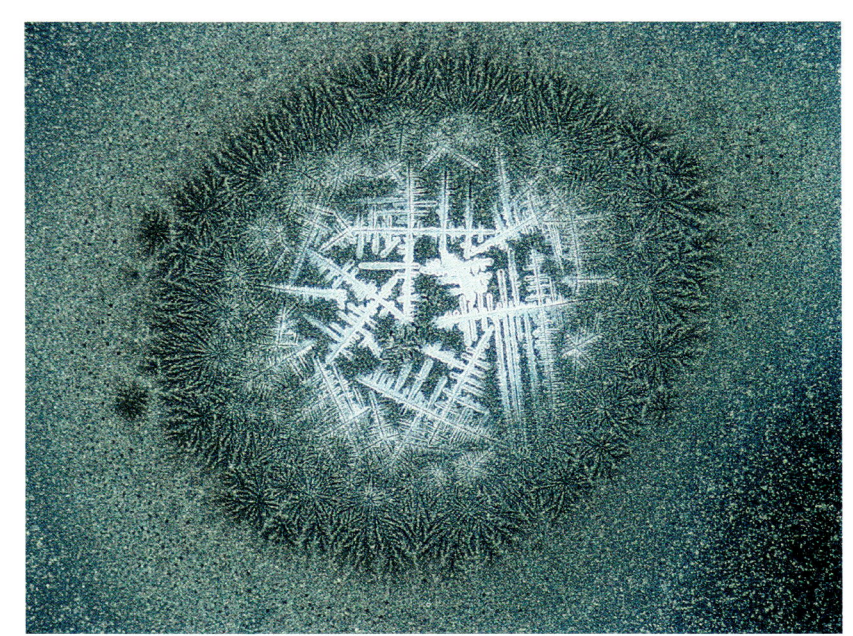

Image 256: Solomon's seal fruct. (400x)

Images 255 and 256 speak of the dark blue berries of Solomon's seal.

In images 257 to 259 (digital camera) we see the drop image of red and black berries of the woolly snowball (*Viburnum lantana*). In all the images "woolly threads" appear. The centers are enchanting and characteristically designed.

Image 257: Viburnum lantana fructus (Wooly snowball) (40x)

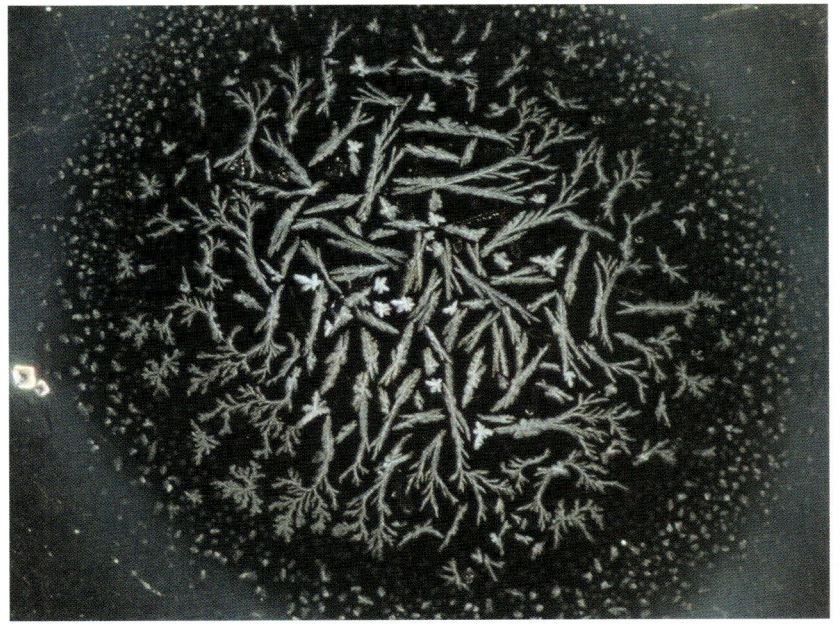

Image 258: Viburnum lantana fructus (400x)

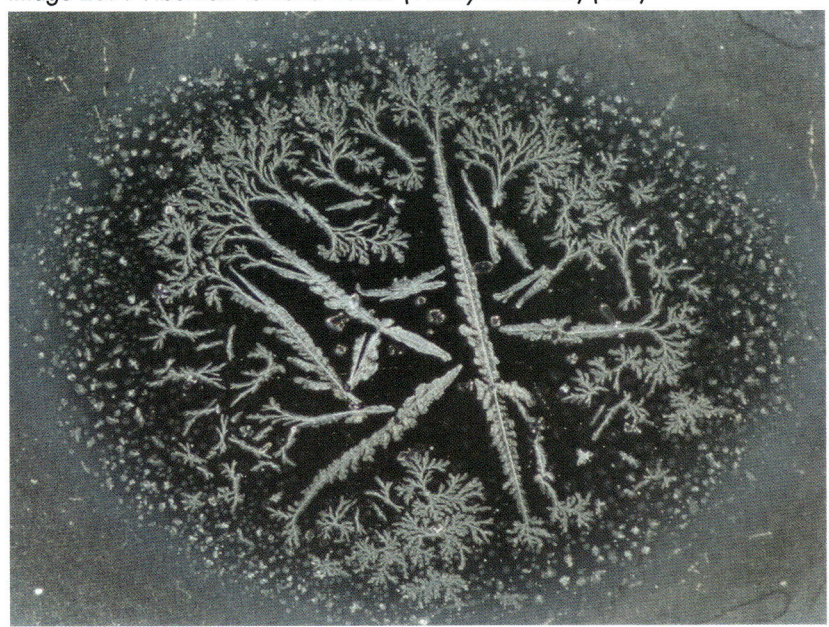

Image 259: Viburnum lantana fructus (400x)

101

THE SILENT LANGUAGE OF LIFE

Image 260: Mistletoe berries (100x)

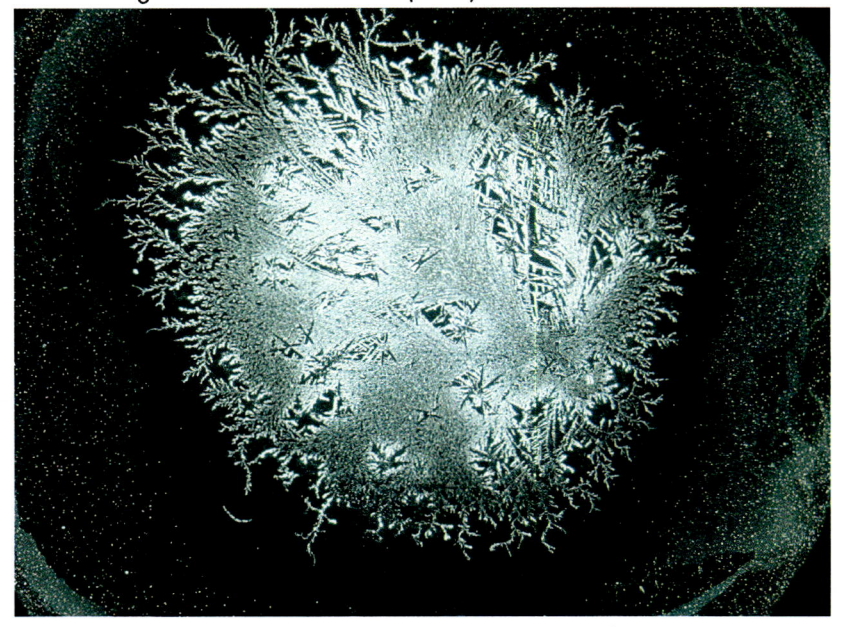

Image 261: Mistletoe berries (preparation diluted 1:3) (100x)

In the following images (260 to 264) we now see plant berries of a different character. These images demand a deeper insight into the plant being itself or into its fruit. This will be dealt with in a modest way.

Image 260 shows mistletoe berries in water.

Inside the mistletoe berries is the flat, triangular seed—called embryo—which lives in the slimy berry substance. Two (there can also be three or four) green horns peep out from the two upper corners. They are the seedlings that later attach themselves to the bark of the tree by means of an adhesive disc. The *sinker* then grows out of the center of this adhesive disc and bores into the cambium of the tree bark.

From Rudolf Steiner we learn a great deal about the nature of the mistletoe plant. It is actually not an earth plant (no root, no metamorphosis into flower...), but has survived from the old moon time of our earth evolution. Its relation to the cosmos is already shown by the spherical shape itself, which we discover high up in the trees. Its breeding ground is not the earth, but the cambium of the tree bark. Mistletoe is thus a parasite. What should be emphasized is its amazing mobility already during the development and anchoring of the seedling as well as its extraordinary vitality.

This is mentioned briefly to show that a deeper look at the objects from nature can give the drop images a background for understanding.

In the mistletoe drop image (260) we do not see a delimited space located in the center. What we see is a huge, spreading, almost sprawling event that stops at the edge of the drop. We think of the small green horns of the flat berry core, which already reveal something of the germinating power.

Image 261: Here the mistletoe berry water has been diluted 1:3. It now contracts more into the interior and reveals its berry character.

Images 262 to 264 show Physalis fructus, here also an uncommonly expanded central area. If we look at the Physalis cherries, we do not see them, for they are enveloped in a reddish-orange husk. It envelops the fruit with a layer of air. They are put into the water with this cover. Filigree, rhythmically curved clusters surround a cluster of playfully arranged "sun figures" with a crystalline halo.

Let us return once more to the typical berry images and their free nucleation in the center to get closer to their mysterious formation.

Plants, Fruits, Seeds, Barks, and Ashes in Water

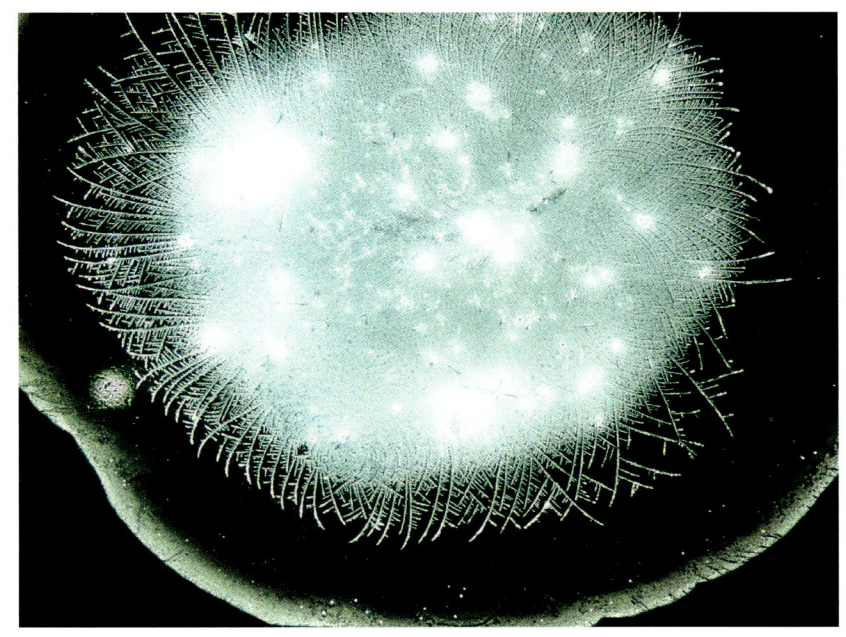

Image 262: Physalis fruct. (100x)

Image 263: Physalis fruct. (200x)

Image 264: Physalis fruct., center enlarged (400x)

THE SILENT LANGUAGE OF LIFE

Image 265: Barberry fruct. (100x)

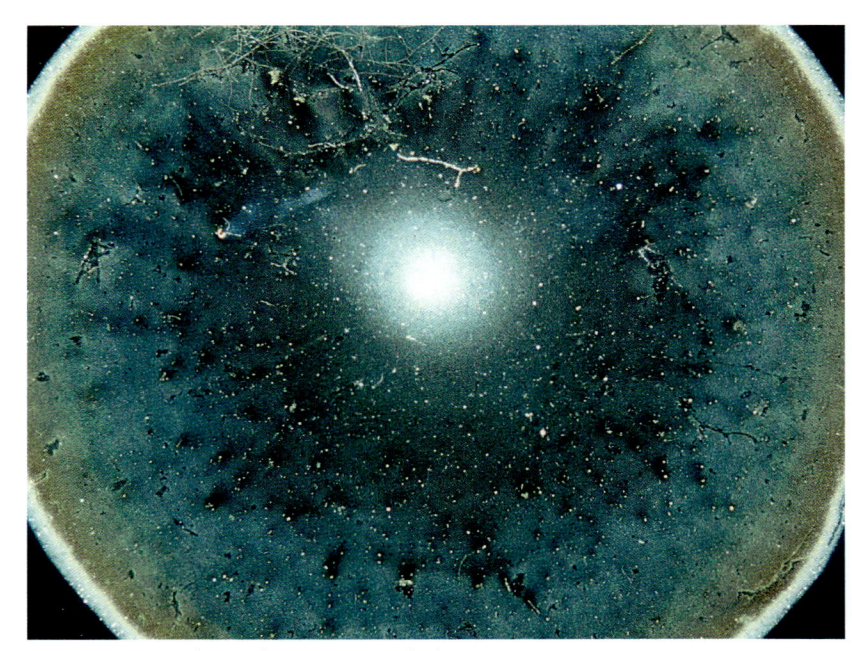

Image 266: Barberry fruct. (1 year dark) (100x)

Image 265: Small red barberry berries lie in a glass of water. After four weeks, the water is dripped on. We see the nucleation in the center and a border typical for these berries.

Now, one day, it was necessary to tidy up the laboratory table. The preparation jars urgently needed to be emptied and their contents returned to their natural conditions. In two black film cannisters, we found barberry berries and hawthorn berries (*Crataegus fructus*). They had been lying in dark water for more than a year.

At the last moment before disposal, it occurred to me to check this water again. There was a surprise that touched me deeply at the same time. The center, as shown in image 265 (barberry water), was still intact (image 266). So the barberry fruits had been in conversation with the water for over a year in the darkness of the water!

Does this batch water change when it comes back into the light? This was tested in the following experiments. We see a restructuring of the core area in barberry berries (I) and hawthorn berries (*Crataegus fructus*) (II).

I. Barberry berries (images 266 to 272):

Image 266: Barberry berries, one year in a dark film cannister in water. This and the following images show the drops from the preparation water from which the berries had been removed! This is also the case with Crataegus fructus, one year in the dark.

Image 267: Barberry fruct. (1 year dark, 2 hours light) (100x)

Image 268: Barberry fruct. (1 year dark, 8 days light) (100x)

Plants, Fruits, Seeds, Barks, and Ashes in Water

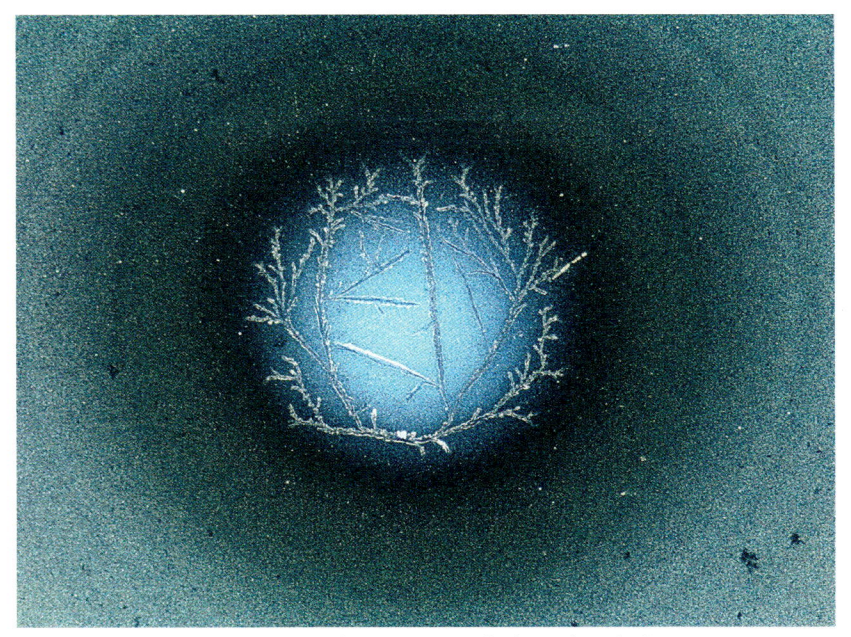

Image 269: Barberry fruct. (1 year dark, 8 days light) (200x)

Image 270: Barberry fruct. (1 year dark, 8 days light) (400x)

Image 267 shows a drop after the water (without berries) has been exposed to light for two hours.

Images 268 to 270 show a restructuring of the center after eight days in the light.

Image 271: After more than eight days, it is still clearly marked next to concentric rings spreading on the drop surface.

Image 272: After eight weeks in the light, the central designs change into harder structures and the circles have become more "physical." Significant is the drop border, which in barberry always appears white on the outside and orange-ocher toward the inside. This border is constant in all the images.

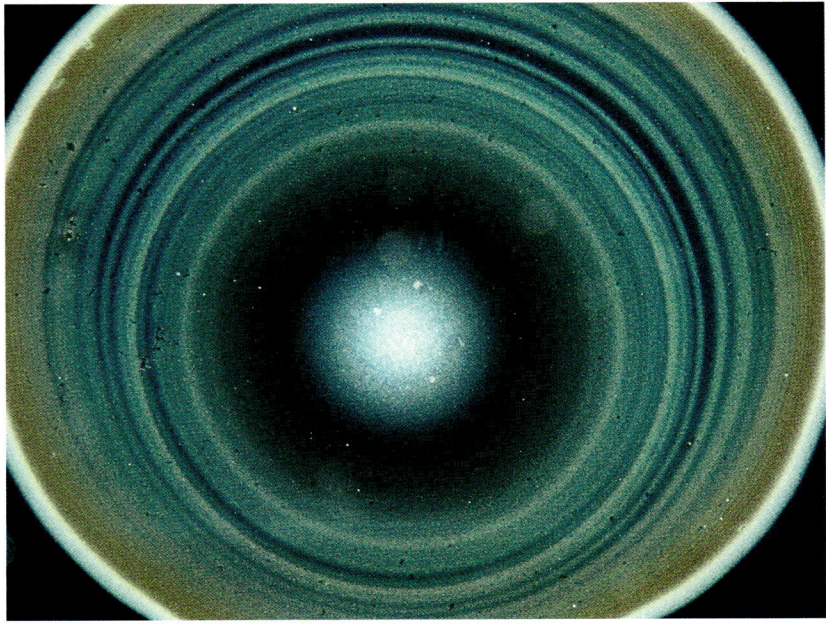

Image 271: Barberry fruct. (1 year dark, over 8 days light) (100x)

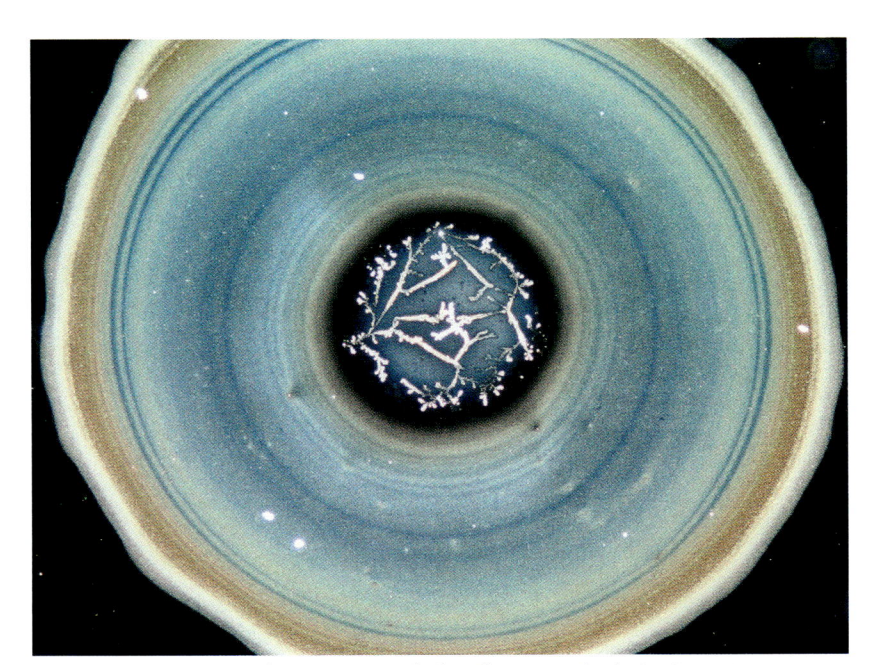

Image 272: Barberry fruct. (1 year dark, after 8 weeks light) (100x)

THE SILENT LANGUAGE OF LIFE

II. Crataegus fructus, 1 year in the dark (images 273 to 278):

Image 273: After one year in the dark, the dripped water still forms a center with a clear border.

Images 274 and 275: We see two different drops after four days in the light. In the enlargement, the delicate center border and a small row of rays appear again.

Images 276 and 277 show the further development after ten days in the light.

Image 278: After five weeks, the center has now turned black — that is, the materiality has "retreated" except for individual lines. The conversation in the water has almost fallen silent because the "interlocutor" was missing. The etheric forces in the water are dying away and can no longer gather and shape the substance in the center. Remarkable is also here the drop border in orange-ocher tinging, which is typical for Crataegus berries. It remains completely intact.

In a particularly impressive way, it is revealed here that water is capable of preserving its imprints over a long period of time.

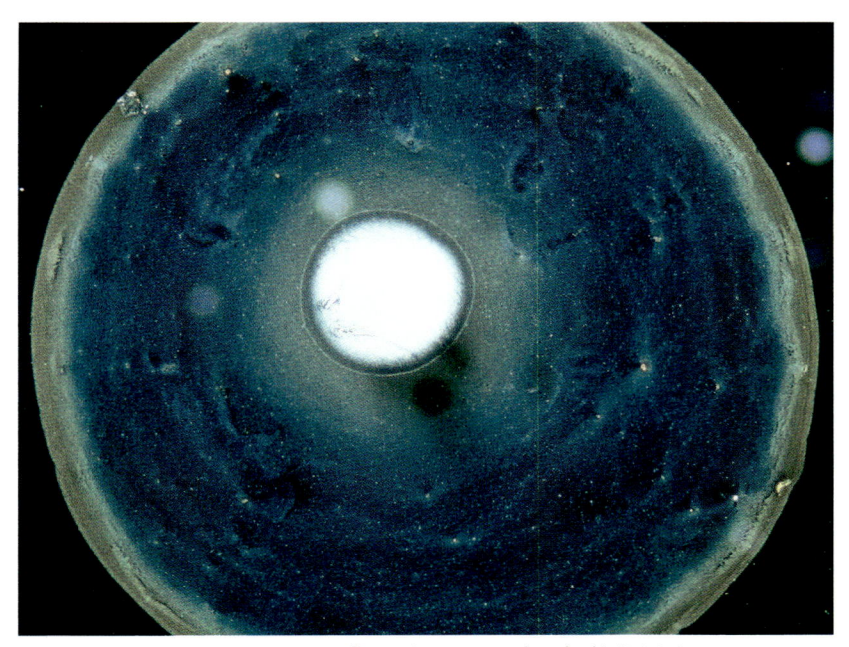

Image 273: Crataegus fruct. (1 year in the dark) (100x)

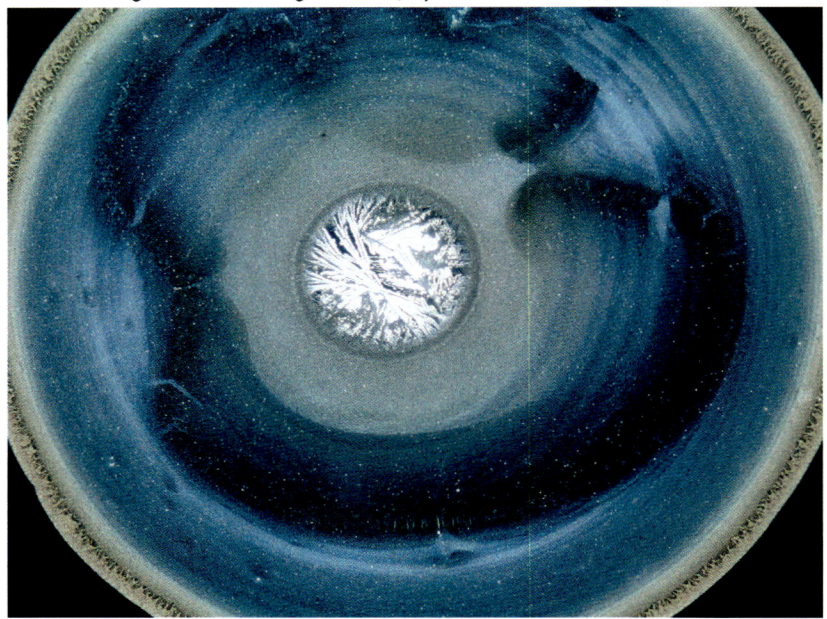

Image 274: Crataegus fruct. (1 year in the dark, 4 days light) (100x)

Image 275: Crataegus fruct. (1 year in the dark, 4 days light) (200x)

Plants, Fruits, Seeds, Barks, and Ashes in Water

Image 276: Crataegus fruct. (1 year in the dark, 10 days light) (100x)

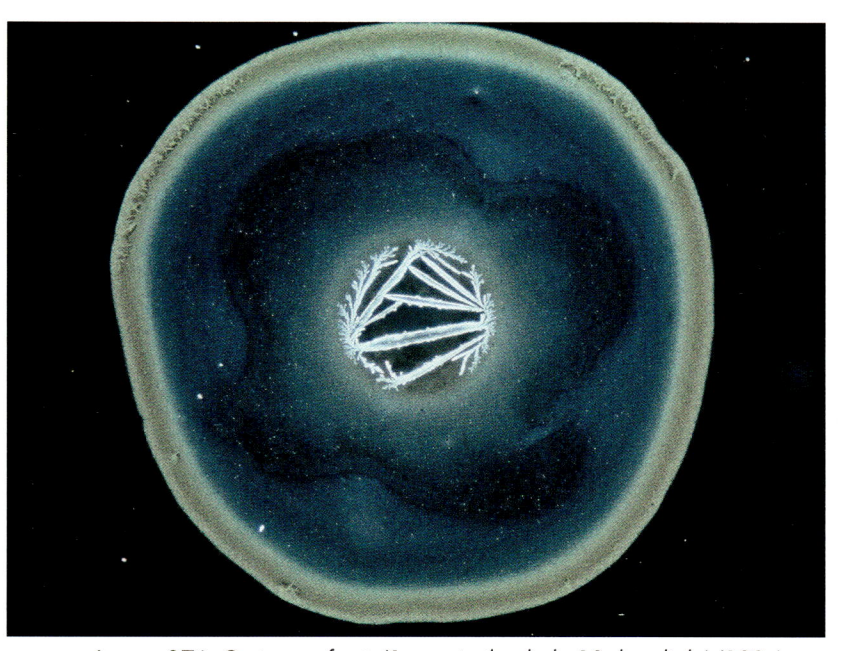

Image 277: Crataegus fruct. (1 year in the dark, 10 days light) (200x)

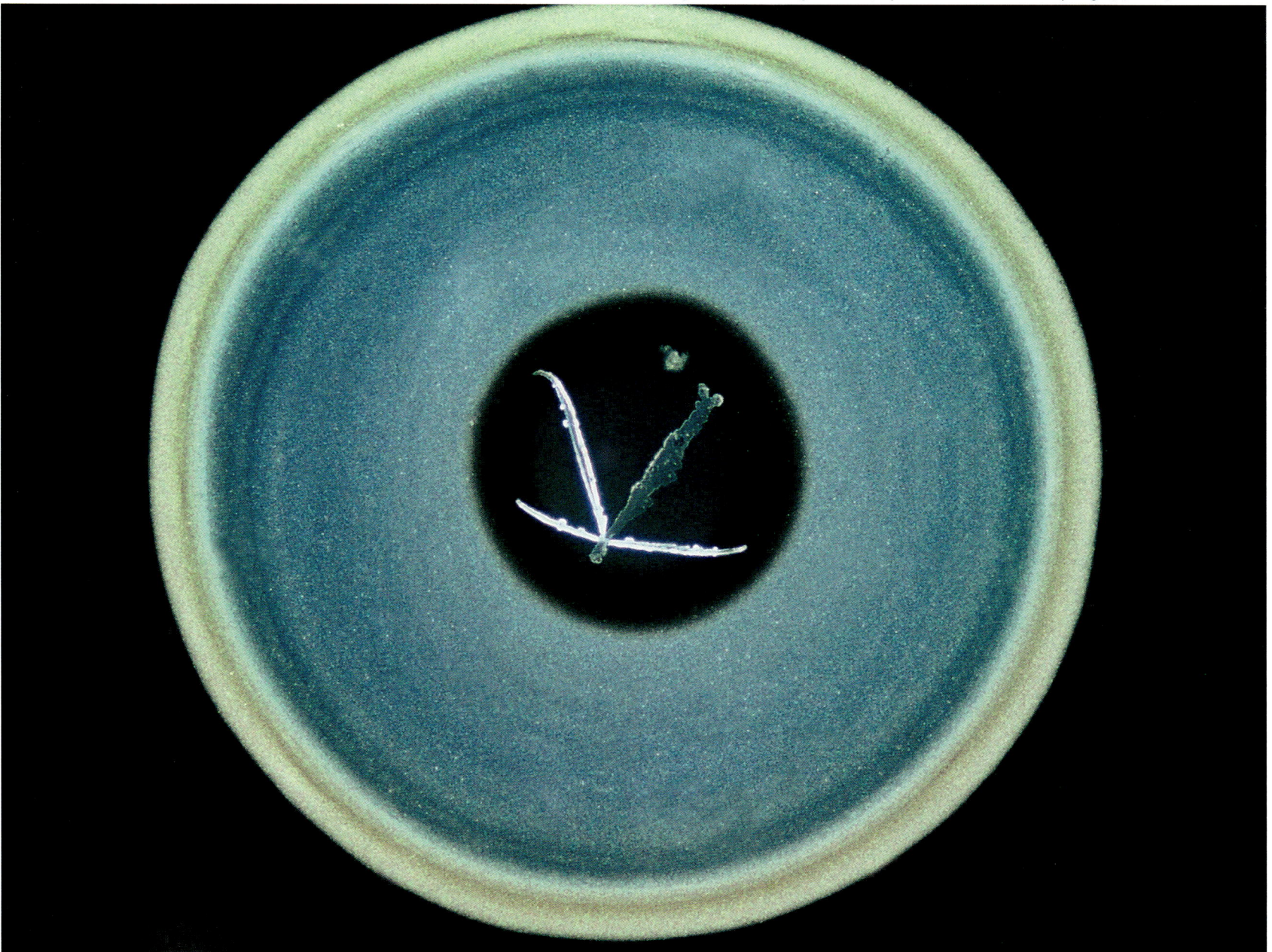

Image 278: Crataegus fruct. (1 year in the dark, after 5 weeks of light) (100x)

Now we want to see whether the environment in which plants grow can influence the thriving of their fruits.

Lily-of-the-valley berries (*Convallaria fructus*) are harvested at three different locations and placed separately in water for three weeks. One time the red berries came from the Black Forest (*Convallaria I*), the other time from the forest of Stuttgart Frauenkopf (*Convallaria II*). In this forest, on the hill, there is a tower with numerous electromagnetic beams. (It is not the television tower—this is 1 km away as the crow flies.) Lily-of-the-valley berries were growing at a distance of about 300 meters from this tower. Convallaria III shows lily-of-the-valley berries at a distance of 1 km from the radiation source.

Convallaria I, lily-of-the-valley berries from the Black Forest (images 279 to 285).

Image 279 shows Convallaria fructus, the early harvest, in its crystalline center formation.

Image 280 as well as all following images of the ripened berries show a rather large center figure with delicate vivification inside. This becomes clearer in the magnifications. The delicate vegetative ramifications spread in two directions from a central nucleus. They border each other to form small fields, which in turn respectfully stop in front of each other.

In image 285 we see a drying process.

The marginal zone remains featureless in lily-of-the-valley berries; sometimes delicate crystalline figures appear above the basic structure.

Image 279: Convallaria fruct., Black Forest, early harvest (100x)

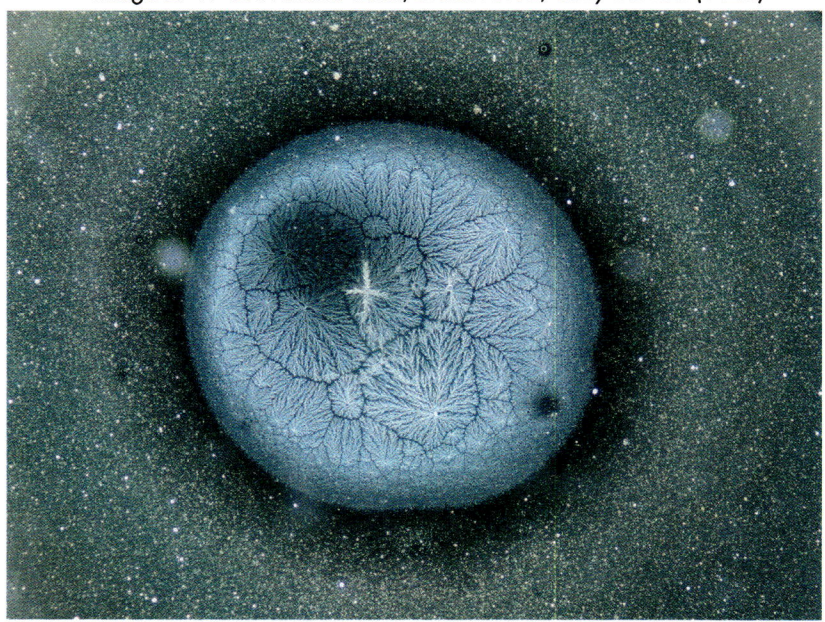

Image 280: Convallaria fruct., Black Forest (100x)

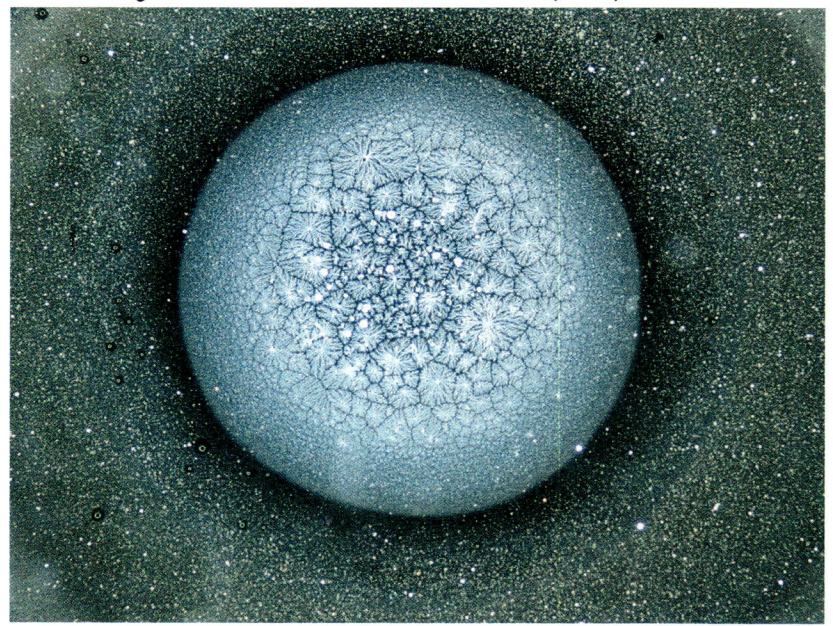

Image 281: Convallaria fruct., Black Forest (100x)

Image 282: Convallaria fruct., Black Forest (100x)

Plants, Fruits, Seeds, Barks, and Ashes in Water

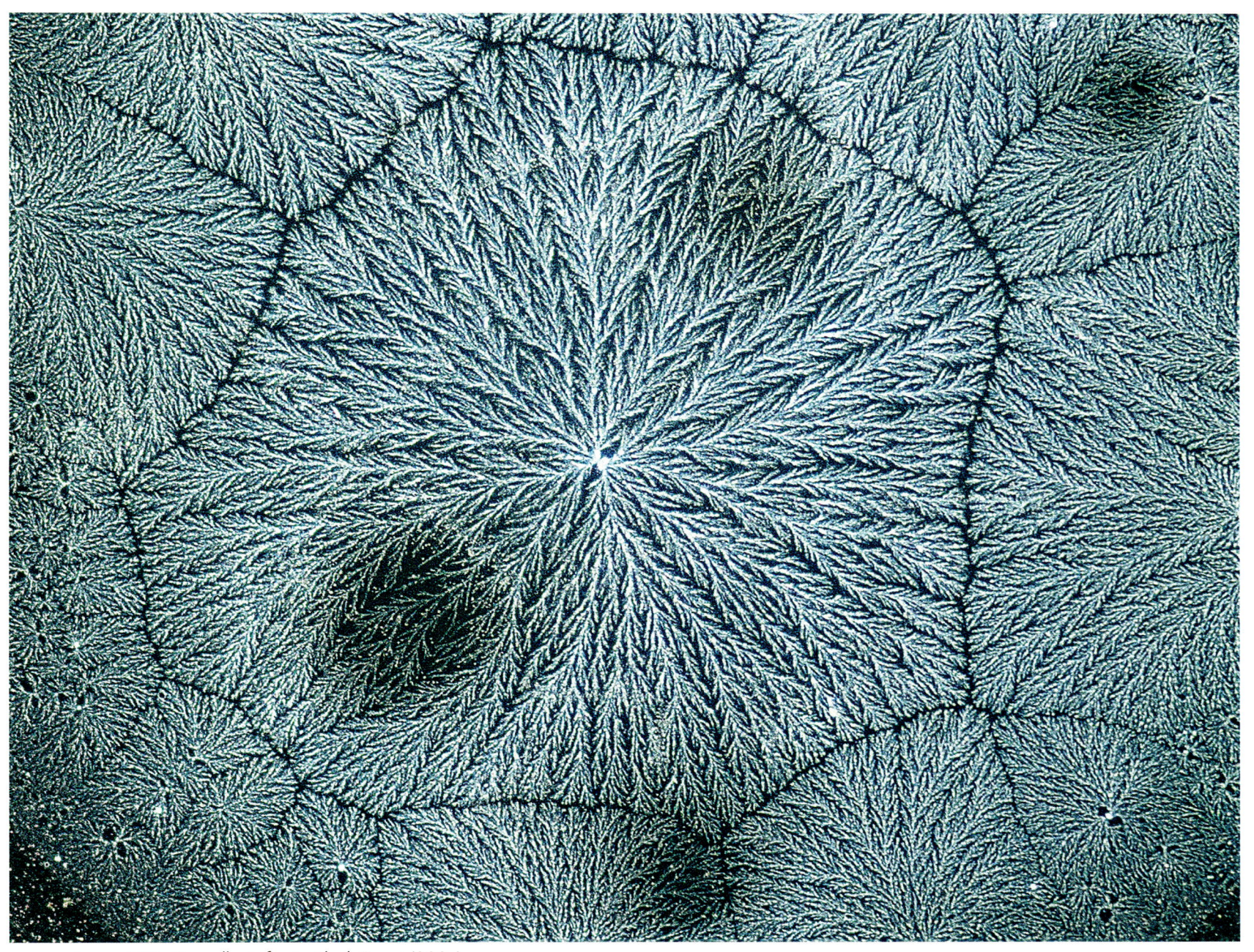

Image 283: Convallaria fruct., Black Forest (200x)

Image 284: Convallaria fruct., Black Forest (200x)

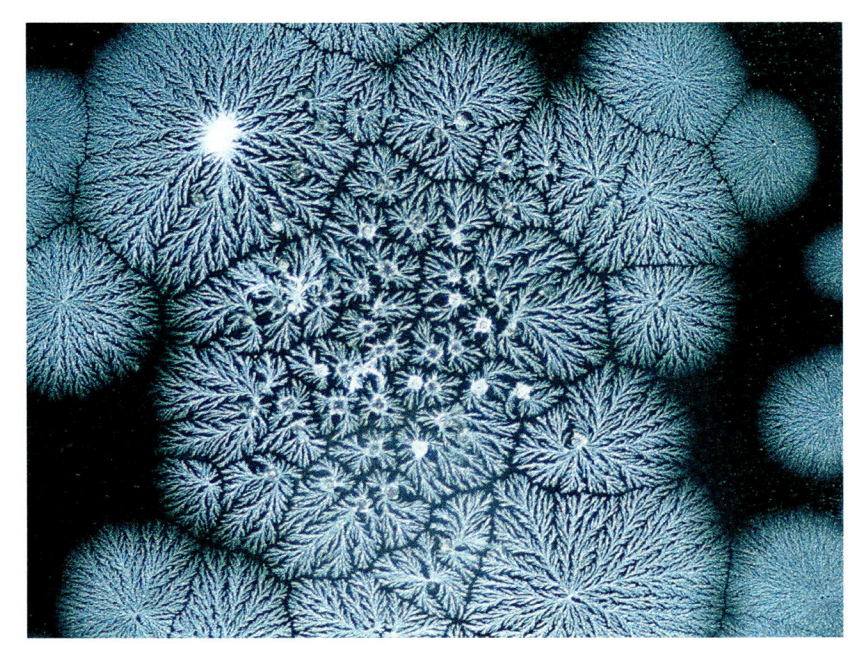

Image 285: Convallaria fruct., Black Forest, drying process (200x)

THE SILENT LANGUAGE OF LIFE

Image 286: Convallaria fruct., Frauenkopf (100x)

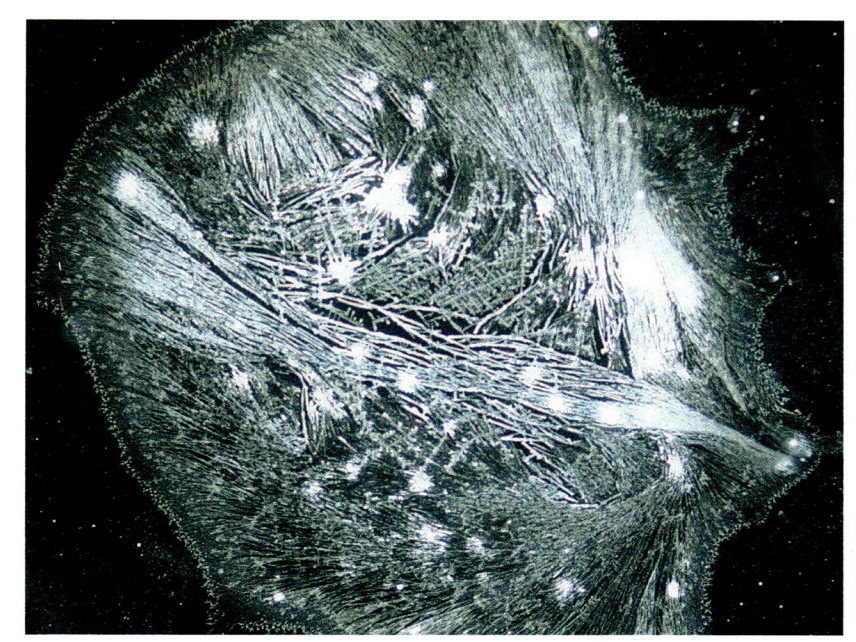

Image 287: Convallaria fruct., Frauenkopf (100x)

Convallaria II, lily-of-the-valley berries from Stuttgart Frauenkopf, 300 m distance from the radiation source (images 286 to 290).

The delicate center designs are lost. A giant event expands explosively over the entire drop surface. We see in the chaotic, torn structures how "the silent language of life" is destroyed.

The images only show the 100x magnification, as there are no finer designs.

How do you think the center of the lily-of-the-valley berries are shaped at a greater distance from the radiation source?

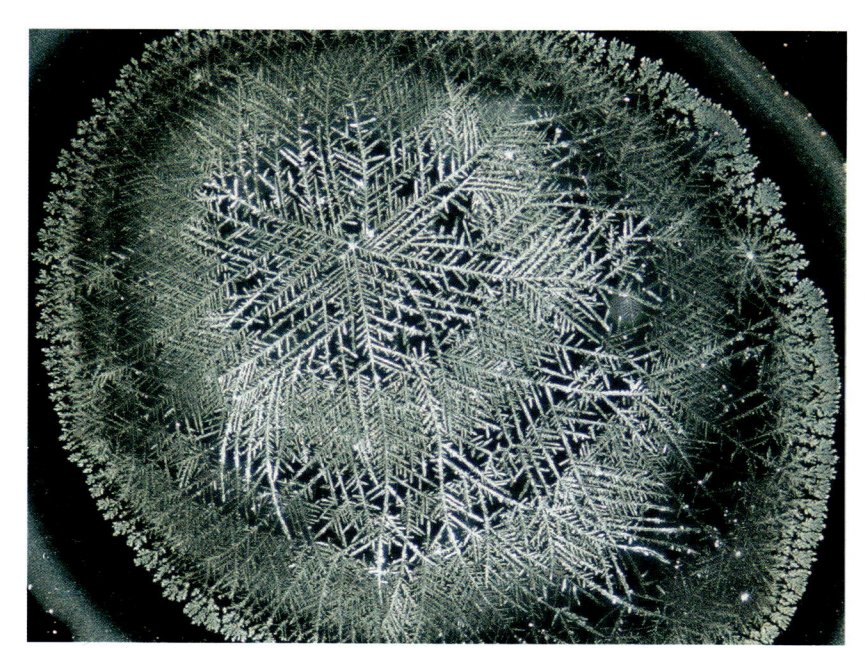

Image 288: Convallaria fruct., Frauenkopf (100x)

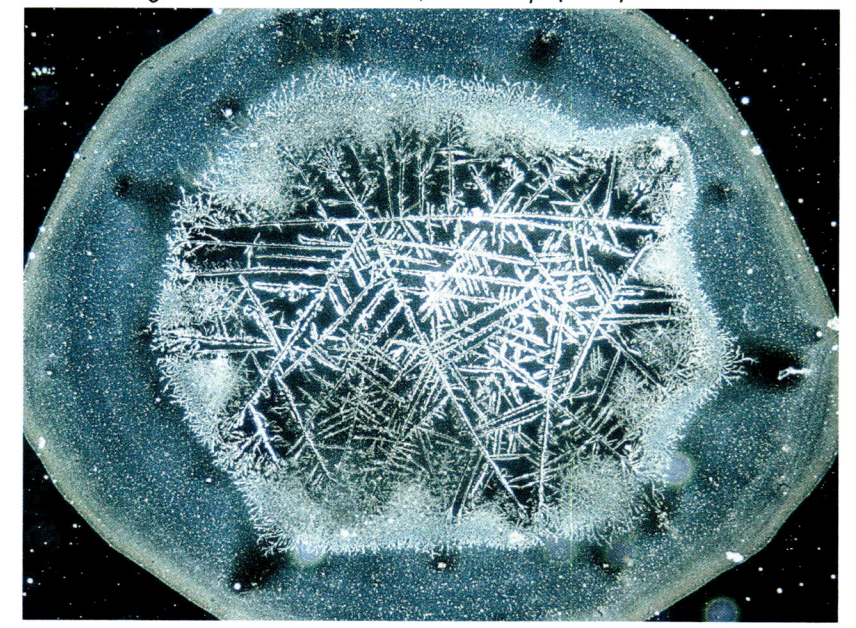

Image 289: Convallaria fruct., Frauenkopf (100x)

Image 290: Convallaria fruct., Frauenkopf (100x)

Plants, Fruits, Seeds, Barks, and Ashes in Water

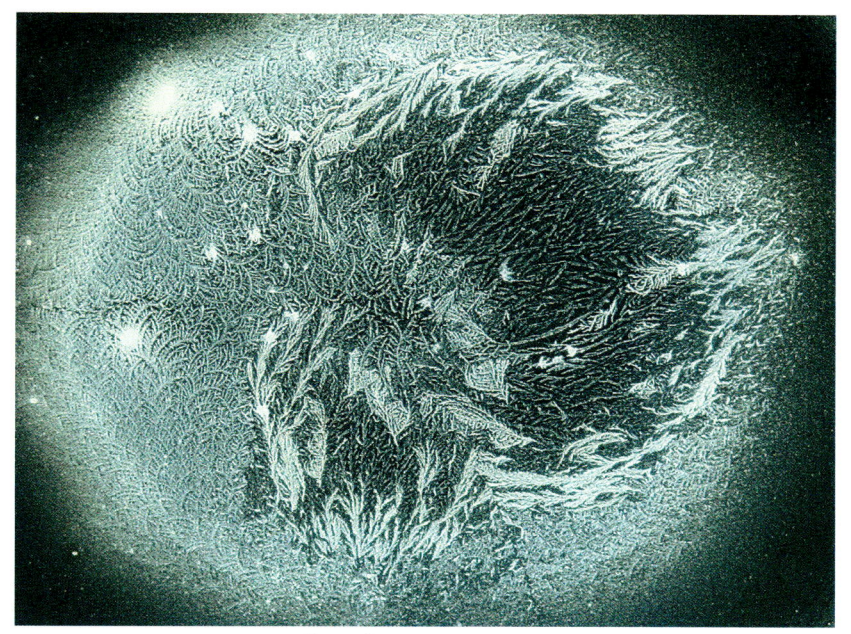

Image 291: Convallaria fruct., Frauenkopf 13 (100x)

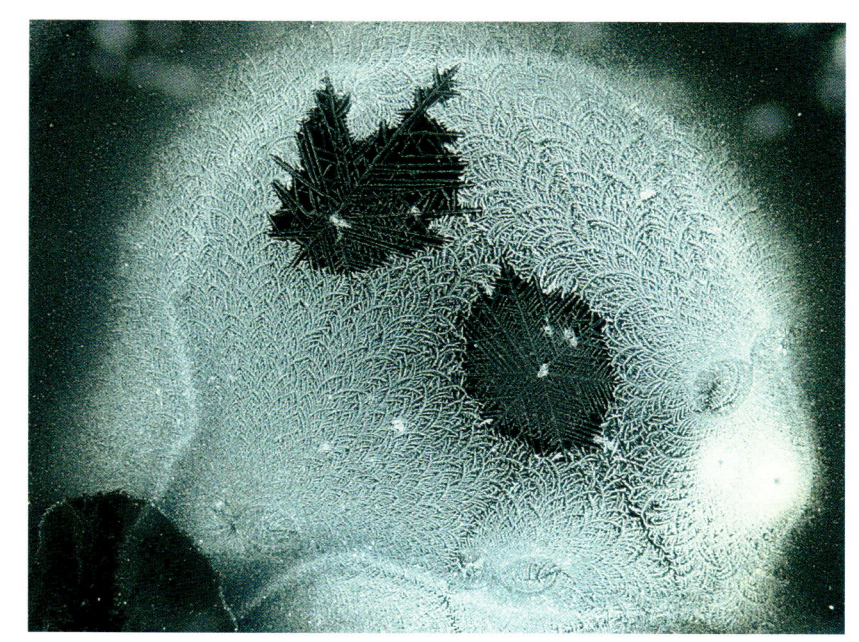

Image 292: Convallaria fruct., Frauenkopf 13 (100x)

Convallaria III, lily-of-the-valley berries from Stuttgart Frauenkopf, 1 km distance from the radiation source (images 291 to 295). They are labeled here as "Frauenkopf 13."

Images 291 to 295 show a mixture of disorderly and harmonious-rhythmic designs. What appears harmonious here, however, does not correspond to the typical lily-of-the-valley berry image!

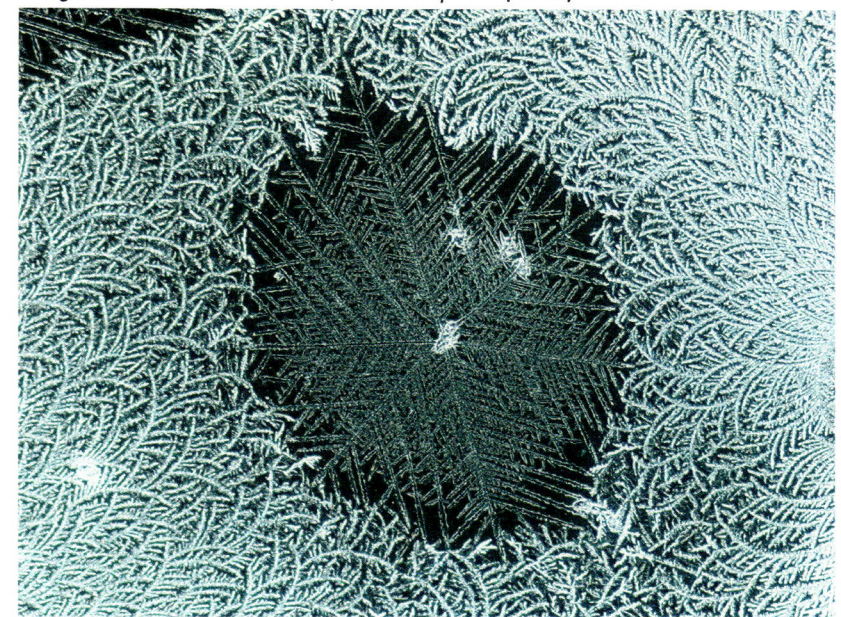

Image 293: Convallaria fruct., Frauenkopf 13 (400x)

Image 294: Convallaria fruct., Frauenkopf 13 (400x)

Image 295: Convallaria fruct., Frauenkopf 13 (400x)

— THE SILENT LANGUAGE OF LIFE —

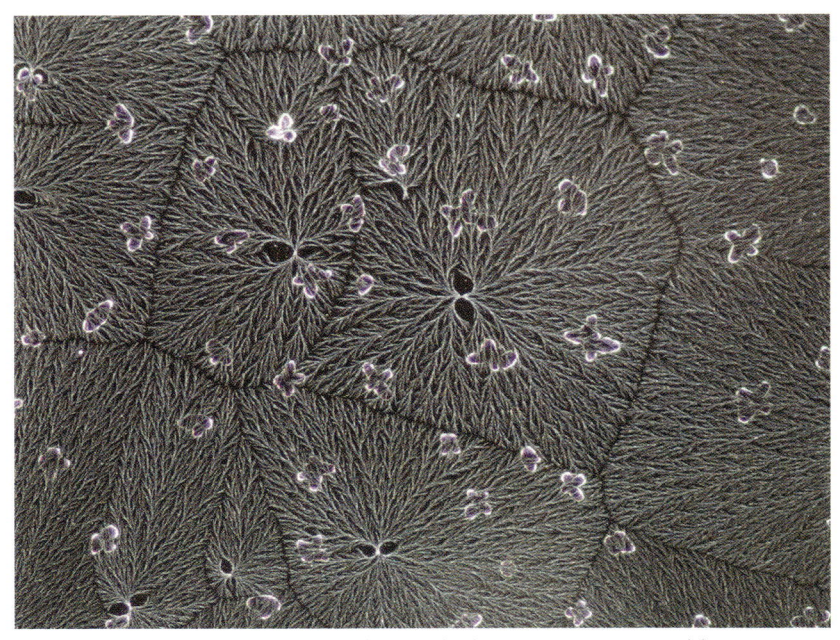

Image 296: Convallaria fruct., Black Forest, 0.5 years old (200x)

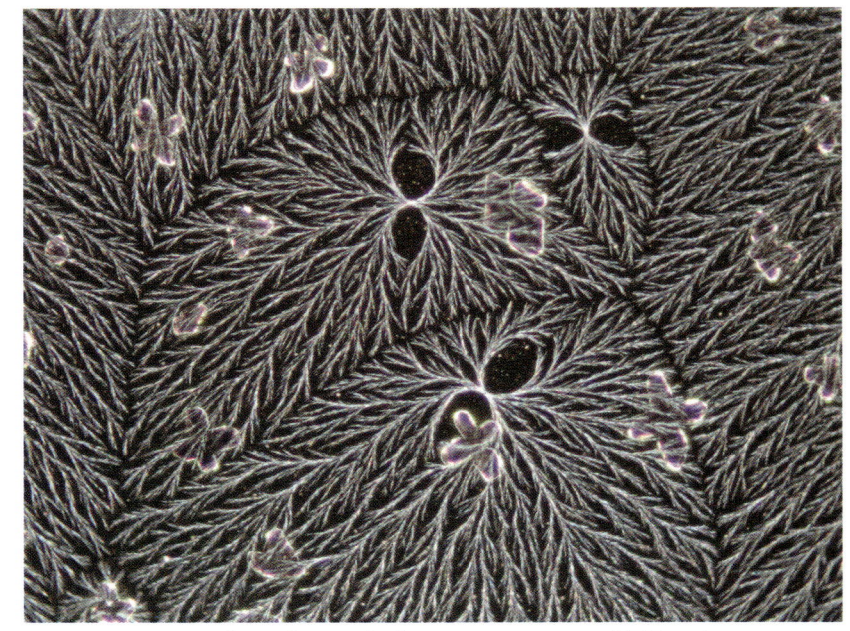

Image 297: Convallaria fruct., Black Forest, 0.5 years old (400x)

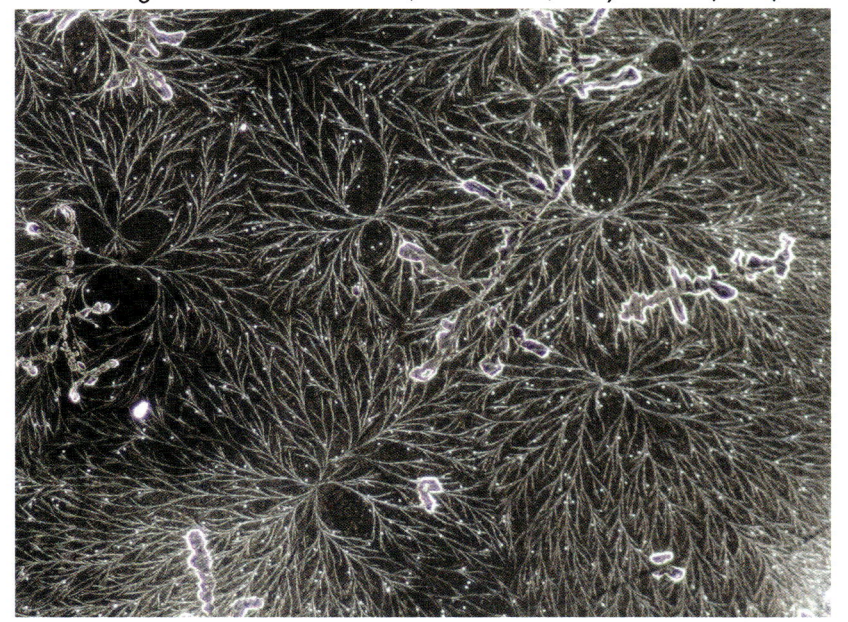

Image 298: Convallaria fruct., Black Forest, 0.5 years old, Stg. W. (400x)

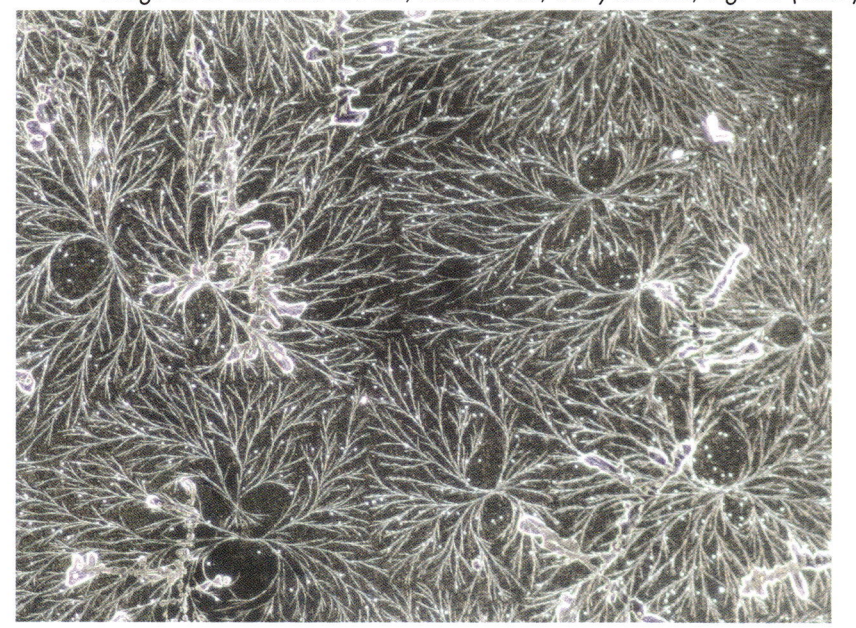

Image 299: Convallaria fruct., Black Forest, 0.5 years old, Stg. W. (400x)

The following four images (296 to 299, digital camera) appear from a different preparation of lily-of-the-valley berries from the Black Forest. The berries were harvested in autumn with stems and left without water until the next spring (half a year). They were shriveled, but of a bright red color. The shapes reproduced here in magnification had spread over the whole drop surface (see above: ripening).

Image 296: Lily-of-the-valley berries in water, six months old (200x).
Image 297: Lily-of-the-valley berries in water, six months old (400x).

The next two images (298 and 299) show the preparations of such berries with Stuttgart water. We see the same Convallaria structure, but of much weaker expression. This is an example of the fact that other waters also produce typical images. Volvic water, however, produces much clearer images owing to its high silicon content (see "Material and Method").

3. Seeds — Kernels of the Plant in Water

In contrast to the berry fruits, the seeds are kernels without a pericarp. So the flesh is removed from the date, likewise from the papaya fruit, sloe berries, grapes, etc. Does the drop image change without the pericarp?

In images 300 to 305 we see seed kernels with just as characteristic a center design as before!

―――― Plants, Fruits, Seeds, Barks, and Ashes in Water ――――

Image 300: Date seeds (100x)

Image 301: Date seeds, other drop center, magnified (400x)

Image 302: Maple seed kernels (100x)

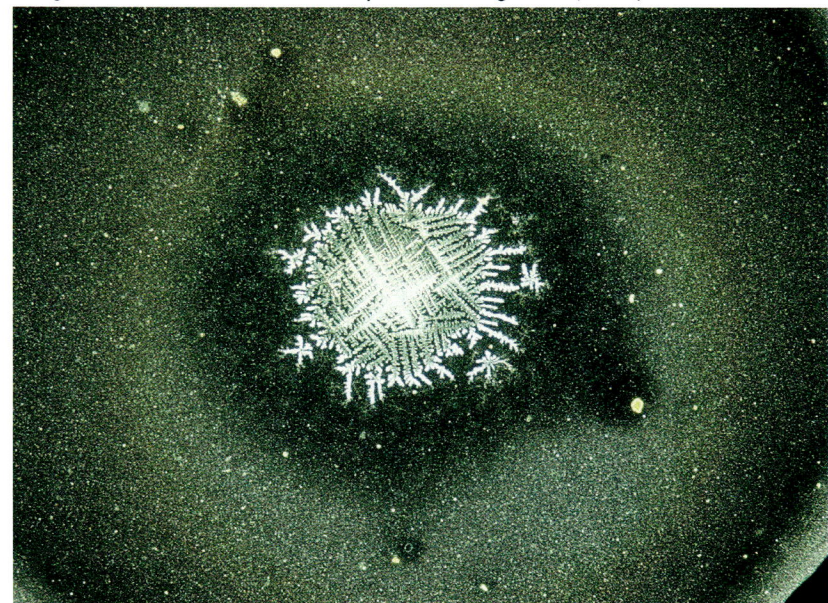

Image 303: Maple seed kernels (100x)

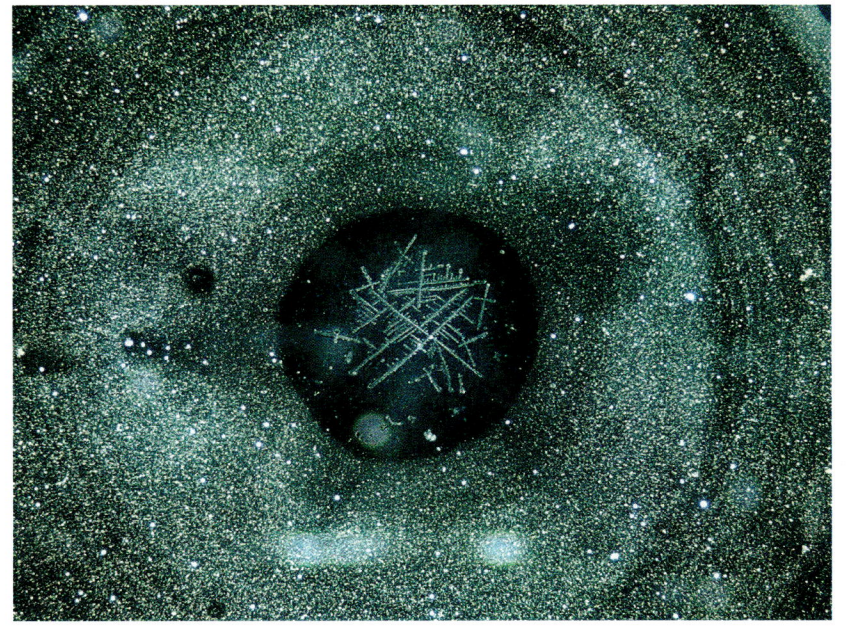

Image 304: Lemon seeds (100x)

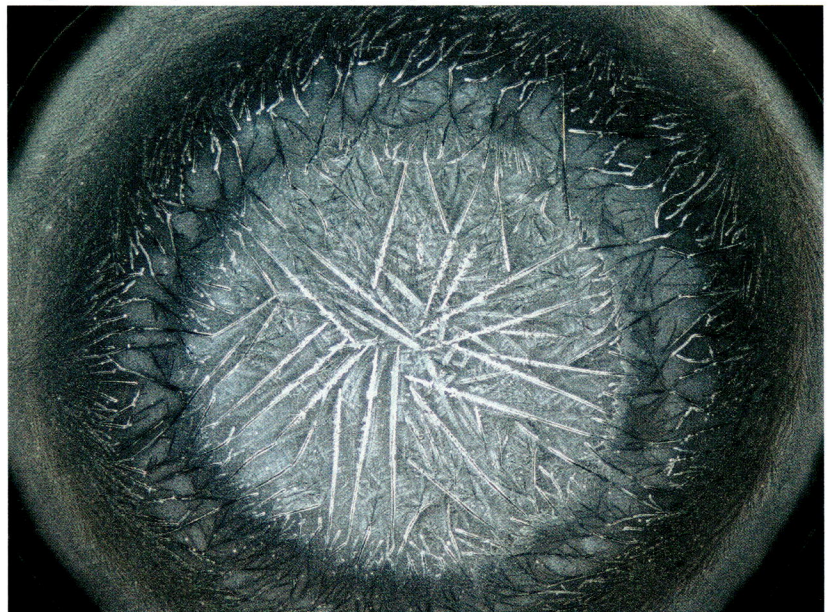

Image 305: Papaya seeds (100x)

THE SILENT LANGUAGE OF LIFE

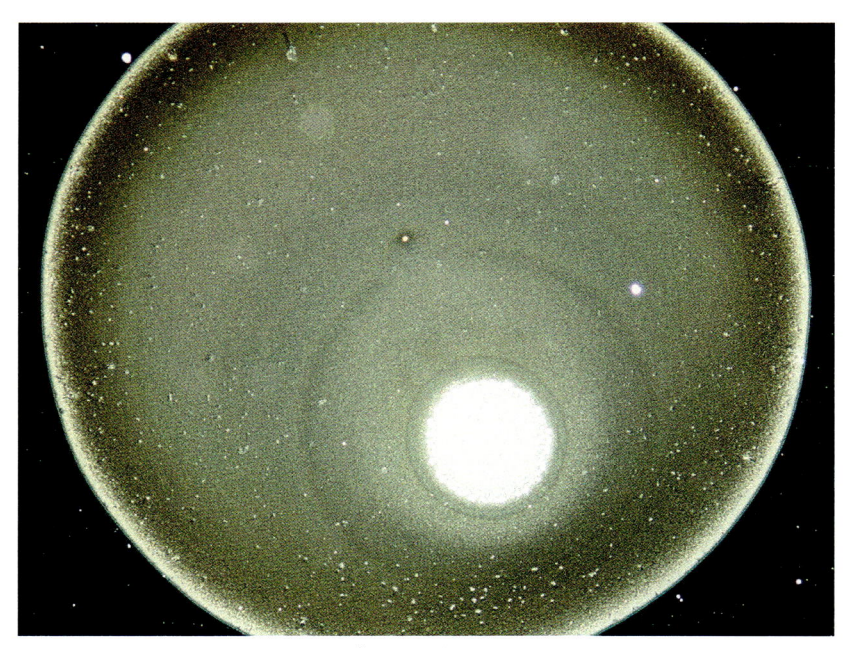

Image 306: Prunus seeds (100x)

Image 307: Prunus seeds, heat drying (100x)

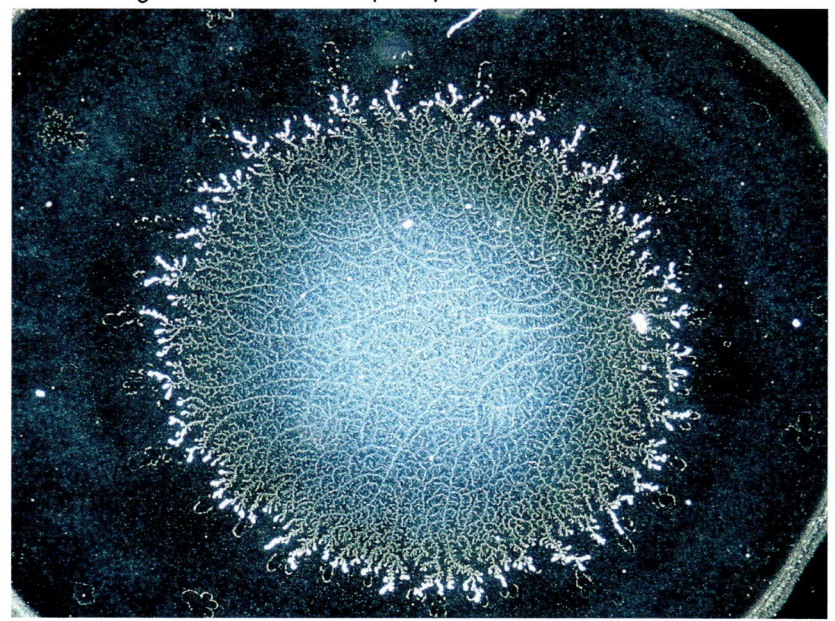

Image 308: Prunus seeds, germinated (100x)

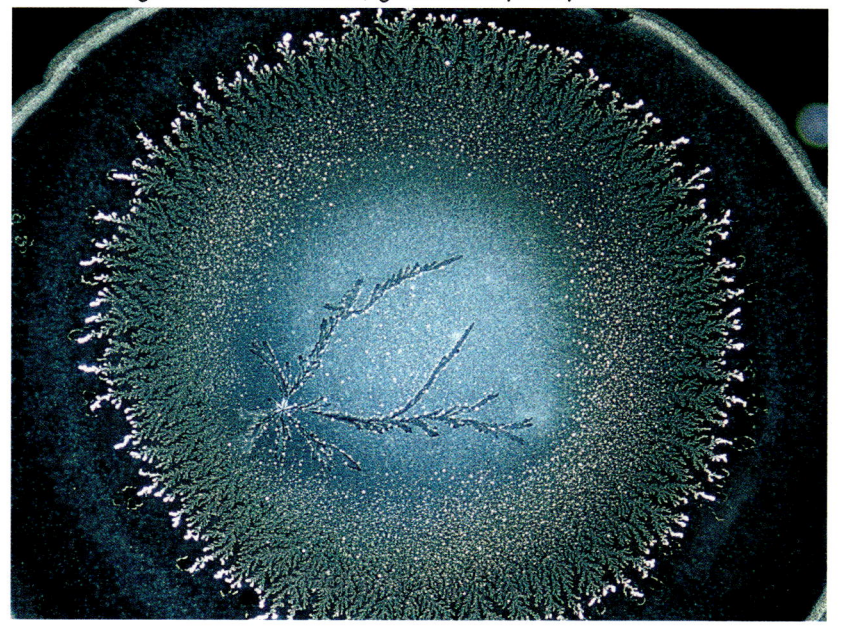

Image 309: Prunus seeds, germinated (100x)

Image 306 shows sloe berry seeds (*Prunus fructus* seeds) in water.

Image 307: the same under heat drying of the drop. The center clearly remains as an agglomeration of matter and cannot be spread over the entire drop surface by the heat. The heat sets the water in motion, which is reflected in delicate rhythmic circles.

Then one day there was a surprise with the seeds. Since the preparations often remained standing for months and some for years, samples were taken from the Prunus preparation of November 3, 2003, on July 20, 2005—after more than 1.5 years, and dripped on. In the glass, the Prunus seeds showed small rootlets and beginnings of germination!

Images 308 to 310: Prunus seeds, rooted and germinated.

The root structure mentioned above can be seen in the magnification in image 310. Depending on the size of the seed, it changes into a "stem structure" in the periphery. At the very beginning of this chapter we had already encountered both structures.

Thus the nucleation in the drop center has revealed itself in its actual potency. What is written here in such sober words actually testifies to an unbelievable event. Is it not beyond our imagination?

How can we think of these central cores of fruits and seeds in terms of location in the water? Are they really contained in it as such, or do they only appear in the circular form of the drop after it has been dripped? And what changes in the water when the seed has taken root and germinated? The result of the rooting and germinating power that appears in the image would then also have to be present in the initial water, otherwise there would be no image. Questions and amazement remain!

Plants, Fruits, Seeds, Barks, and Ashes in Water

Image 310: Prunus seeds, germinated, root structure (200x)

THE SILENT LANGUAGE OF LIFE

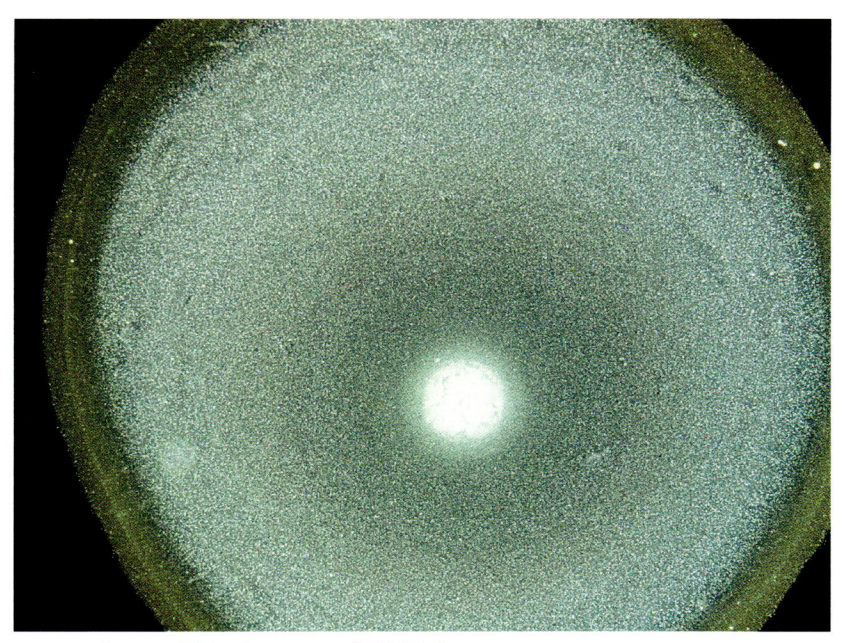

Image 311: Grape seeds (100x)

Image 312: Grape seeds, 2nd drop of water (100x)

Image 313: Grape seeds, 2nd drop of water (100x)

Image 314: Grape seeds, 2nd drop of water (40x)

Image 315: Grape seeds, 2nd drop of water (100x)

Image 316: Grape seeds, mirrored center magnified (400x)

Plants, Fruits, Seeds, Barks, and Ashes in Water

Grape seeds in water with new surprises follow.

Image 311: Grape seeds in water. In sparse expression, they show a central core. As frequently before, it was tested for its longevity, since according to the previous one, a special "place of power" was assumed in this one.

Images 312 to 316: A drop of water (not a batch of water) is deposited at the edge of the already dried drop. After the drying of this second drop, the center of the first drop has clearly increased in its expressiveness (image 312), this is also visible in image 313. Here, circles also appear around the stimulated center. What communication between the dried center and the fresh water at the edge of the drop!

In the following three images (314 to 316), it is not just communication that we observe. The water drop is now added to the dried drop, so that it overlaps the center (images 314 and 315). What happened during the drying of the second drop? The center is extended without disappearing, as if "mirrored."

Image 316 shows a magnification of the "mirrored center."

The numerous images of this series of experiments are impressive and enigmatic. By soulfully feeling into this process, one could ask oneself whether the dried central nucleus *"longs"* for the water. Then it would find its fulfillment in the following image.

Images 317 and 318: After months, the grape seeds in the water had also rooted and germinated. The center has dissolved. A tender, central root structure as well as a peripheral stem structure show up.

Image 319: In another preparation without germination, it holds its own in the freezer compartment of the refrigerator for ten days.

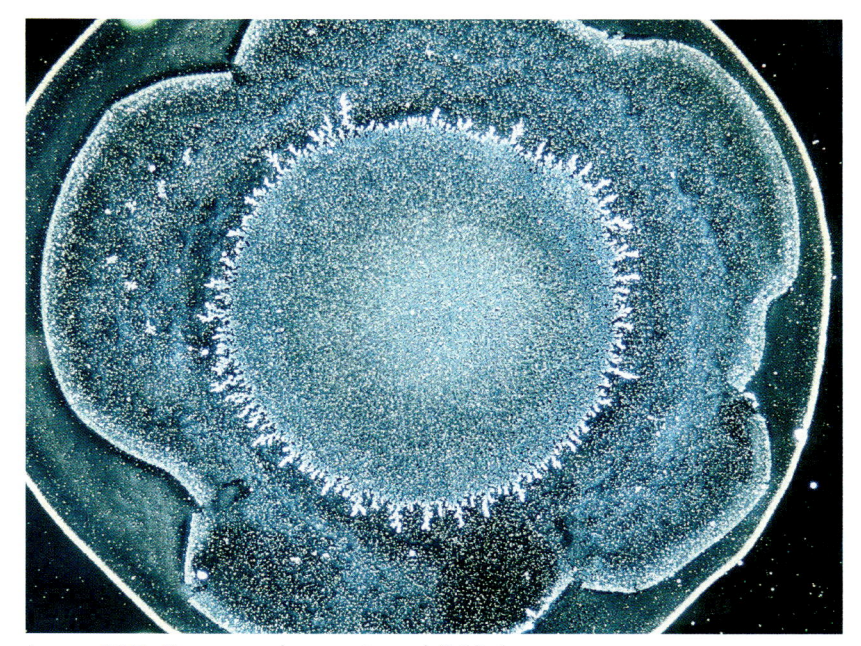

Image 317: Grape seeds, germinated (100x)

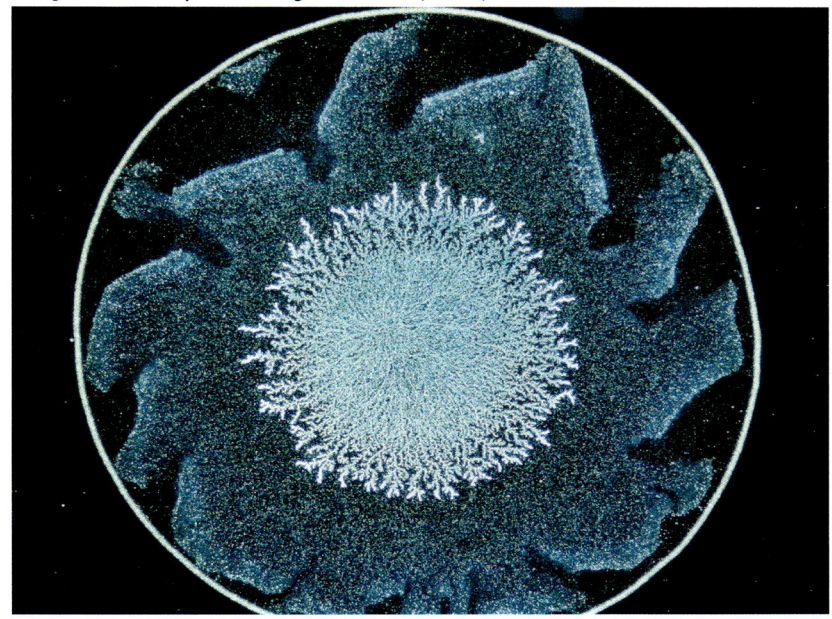

Image 318: Grape seeds, germinated (100x)

Image 319: Grape seeds, 10 days in the freezer (200x)

— THE SILENT LANGUAGE OF LIFE —

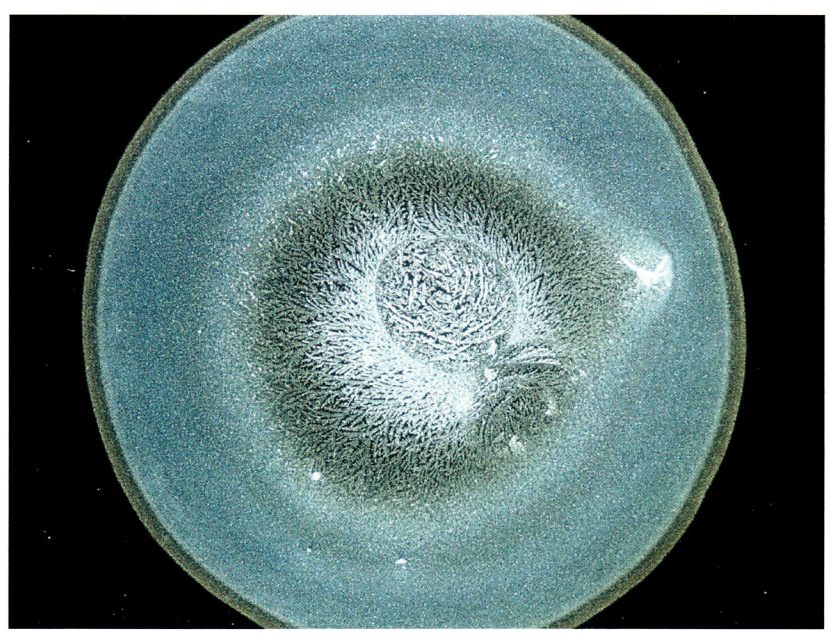

Image 320: Beechnuts (100x)

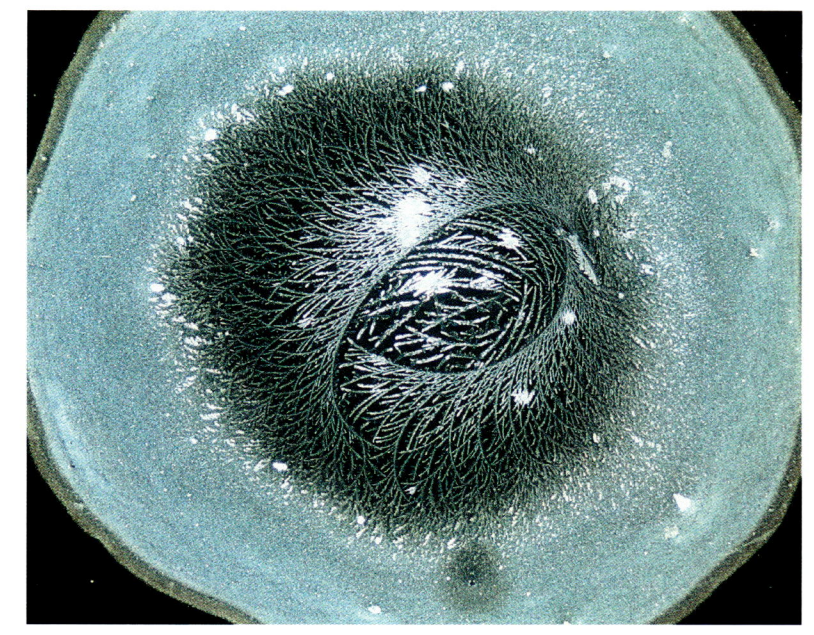

Image 321: Beechnuts (100x)

Images 320 to 322 show us beechnuts after three weeks in the water. The center of these quite elastic, triangular seeds is always very pronounced.

What does the seed, which can also be eaten in moderation, show when its shell is removed?

The shell is removed, and the bare triangular seed kernel is placed in the water.

Image 323 shows its drop pattern after three weeks. The rhythmically shaped center has disappeared; the imprinting of the marginal zone has changed. This points us to the importance of the seed shell. We will encounter these "shell forces" of fruits and seeds in the rice in a moment.

Image 322: Beechnuts (400x)

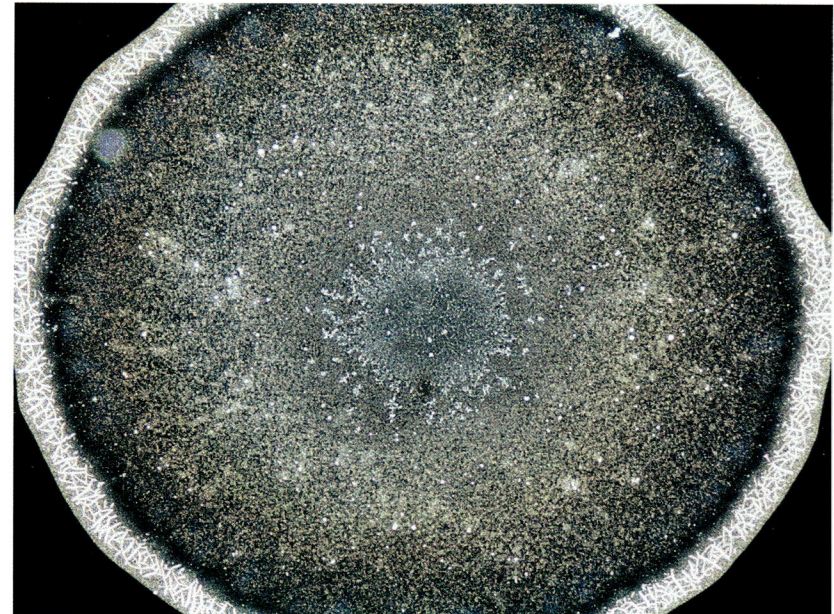

Image 323: Beechnuts without shell (100x)

118

Plants, Fruits, Seeds, Barks, and Ashes in Water

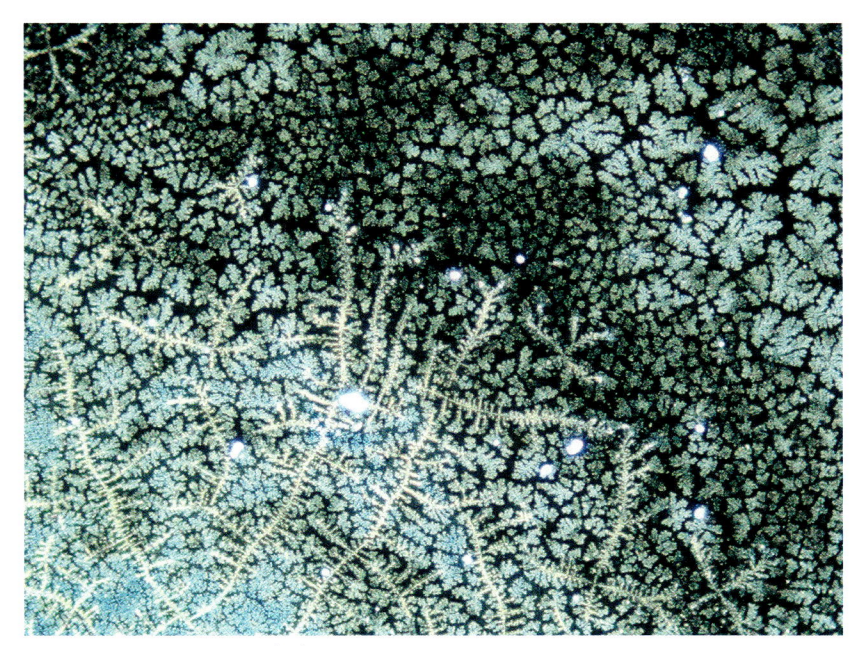

Image 324: Whole rice (200x)

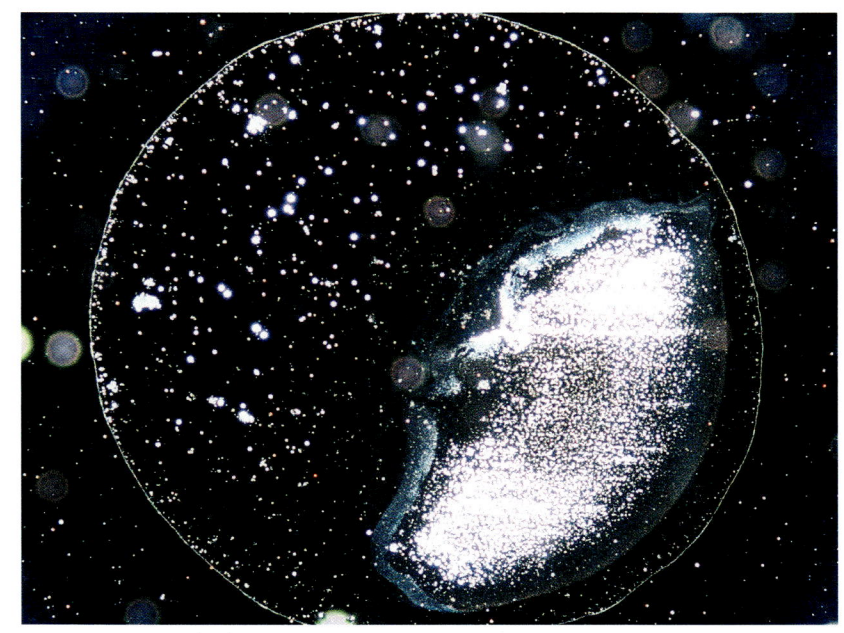

Image 325: Husked rice (100x)

4. Seeds of the Plant in Water: The Importance of the Seed Shell

Even the small seeds of annual plants form a center in the drop center. It is different with rice, wheat, and rye. These spread their figurations over the whole drop surface.

Image 324 shows a section of whole rice in water.
Image 325 shows rice without its husk, white rice—
 a striking difference!

When a Dutch doctor in Dutch India was in charge of the care of Beri Beri patients, he noticed that chickens fed with husked rice were wasting away from symptoms similar to those of Beri Beri patients (dissolution of the muscle structure and nerve degeneration).

When the chickens were fed rice *bran* (a rice product made from fruit and seed husks), the symptoms disappeared. This was the decisive observation that led to the explanation of Beri Beri disease as a vitamin deficiency disease.

Dr. Rudolf Hauschka (1891–1969) reported this in his book *The Nature of Substance: Spirit and Matter*[13] and described vitamin B, a deficiency of which is the basis of the disease, from a spiritual-scientific point of view, in connection with the shells of fruits and seeds.

Thus we see in images 323 and 325 a significant change in the drop pattern. The typical seed structure is lost after the removal of its shell.

Images 326 to 329 show wheat and rye in their black shape sowing over the whole drop area.

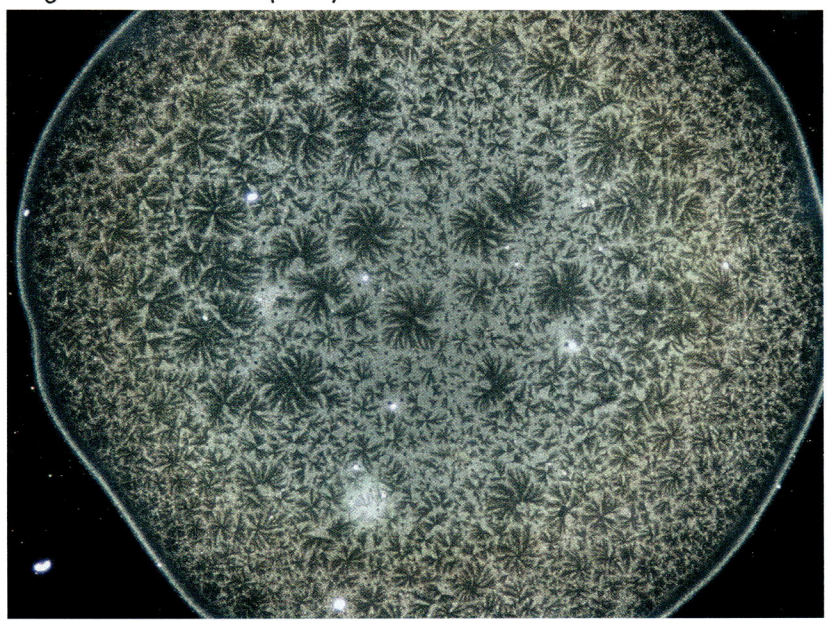

Image 326: Wheat, 3rd week (100x)

Image 327: Wheat, 3rd week (200x)

THE SILENT LANGUAGE OF LIFE

Image 328: Rye, 3rd week (100x)

Image 329: Rye, 3rd week (200x)

Plants, Fruits, Seeds, Barks, and Ashes in Water

The other images (330 to 341) may remain without comment. They give us a small insight into the different designs of the drop centers of seeds.

Thus, with few exceptions, we generally see a delimited center area in fruits, seeds, and the small seeds.

Image 330: Taraxacum seeds (dandelion seeds) (100x)

Image 331: Taraxacum seeds (dandelion seeds), magnified (400x)

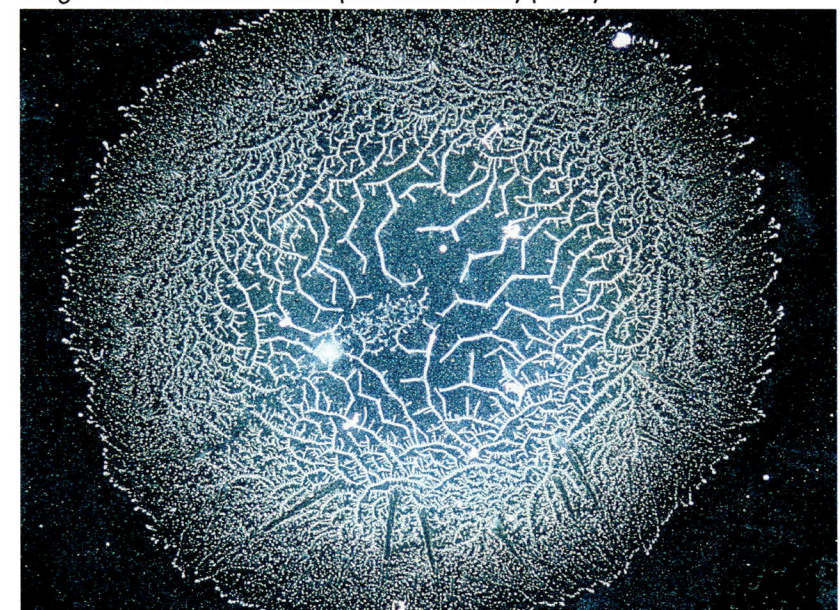

Image 332: Taraxacum seeds (dandelion seeds), rooted (200x)

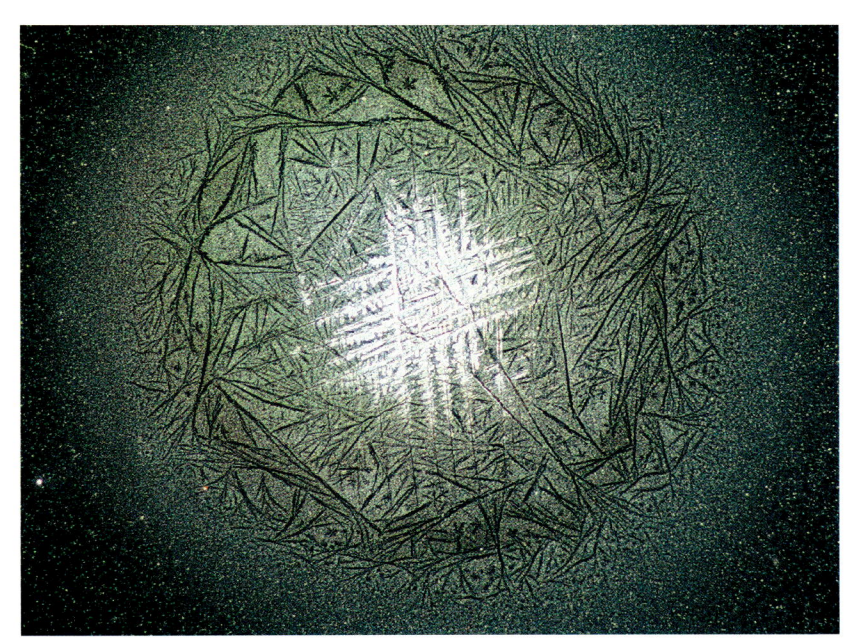

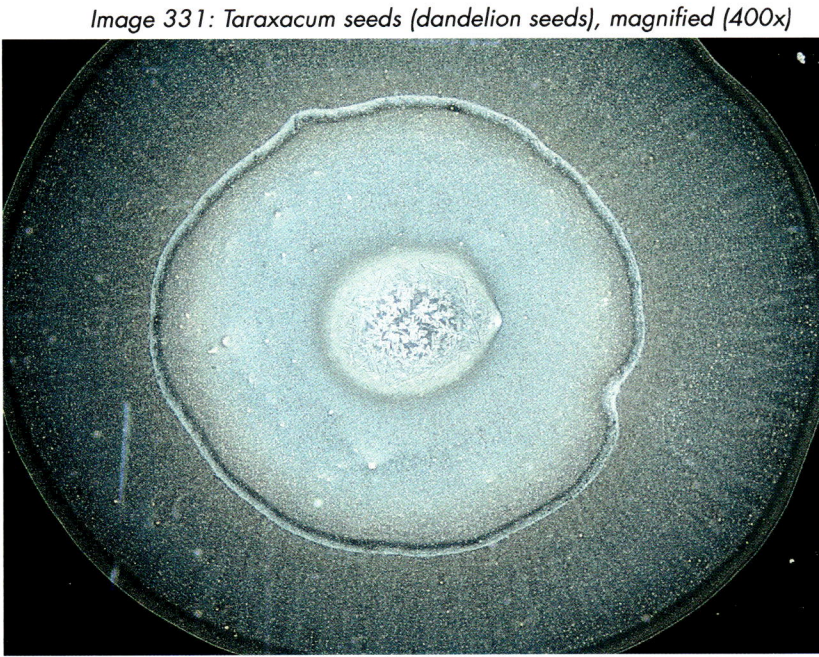

Image 333: Coltsfoot seeds (100x)

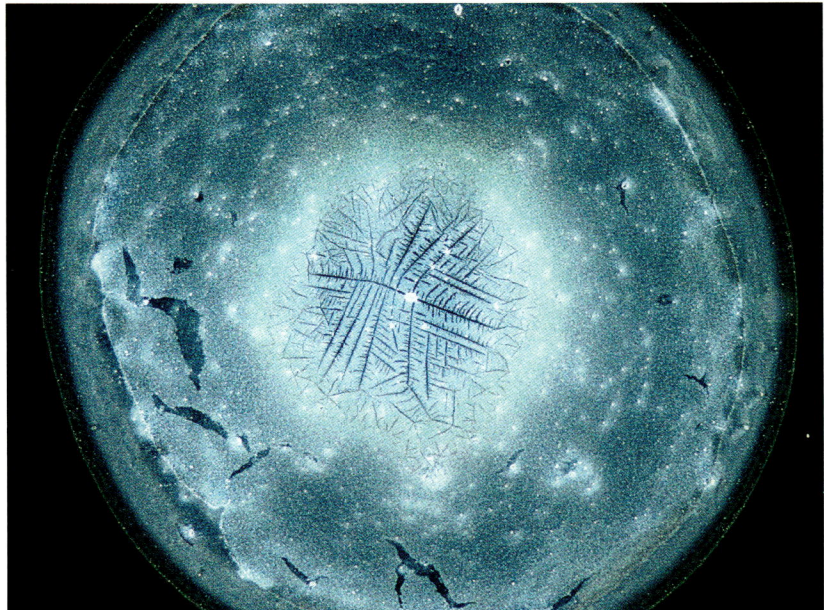

Image 334: Euphrasia seeds (eyebright seeds) (100x)

121

THE SILENT LANGUAGE OF LIFE

Image 335: Aconite seeds (40x)

Image 336: Water lily seeds (100x)

Image 337: Corn poppy capsules (200x)

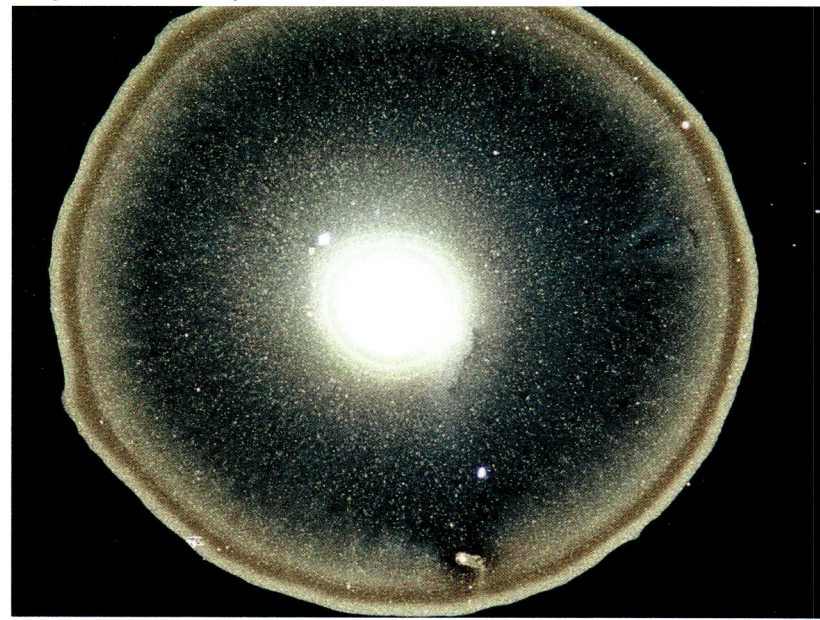

Image 338: Blackberry seeds (100x)

Image 339: Red seed (N. N.) (100x)

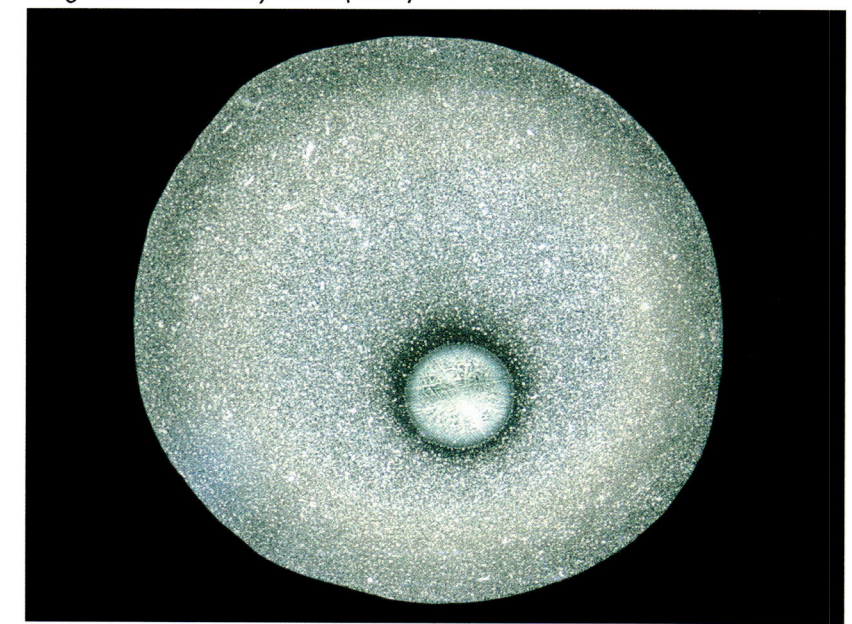

Image 340: Peony seeds (40x)

Plants, Fruits, Seeds, Barks, and Ashes in Water

Image 341: Star moss seeds (100x)

5. Barks of Young Twigs and Trees in Water, the Cambium

This topic was initially outside of this research. In the context of medical practice, I had dealt with Aesculus cortex, the chestnut bark, and with the substance aesculin, which is extracted from the bark of young twigs. Then this topic also flowed into this research and led to the question of the cambium of the tree.

Before that, however, some images of the outer branch formation should be shown.

Although the "rose thorns" are actually spike formations, since only epidermal and no deeper layers are involved in their formation, the term "rose thorns," which is common in linguistic usage, is used here.

Images 342 and 343: Rose thorns after three weeks in water (digital camera).

In the differentiated abundance of these formations, right down to the curling figurations that look like flower buds, an image of the plant being itself can appear to us.

We come to the barks of the plant branches with their cambium.

The horse chestnut (*Aesculus hippocastanum*) appears to us as a mighty, impressive tree. It delights us in spring with its light brown, sticky buds and lush flowering candles. In autumn we like to collect its smooth brown, shiny chestnuts. Its healing properties, primarily those of its fruits, are known in homeopathy, especially for vein problems and fluid congestion in the tissues. In the anthroposophical medical context, however, it is not only the seed fruit of the horse chestnut that finds its use, but also its bark: Aesculus cortex.

Following the lively suggestions by Wilhelm Pelikan in his third volume of *Healing Plants*,[40] the bark of young, strong branches of the tree was peeled off in spring, cut into small pieces and shaken with water in a screw-top jar. The bark is peeled down to the wood so that the cambium, the innermost layer of the bark, comes to the fore. In the light of day, the solution shows nothing special, but "glows and shimmers in the most beautiful sea blue in the striking light"[40] (Wilhelm Pelikan).

This fluorescent effect was experienced in the dark with a torch. Aesculin, which is known as a substance and the cause of fluorescence, has a special relationship with light in that it attenuates UV radiation and can, for example, protect the skin from its effects. Aesculin is found abundantly in the bark of young twigs, especially in spring, but not in seeds!

Dr. Rudolf Hauschka has impressively demonstrated this relationship to light in *The Nature of Substance*[13] and his experiments with the light spectrum. The chemical part of the spectrum (UV light) is swallowed up by aesculin solution. This can also be found in Rudolf Steiner's second natural science course.[15]

Aesculin, potentized, is also known as a remedy (e.g., Aesculinum D30).

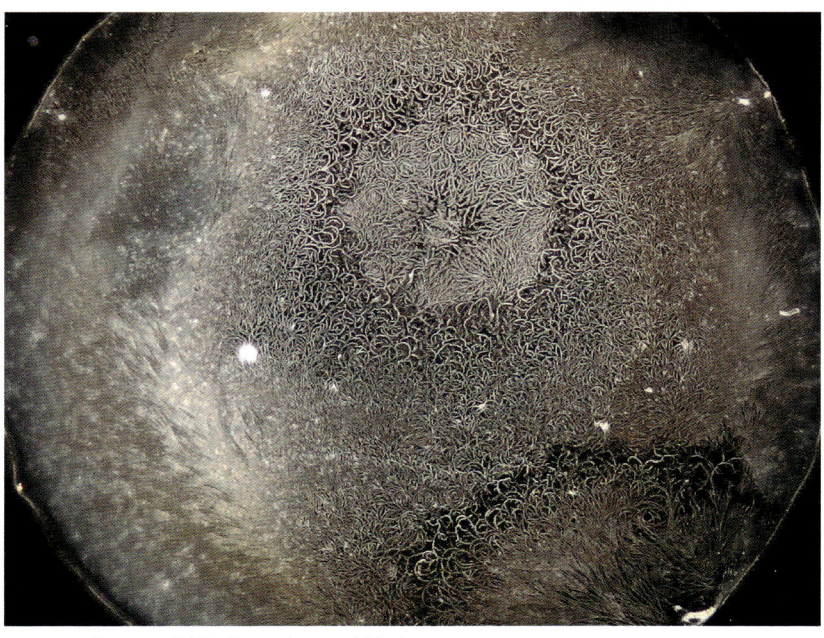

Image 342: Rose thorns (40x)

Image 343: Rose thorns (200x)

Plants, Fruits, Seeds, Barks, and Ashes in Water

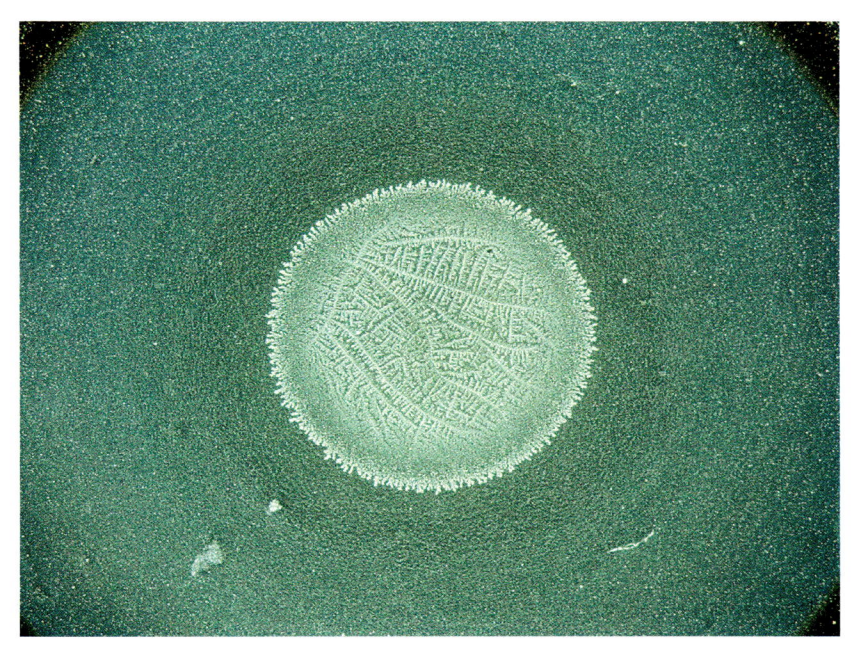

Image 344: Aesculus bark (100x)

Image 345: Aesculus bark (10 days dark) (200x)

This solution with the Aesculus bark was left to stand for a few weeks and then dripped on.

Image 344: Aesculus bark in the water. A clearly structured, circular formation can be seen in the center of the drop! This was a big surprise and in no way expected. What does the bark have to do with the fruits and seeds? Does this center disappear when the batch is placed in darkness for ten days?

In image 345 we see the result after dripping. The center has faded, become materially impoverished, shrunk, and slipped out of the center of the drop.

Image 346 shows the enlarged center of another drop after darkness.

Can it be awakened by the light in the same way as the plant fruits? The preparation solution from the dark is dripped on again and the slide with the drops is dried on gold leaf in bright sunlight.

Image 347 shows us the center magnification of a sample drop under the influence of sun and gold leaf. A truly amazing revival through light!

The first stanza of six from Rudolf Steiner's *Truth-Wrought-Words* is quoted here. It can also be a companion for us in many other places.

> Sun, you radiance-harborer,
> your pure light's material power
> Conjures life from out of the earth's
> Immeasurably fertile depths.[41]

Image 346: Aesculus bark (10 days dark, other center, magnified) (400x)

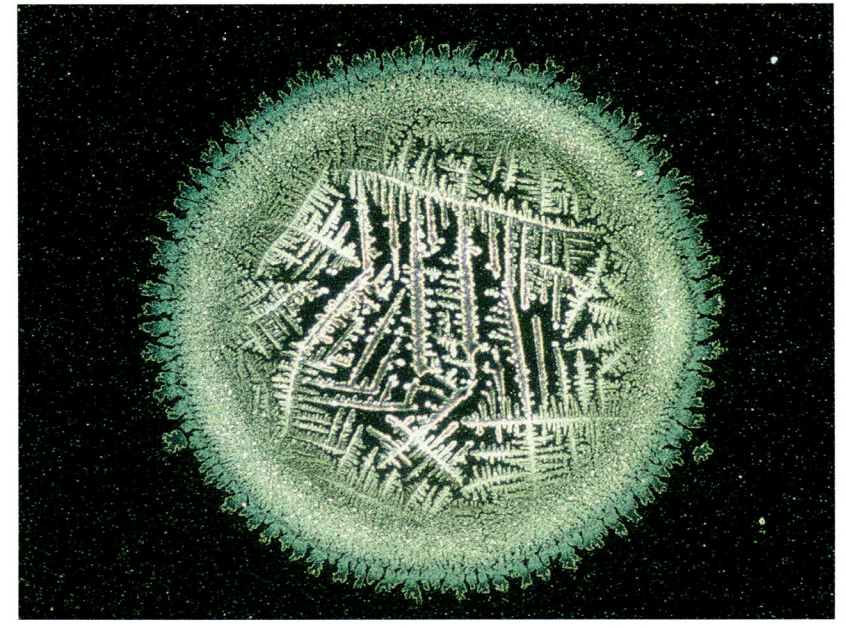

Image 347: Aesculus bark, gold leaf, sun, center magnified (400x)

Image 348: Aesculus bark, 8 weeks in syringe (100x)

Image 349: Aesculus preparation water, 7 months in syringe (100x)

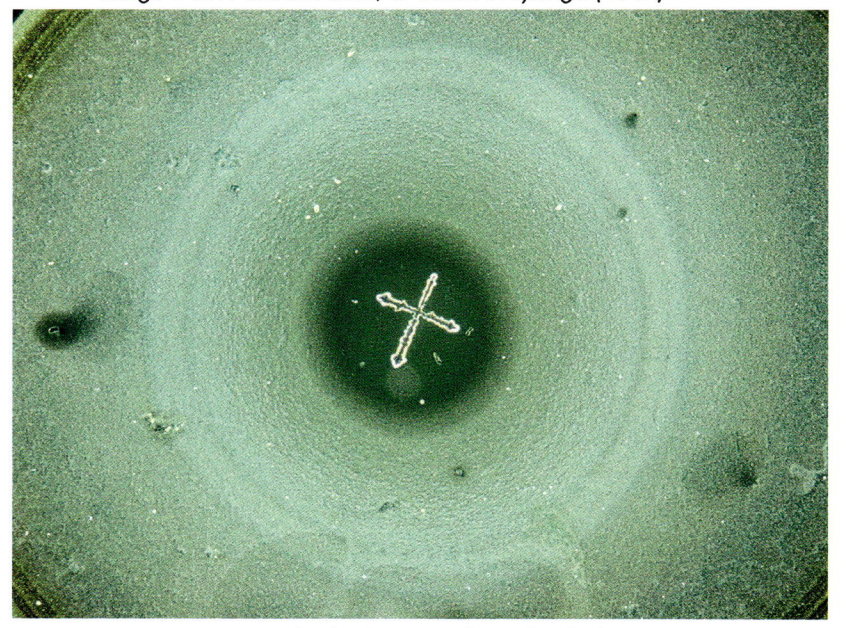

Image 350: Aesculus bark, 2 months in the dark (100x)

Image 351: Aesculus bark, 10 days in the freezer (100x)

The question was now what would be seen when the bark was removed from the water. Bark preparation water is taken up in a 5 ml syringe and dripped on again after eight weeks. Image 348: Here is the result.

Image 349 shows Aesculus bark after seven months in the syringe, dripped on. Even without bark, the center is preserved in the water.

Now this water syringe is placed in the dark for two months. Image 350: We see a preserved but black center area with a bright cross!

The cross is not an isolated phenomenon in this research. Under certain circumstances, which cannot be explained, it appears again and again. The salt experiments are excluded from this.

Now a batch of water (over the years—as in all experiments—new batches have been made again and again) was placed in a freezer for ten days. Image 351: Aesculus bark water from the icebox, dripped on. The center asserts itself, as we can see.

Of course, the urgent question arose whether this Aesculus bark imprinting in the water is due to the aesculin effect alone, or whether other conditions come into play for this. Therefore, the young twigs of many different trees were peeled off and attached. All of them showed a clearly pronounced center! Here are two examples:

Plants, Fruits, Seeds, Barks, and Ashes in Water

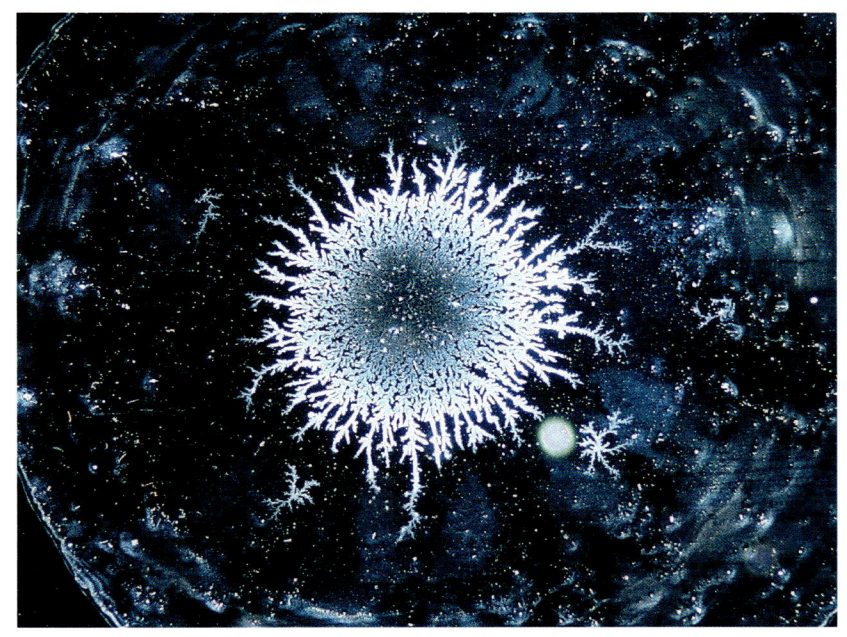

Image 352: Birch bark (100x)

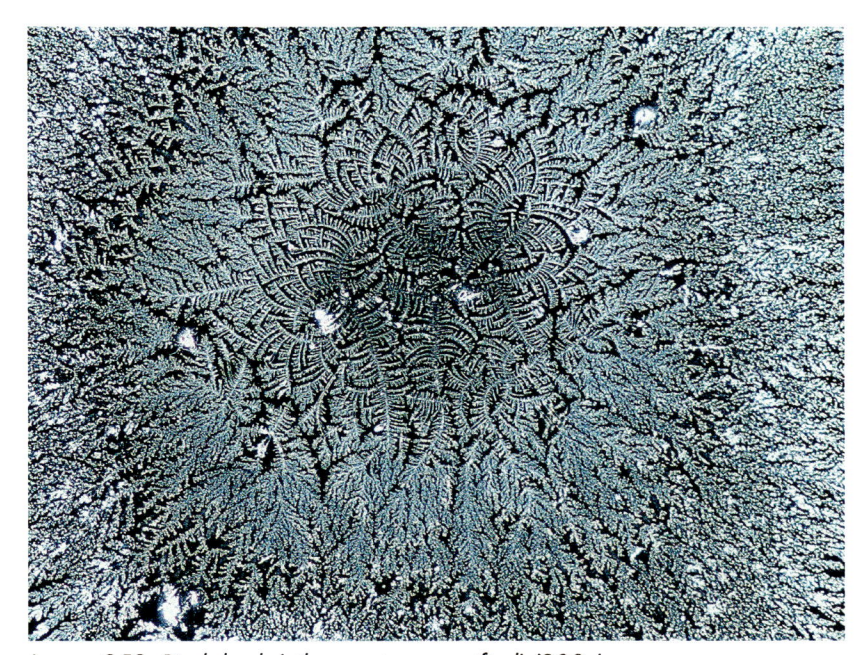

Image 353: Birch bark (other center magnified) (200x)

Image 352: Birch bark in water.
Image 353: Birch bark in water, other center magnified.
Image 354: Alder bark in water.
Image 355: Alder bark in water, center magnified.

Now there was another surprise! At the beginning of chapter 1, I mentioned that the images are expressed only in the closed circular form of the drop (image 15, rose thorns with drop; image 16, rose thorns without drop).

This was confirmed again and again by the fact that the preparation water spread over a large area on the slide did not produce any typical images.

This was then tested again with the alder bark preparation. Macroscopically, toward the end of the surface drying, a clear accumulation of water was seen at one point on the slide.

Image 354: Alder bark (100x)

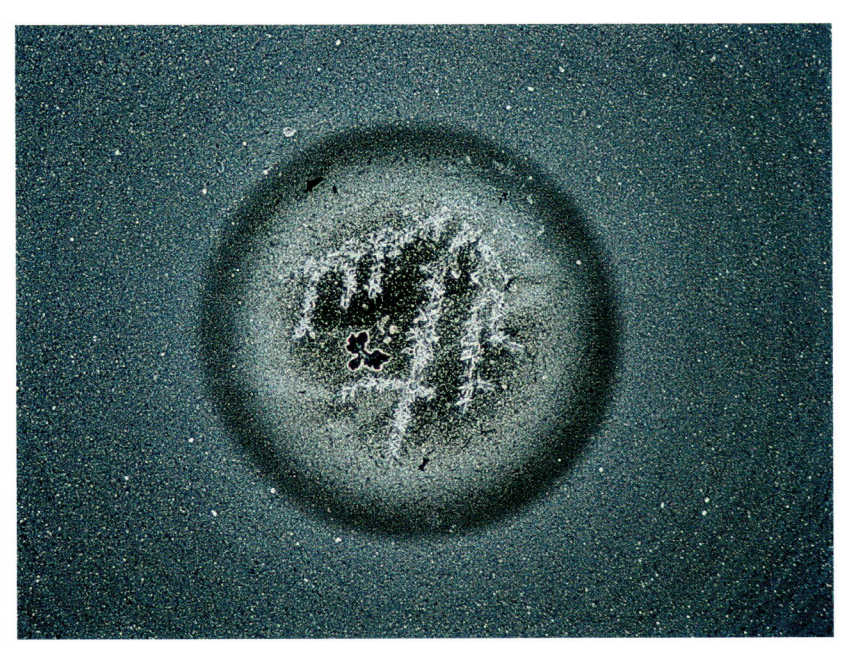

Image 355: Alder bark, center magnified (200x)

127

THE SILENT LANGUAGE OF LIFE

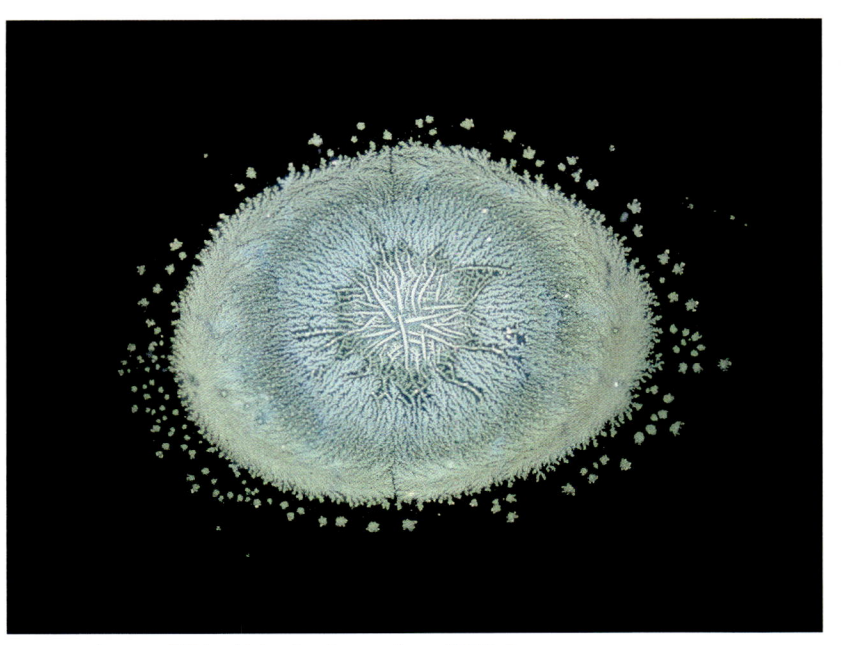

Image 356: Alder bark, no drop (200x)

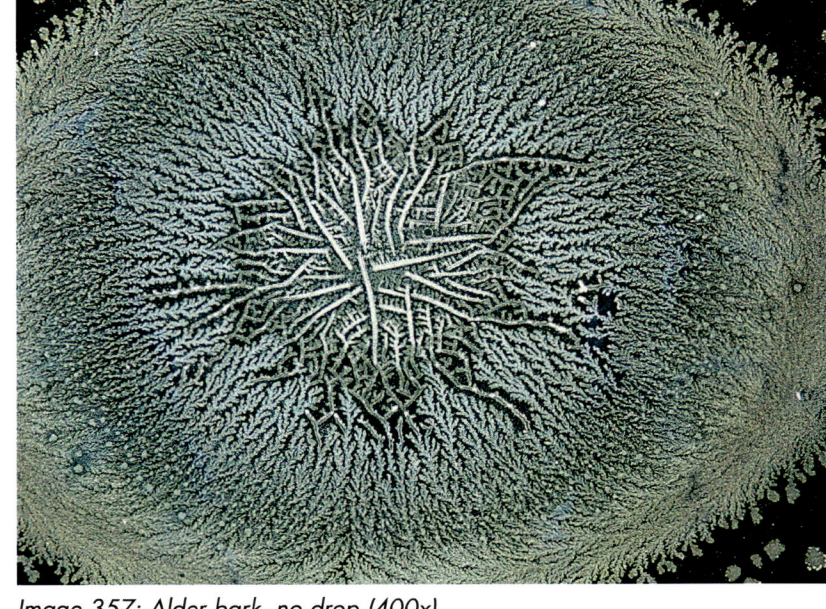

Image 357: Alder bark, no drop (400x)

After drying, images 356 and 357 do not show us an actual "drop image," but rather what actively gathered in one place. Thus, a special formative will power can be seen here in the preparation water of the alder bark, which has been spread over a wide area: This *will* forms its "own drop" from a right-angled water surface!

A crystalline, circular center forms in the lively oval of the perimeter, into which vegetative branches grow inward from the central zone. An enchanting formation!

This raised the question of whether this peculiarity of alder bark—to form its own center on the spread-out surface—can also be found in the fruits. We return to these for a moment. Crataegus fructus (hawthorn berries) from a four-month-old batch was dripped on and also spread out on a surface (digital camera). We first see the physical arrangements in the image.

Image 358: Preparation jar with hawthorn berries.

Image 359: Glass and dripping; the centers are visible in the dried drops.

Image 360: "Giant drop" and surface area. Here we can already see a macroscopically indicated wrapping spiral that leads into the newly formed "center."

The actual drop images now follow:

Image 361: Crataegus fructus, drop image.

Image 362: Crataegus fructus, "giant drop."

Image 363: Crataegus fructus, no drop; a new yellow-orange "center" forms on the surface area.

Image 358: Crataegus fructus seed jar

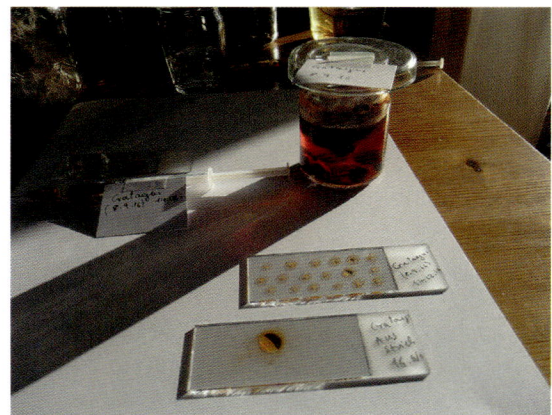

Image 359: Crataegus fructus seed jar and drip-on

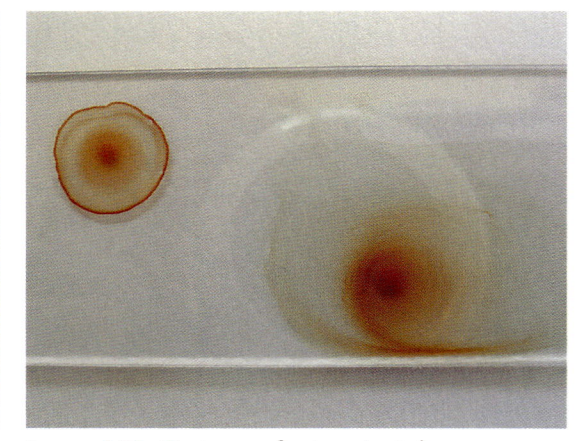

Image 360: Crataegus fructus giant drop and surface area

Plants, Fruits, Seeds, Barks, and Ashes in Water

Image 361: Crataegus fructus (100x)

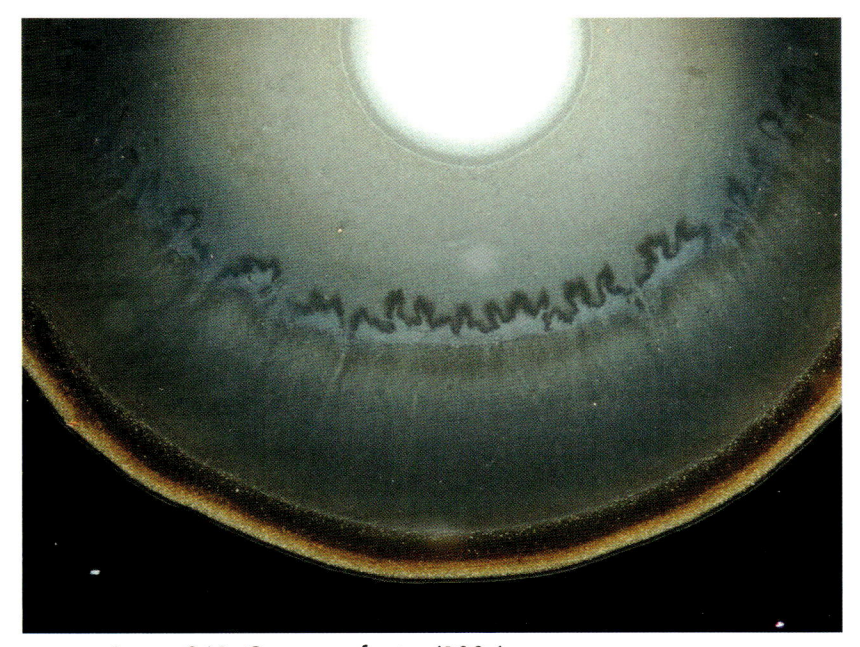

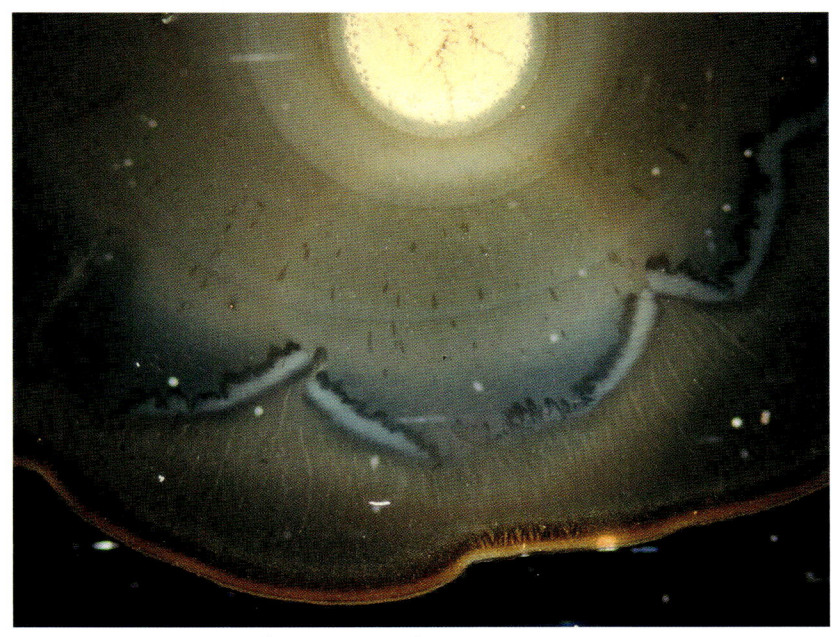

Image 362: Crataegus fructus, giant drop (40x)

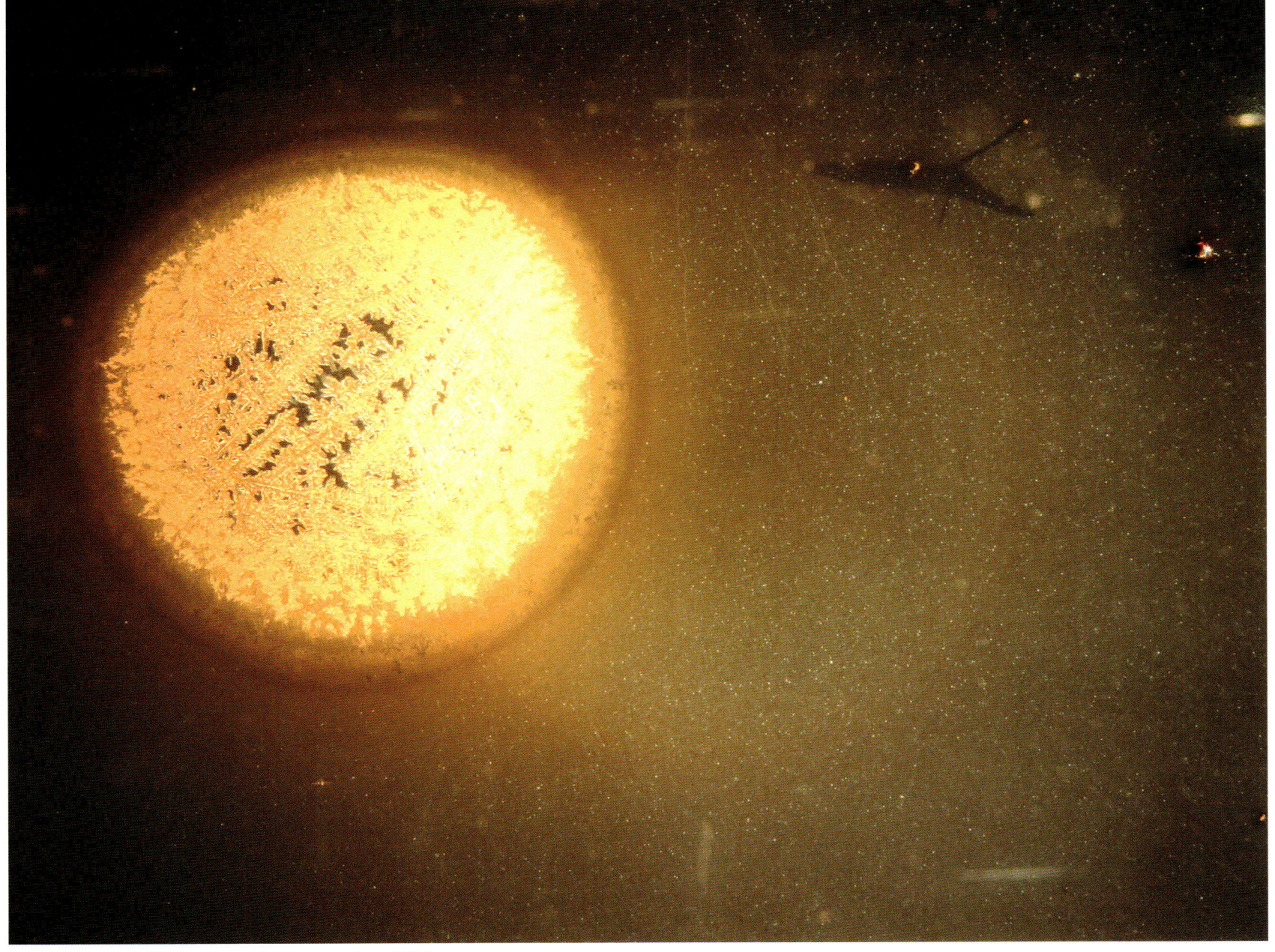

Image 363: Crataegus fructus, no drop (40x)

— THE SILENT LANGUAGE OF LIFE —

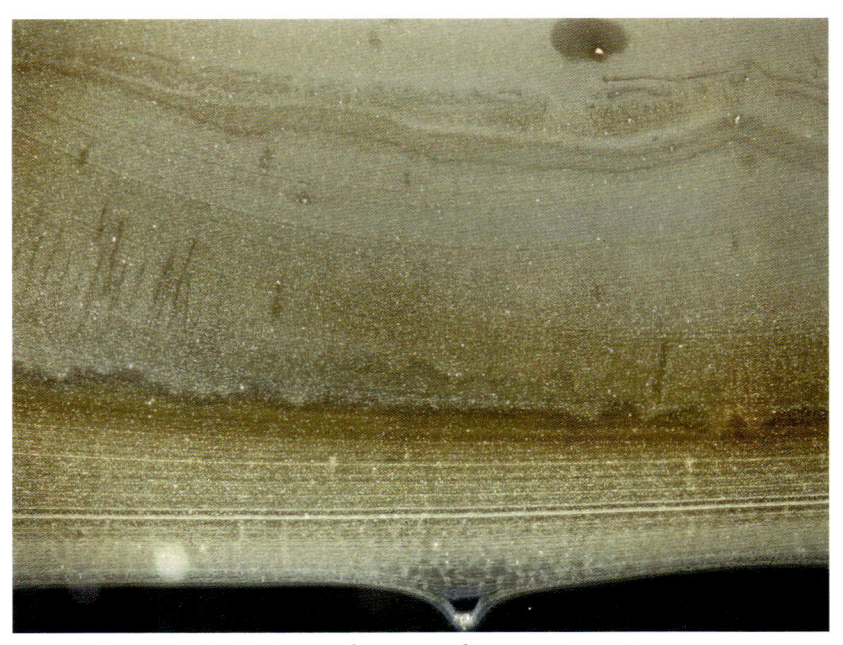

Image 364: Crataegus fructus, surface area (200x)

Image 365: Crataegus fructus, surface area (200x)

Images 364 to 367: Here we see sections of the rectangular surface area of Crataegus fructus. Parallel, delicate water currents come into the image in blue and ocher structures. They are an expression of an extraordinary dynamic, caused by the central suction process.

Thus, both fruits and rinds develop a strong, centering power on the water surface. Here we encounter in an impressive way the formative forces which, starting from the fruits, set the water in motion.

We return to the bark and want to look at how the oak bark forms its center in the water drop in small steps. A drying process of oak bark is given in the images 368 to 372.

Image 368 shows in the fluid, black circular form the activity of the "sparks of light" that, as described above, criss-cross a plane, leaving their traces.

As they continue, a ring forms and contracts into a circular disc in the center.

Image 373: Oak bark, different approach.

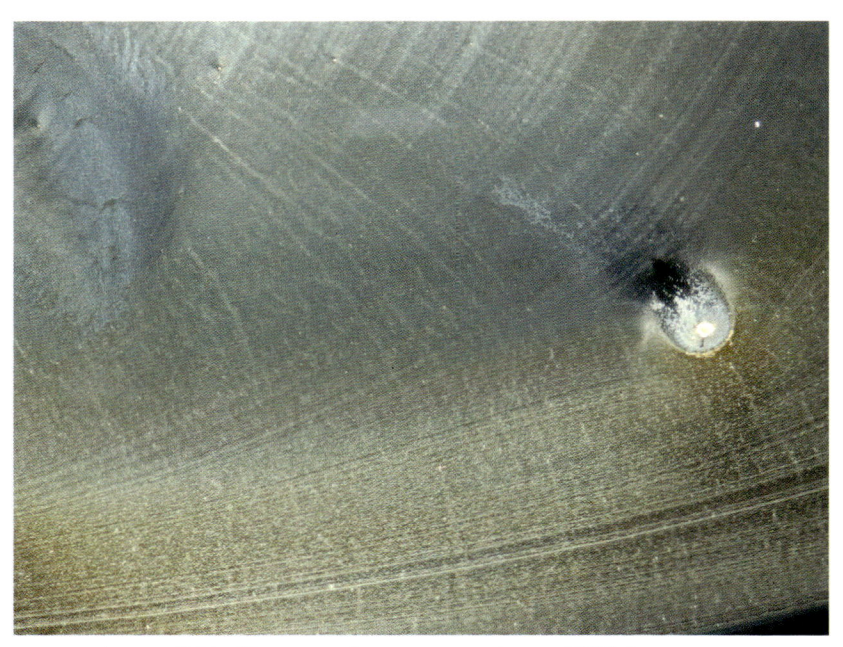

Image 366: Crataegus fructus, spread area (200x)

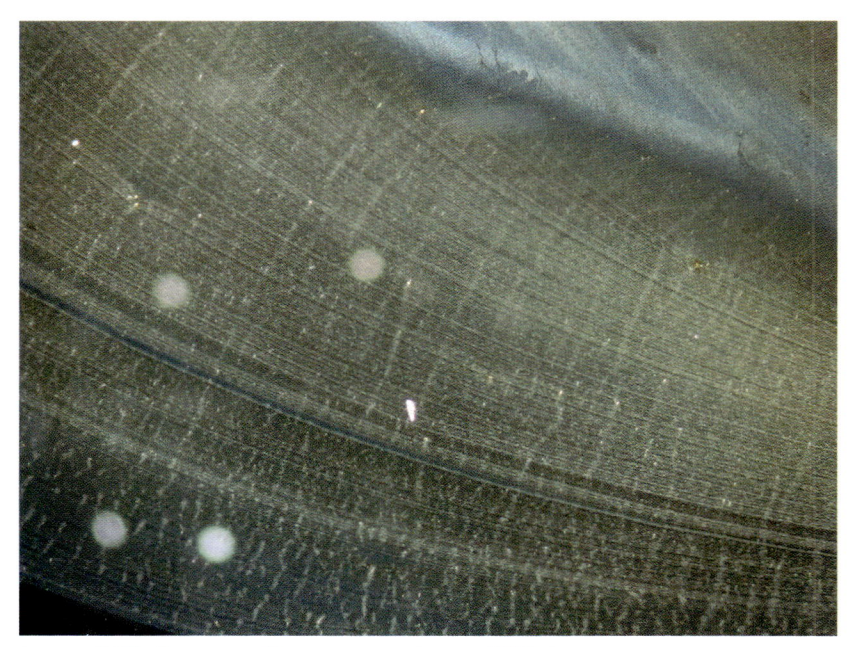

Image 367: Crataegus fructus, spread area (200x)

Plants, Fruits, Seeds, Barks, and Ashes in Water

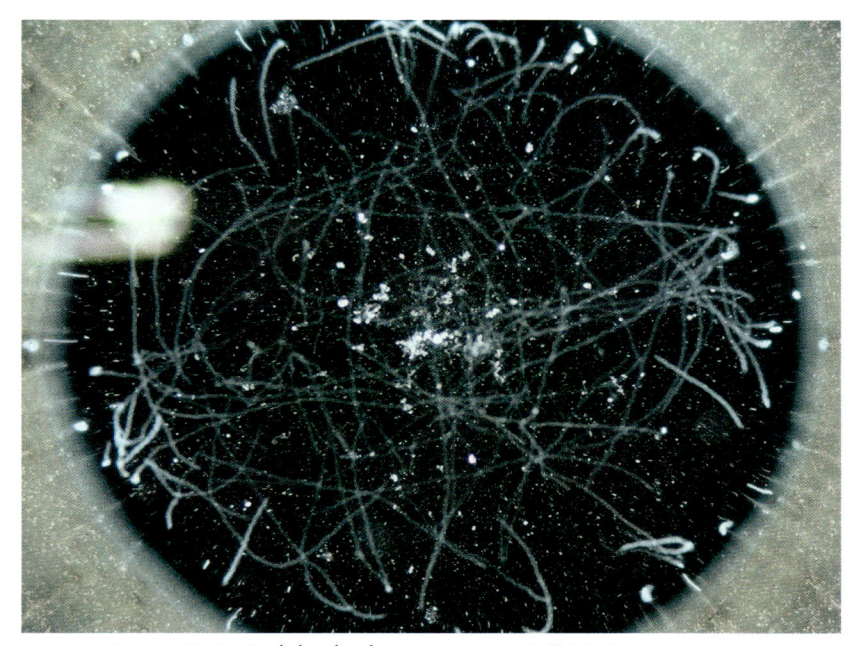

Image 368: Oak bark, drying process 1 (200x)

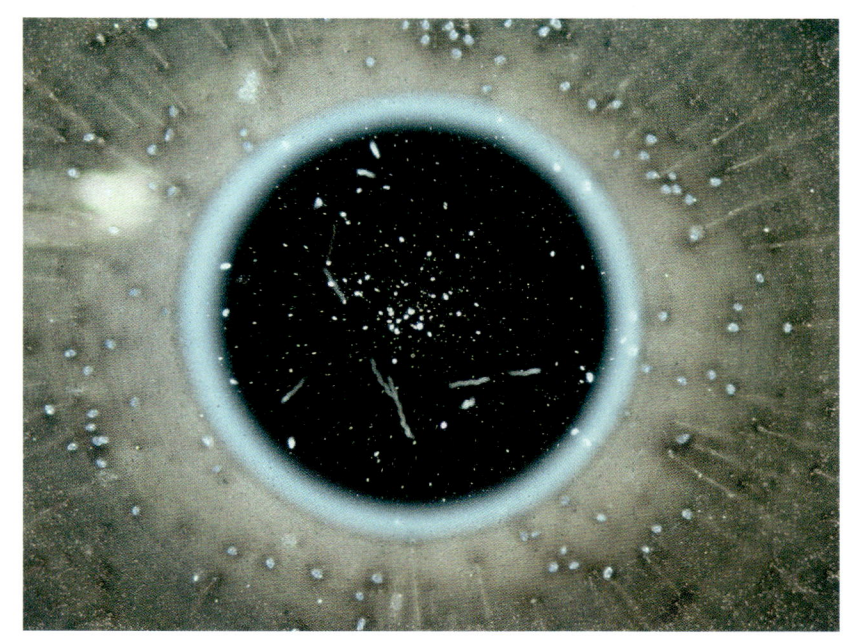

Image 369: Oak bark, drying process 2 (200x)

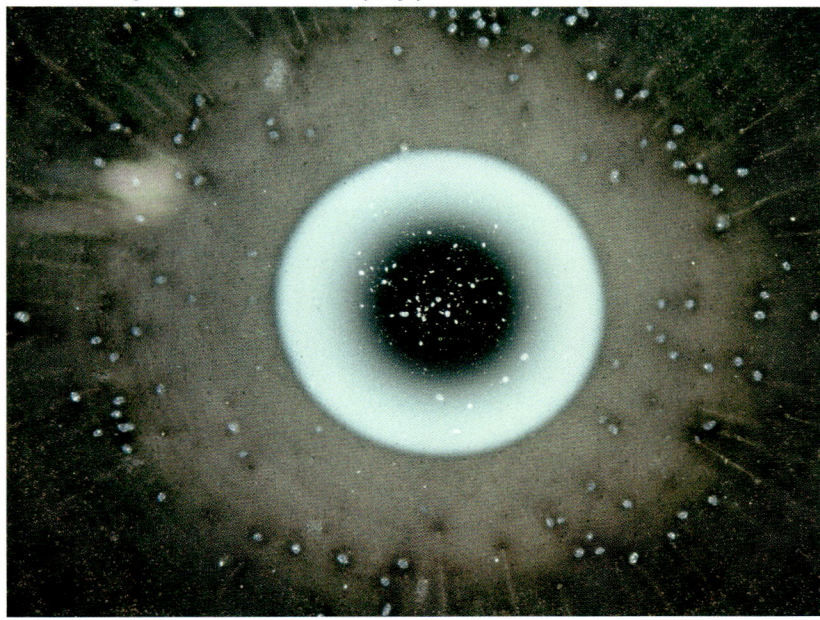

Image 370: Oak bark, drying process 3 (200x)

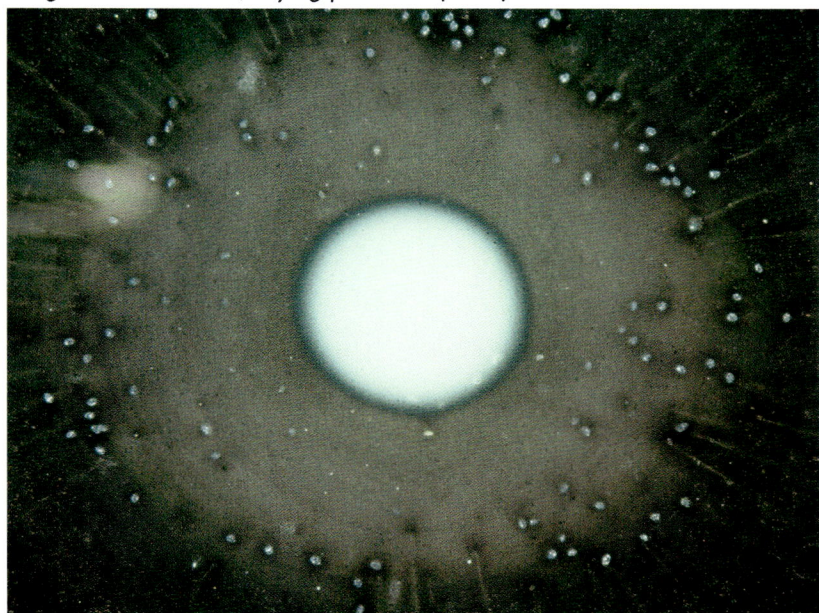

Image 371: Oak bark, drying process 4 (200x)

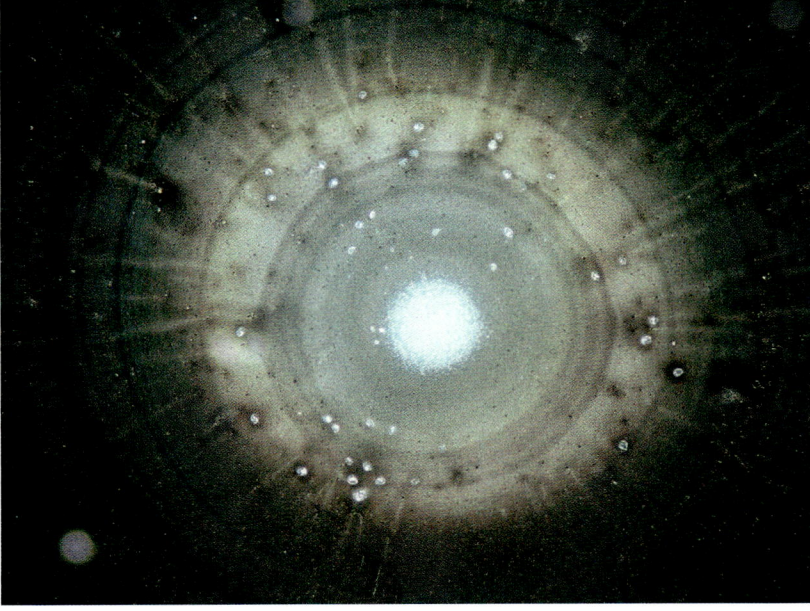

Image 372: Oak bark (200x)

Image 373: Oak bark (100x)

THE SILENT LANGUAGE OF LIFE

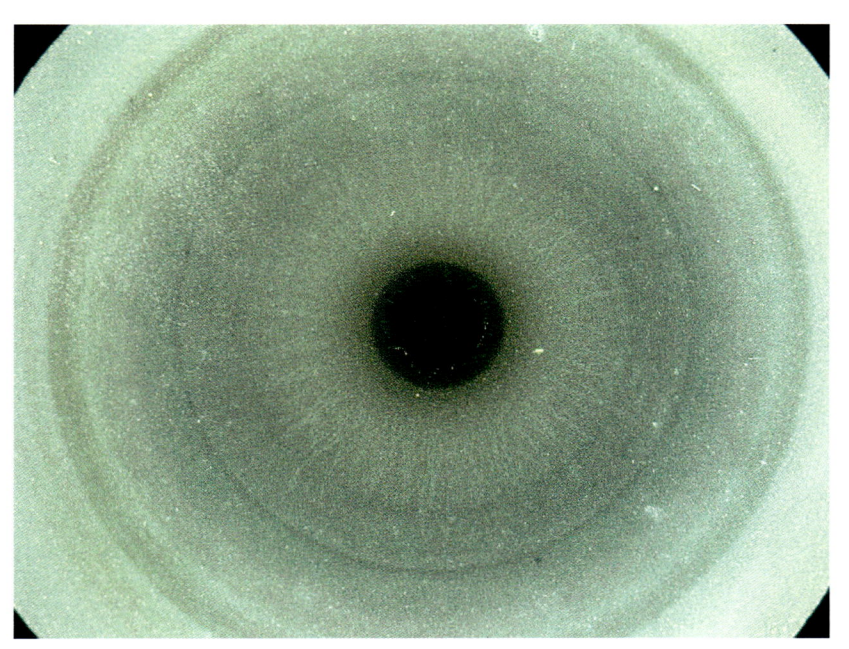

Image 374: Beech bark, cambium dead (100x)

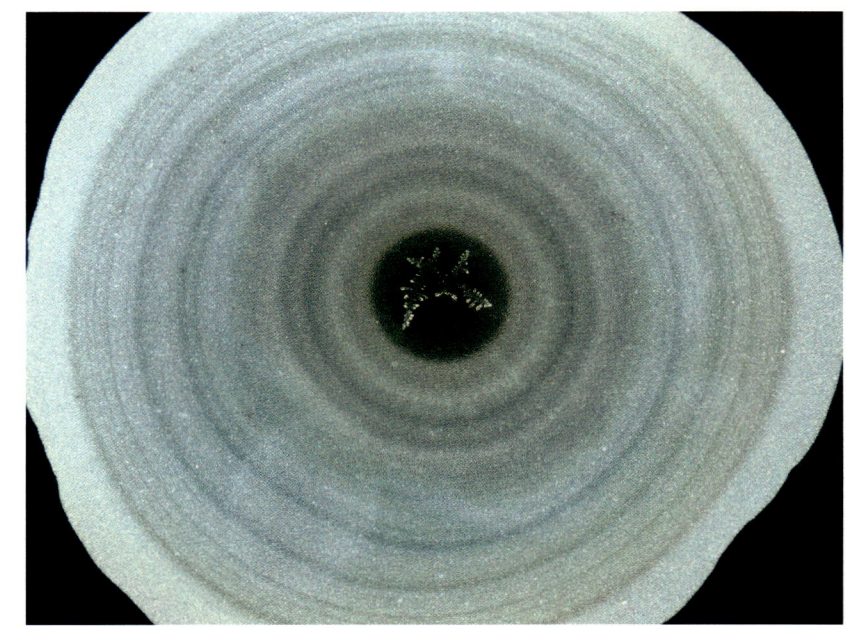

Image 375: Beech bark, cambium dead (100x)

So far, we have seen the peeled bark of the young branches in the water. What does the bark of a thick beech trunk show? The tree had been felled and the bark had already come off the trunk in fragments. The innermost layer, the cambium, seemed to have dried out.

Image 374: Beech bark in water; here we see a black center—i.e., there is no longer any nucleation. In many other drops the center was still rudimentarily "labeled." This is shown in images 375 to 377, where we again see crosses in the center. The formative force of the cambium has been extinguished.

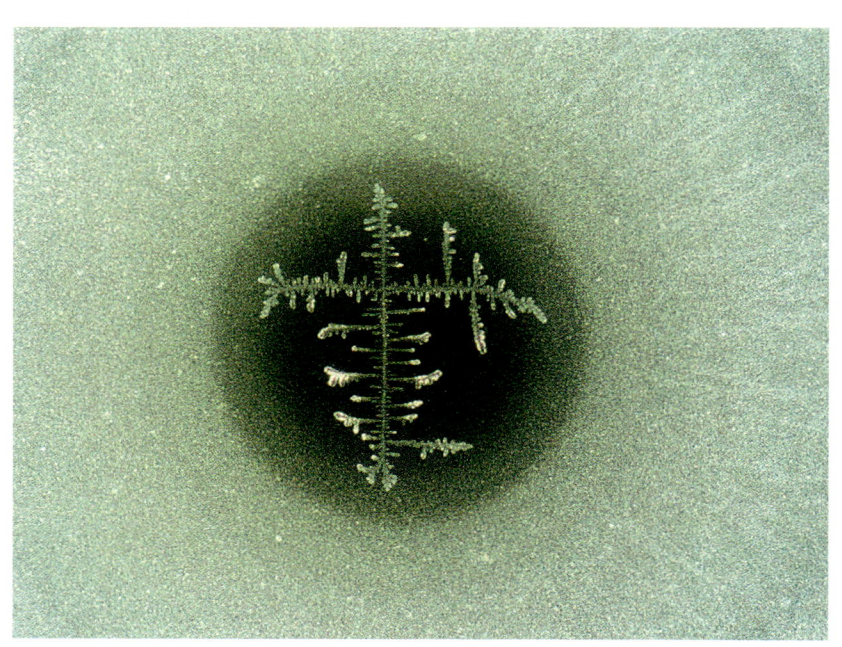

Image 376: Beech bark, cambium dead, center magnified (200x)

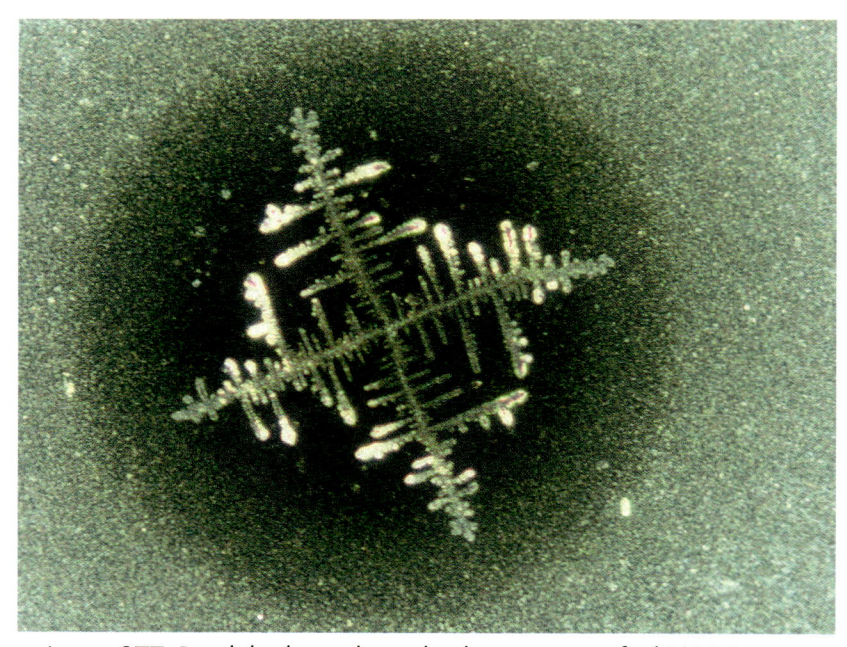

Image 377: Beech bark, cambium dead, center magnified (400x)

Plants, Fruits, Seeds, Barks, and Ashes in Water

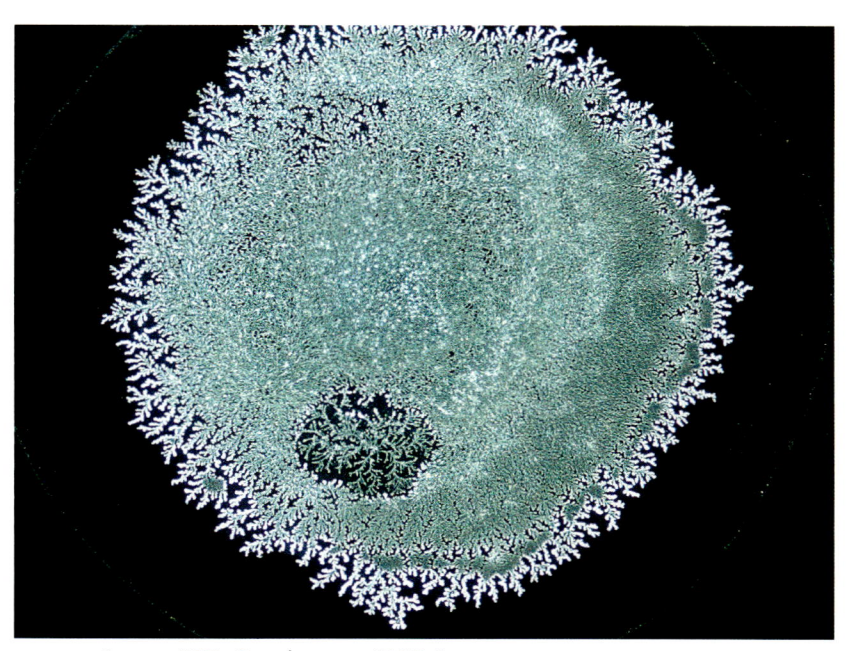

Image 378: Beech tumor (100x)

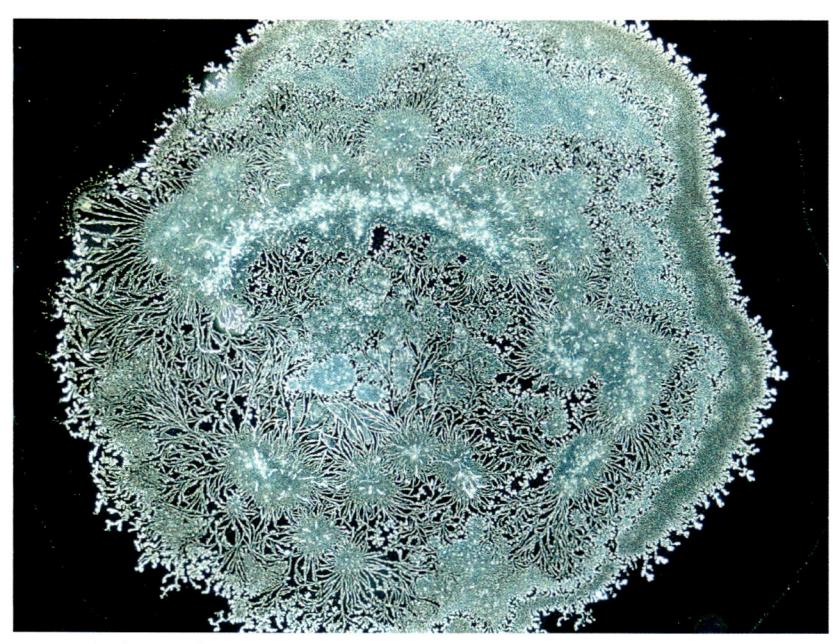

Image 379: Beech tumor (100x)

In this forest there was a beech tree with a monstrous growth. The question was what a piece of bark of this enormous, partly smooth-walled formation would show in the water. Small pieces were peeled off and placed in the water.

The images 378 to 381 show a dense network and give us an impression of the enormous etheric power of the tumor. A core center did not form.

Rudolf Steiner speaks in connection with mistletoe of the bark and trunk outgrowths of trees that one can encounter in the forest:

> If one studies it more closely, one finds that when the tree gets such an outgrowth, the following then occurs: somehow the physical body of the tree is inhibited. There is not enough physical matter everywhere to enable the etheric body to keep up with its power of growth.... The etheric body, which otherwise endeavors to hurl physical matter centrifugally out into the universe, is, as it were... from then on left alone for a certain part.... The consequence of this is that the etheric body takes the turn down to the lower part of the tree, which is equipped with stronger power. So it is essentially again the etheric body that becomes strong.[42]

The intense, rather mixed structuring of the drop image seems to be an expression of this strong etheric force.

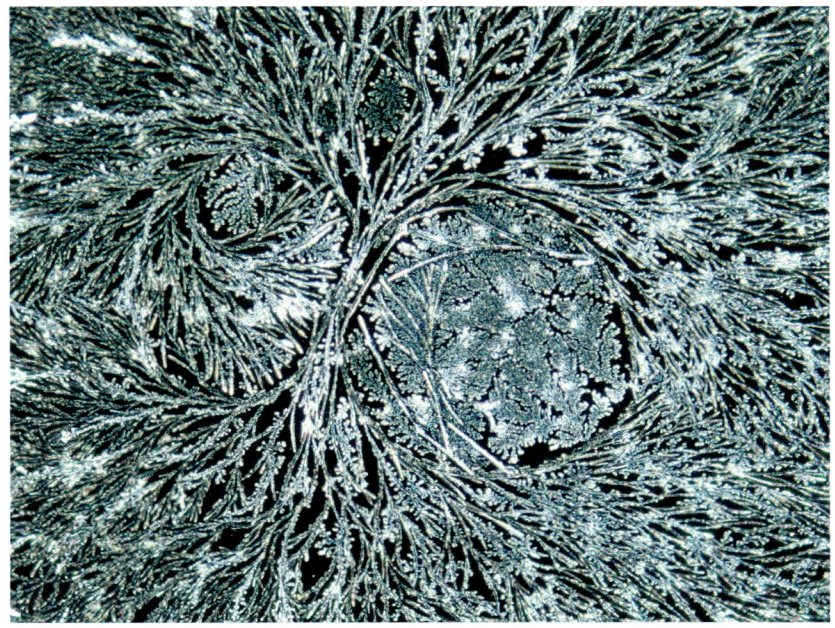

Image 380: Beech tumor, magnified (400x)

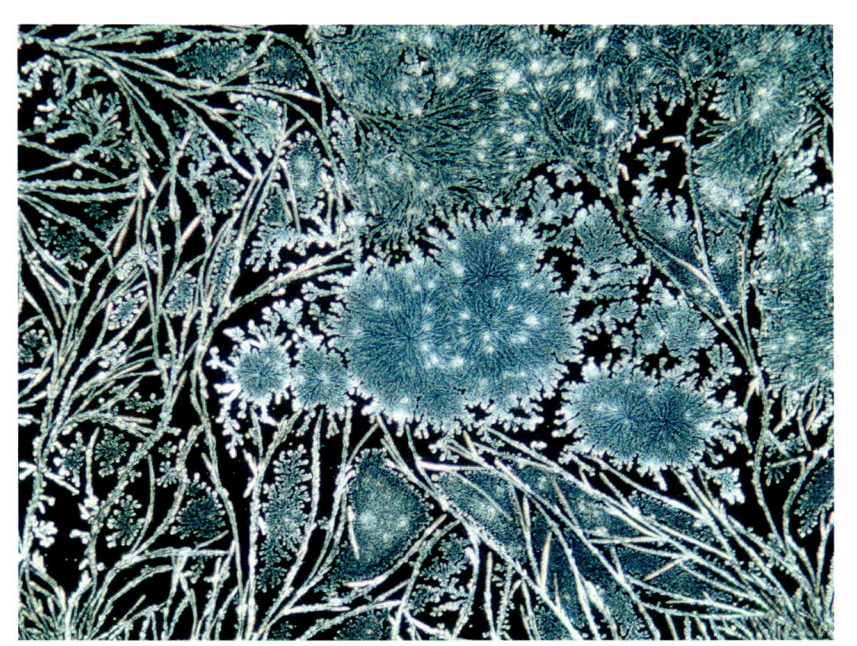

Image 381: Beech tumor, magnified (400x)

6. Reflections on the Seed in the Center and the Cambium

The Center

We had previously gone out with the flowering plant into the astral realm of the earth's orbit and gained a little insight into the intermediate realm of flower and bee. Agitated "whirling wheels" showed us the flower pollen in the water. They could be an expression of fiery, active beings, because the bee lives in warmth. The "butterfly blood" in the drop image aroused completely different sensations. In delicate lines, something from the lifeblood of the butterfly was revealed, which lives entirely in the light.

Now, after the flowers have unfolded, the plant pulls itself completely inward and forms the ovary in which the seeds ripen. In them, the plant form is contracted to the utmost. Whether fruits, kernels, or small seeds—they all create an image of their future becoming in the drop center. The centers appear as individual structural formations typical of the plant. Through these and the drop border, the seeds of the plants can be distinguished from each other and arranged. They give us something of the inner idea of the plant in our experience.

The substance that gathers in the center of the drop images of fruit, seed, and bark shows peculiarities compared to the materiality of other drop images:

1. It asserts itself when exposed to heat; it cannot be dissolved and carried to the periphery (e.g., images 249 and 307).
2. It resists the cold in the freezer compartment of the refrigerator for more than ten days (images 319 and 351).
3. It does not fade even if the plant fruits spent more than a year in the dark (images 266 and 273).
4. It can be revived by the light and even enhanced by the sun and gold leaf (image 347).
5. It shows us a relation to time, in that the still immature seed has a crystalline or also a delicately lively design in the center, and, as it matures, it grows in a richly shaped expanse over the surface of the drop (images 239 to 242).
6. It can be stimulated by a second drop of water at the edge of the first, and can even be drawn into it when the second drop of water reaches the center (images 312 to 315).
7. The seed power of the preparation water forms its own circular center when the water is spread out as a surface on the slide (images 356, 357, 363).
8. The central area is marked—but becomes black—when the cambium of the bark has died (image 374).
9. The center is lost when the seed is stripped of its enveloping shell (images 323 and 325).
10. The center dissolves when the seed dies into the rooted germ (images 308 to 310, 317 and 318, 332).

In winter, the sun's power rests in the many fruits and seeds of the earth's soil. Walking through the winter forest, this is a beautiful image.

The reference to time was mentioned above. The development and expansion of the central configuration that we encounter during the ripening of the seed are not shown in the drop images of root and stem. These images show structures that have come to an end and do not transform. This is different in the fruit and seed images. Something rests in them that harbors forces that are becoming. After germination, the center is lost. The typical shapes of root and shoot then appear in the drop image.

> What summer can become takes place later under the surface of the earth during the winter. And the consequence of this is...that the plant, when it now grows out of the earth in its yearly course, actually first grows with the forces that the sun has given to the earth at least in the previous year, for it draws its power from the soil.[43]

Thus, for many weeks and months, the water preserves a special imprint of this solar power in the center of the drop and shows us an image of it again and again, tireless and typical in its kind—even when the fruits and seeds are removed from the water.

The Cambium

The surprise that the circular center of the Aesculus bark showed in the water became a fascinating question. The center of the fruits and seeds could be recognized as a germ for future life. To what extent should the barks of the young twigs also harbor a germinating power?

The answer to this question is from Rudolf Steiner. In lectures to the workers at the Goetheanum in October 1923, he speaks about the three saps of the tree:

1. The wood sap rises from the earth in spring as living sap and then dies into pure chemistry.
2. The tree unfolds its leaves in the moist and airy surroundings and experiences new powers through the life sap. "Life continuously dies in the trunk; it renews itself in the leaf."[44]
3. "But this is not yet over; but now, while this is happening, a new layer of plant forms between the bark, which is still permeated by the sap, and the wood; I can no longer say that a sap forms...because what forms there is quite thick. It is called cambium."[44]

Steiner goes on to speak about the cambium, which will bring us closer to the solution of the riddle:

> But the cambium—this brings the plant into connection with the stars, with what is above. And it is in this cambium that the shape of the next plant already develops. This then passes on to the seed, and through it the next plant is born, so that the stars, through the cambium, produce the next plant. So, the plant is not merely

produced from the seed—that is, it is of course already produced from the seed, which first needs the influences of the heavens.[44]

Rudolf Steiner writes on the blackboard as an overview:

Wood sap : Earth : Chemical
Life sap : Earth's periphery : Living things
Cambium : Stars : Spirit[44]

So we can understandably encounter this astonishing central imprinting of fruits and seeds also in the barks of young branches, whose cambium preserves the archetype of the plant from the starry world.

The mistletoe plant bores its sinker into the cambium of the tree and unfolds its spherical shape from this region of power.

We see in the forest—more and more frequently in recent years—how new plants grow up from the cambium of felled tree stumps. They grow to considerable size and then die.

In this way, we are led from the flowering plant through seed fruit and cambium even further out into the stars. Fruits, seeds, and barks speak from this sphere.

The etheric circle of the earth was described above.

In four ways the etheric forces "thread" themselves into the work of the elements and appear in all growth and regeneration processes of the living world. *Specific* forms do not yet appear at this level. Etheric forces are pure life forces and not form-giving formative forces. However, they are the basis on which the star forces can manifest their formative images.

It was indicated above that the actual, species-specific formative forces from higher regions—the planets and fixed stars— work into this sea of ether. Only through this does the violet become a violet, the sunflower a sunflower.

And what do we see in the images of the fruits, seeds, and barks? We can distinguish them from each other and classify them according to their origin. We see an individually designed drop image. This was not the case with roots and plant stems. In these images, an individual assignment was not possible. Species-specific, cosmic forces seem to shape the drop image of the seed and the bark in its indestructible expressiveness.

Let us dare to relate the fruit, seed, and bark images to the ethers. The circular center gives us an image of the ovule of the plant. The ripening of the seeds takes place in warmth and in time. In this way we can find a reference to the warmth ether, even if we consider that the seed must pass into chaos so that the world of archetypes can be "inscribed" in the seed.

Ernst Lehrs describes these warmth-ether forces in his book *Man or Matter*[45] and calls the warmth ether "chaos ether." Goethe speaks in this respect—when the plant passes into blossoming and fruiting—of the "spiritual anastomosis." Rudolf Steiner describes to us how, after the transition into seed, the plant dwells "for a moment" in the divine-spiritual world and is refreshed by it.[46] The primordial images are mediated by the warmth ether, and the seed, fruit, and bark images show a species-specific image expression, through their central circular design, which is in harmony with the edge of the drop.

Etheric forces never work on their own, and so here it is probably the life ether that is added in liquid form. Life ether creates out of an individual, always in relation to wholeness. The life ether penetrates the substance due to its more centripetally directed development of forces and enables an individualized wholeness. Skin formation is suited to the life ether as an expression of the inner being.

7. Charcoal and Ash of the Plant in Water: Mistletoe Charcoal (Viscum carbo) in Water

Charcoal and ash preparations, potentized, are often used as remedies. What does the drop pattern show when charcoal or ash is added to water?

Mistletoe (*Viscum album*) was mentioned among the berry fruits. Its whole plant being shows us that it is actually not an earth plant. It grows out of the cambium of the tree as a parasite in spherical form and lacks much of what we perceive in the other plants. In its vegetation cycle, which is contrary to the course of the year, we also see a strong force of its own. Here we shall now look at its charcoal.

A tree with mistletoe had died and been felled. A mistletoe bush had withered and was golden-yellow in color. Its branches were broken off (they are connected like little joints) and the stems were cut into small pieces.

The whole thing was burnt in an open fire.

Now no ash was produced, but the plant parts were charred (*Viscum carbo*).

This was pulverized in a mortar and the black charcoal was put into a test tube with water. These preparations showed the same images for months. Each time they aroused great astonishment and joy because of their differentiated, rhythmic richness of form on the one hand and their coherent consistency on the other.

THE SILENT LANGUAGE OF LIFE

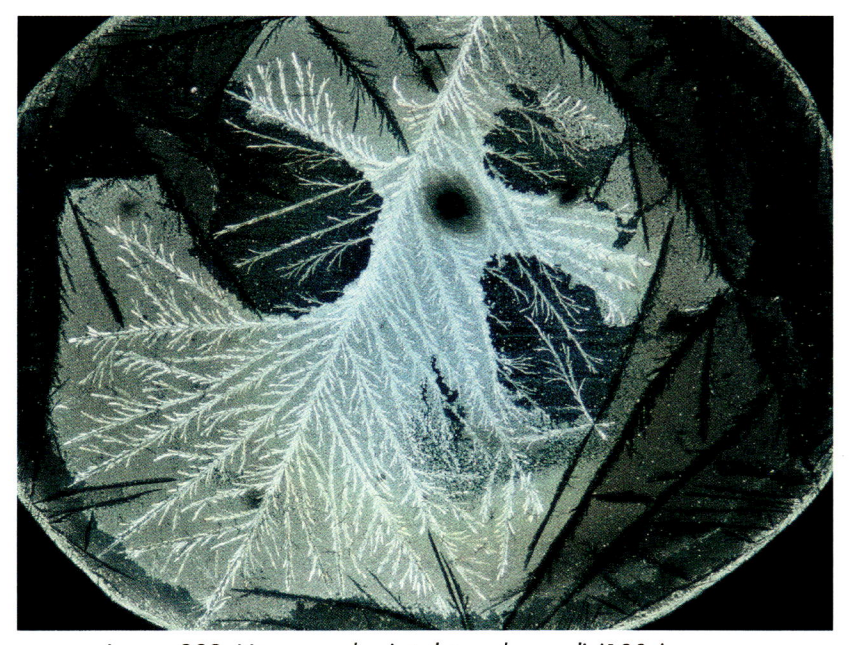

Image 382: Viscum carbo (mistletoe charcoal) (100x)

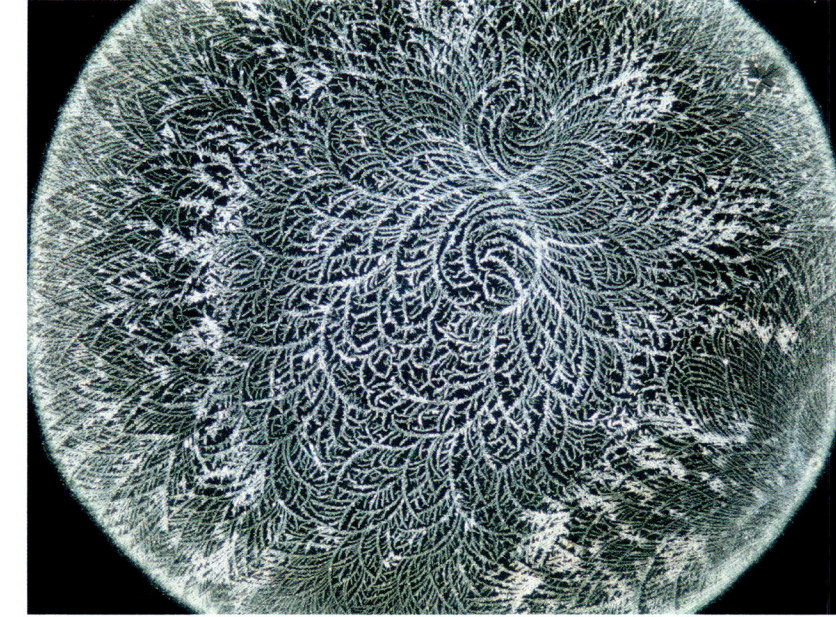

Image 383: Viscum carbo (100x)

Image 384: Viscum carbo (200x)

Image 385: Viscum carbo (200x)

Image 386: Viscum carbo (400x)

Image 387: Viscum carbo (200x)

Plants, Fruits, Seeds, Barks, and Ashes in Water

In these images (382 to 391), we see, besides dense tissue structure, above all lemniscates again and again in the most whimsical variations. It twists and spreads and swirls into itself or grips the circumference through large-arched figures of eight.

A very moving, "willful" event; again and again, a duality is experienced in it, which wants to form itself into a unity. A duality also appears in the figures characterized by light and darkness, which often appear in the image as stark contrasts.

Mistletoe is characterized by strong astral forces. We also saw lemniscates and rhythmic arch figures (interior formation!) in flowering lavender. The flowering plant touches the astral realm and meets the world of insects.

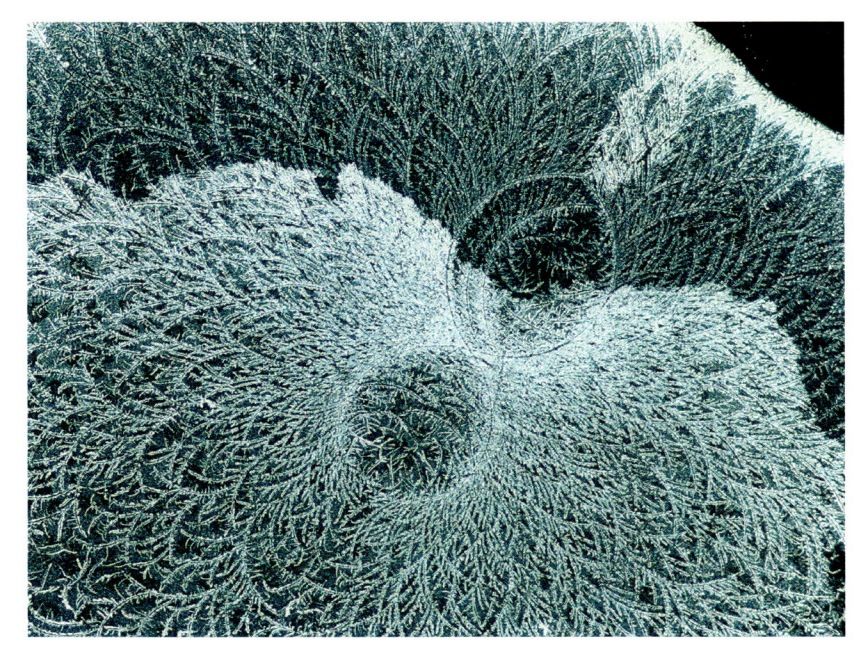

Image 388: Viscum carbo (200x)

Image 389: Viscum carbo (200x)

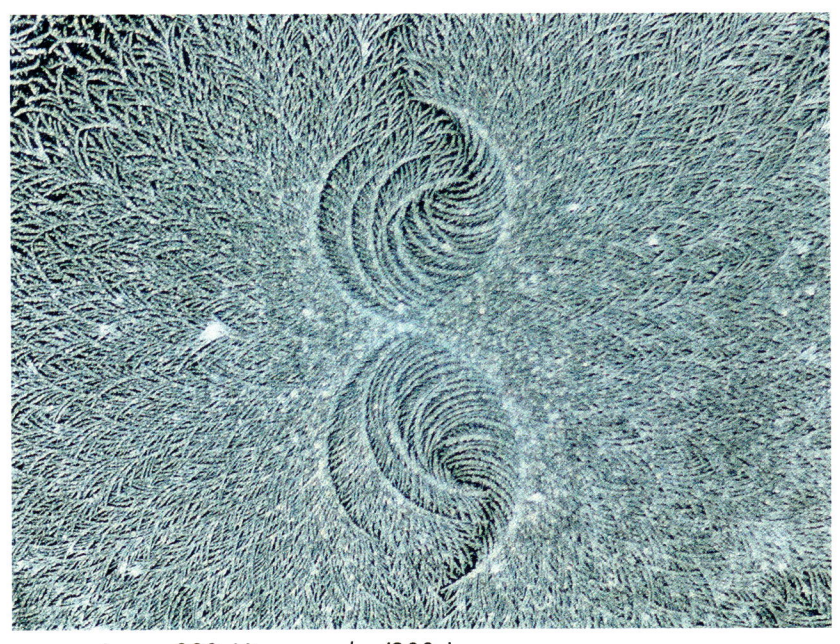

Image 390: Viscum carbo (200x)

Image 391: Viscum carbo (200x)

Image 392: Viscum carbo, batch 5 years old (200x)

Now there was a mistletoe charcoal preparation that had stood for over five years. We see its "water language" in image 392 to 395. Something new appeared, insofar as the duality had now separated. Elongated, arching branches or even clusters are dotted with small "blossoms." These images are enchanting because of their playful beauty.

Image 393: Viscum carbo, batch 5 years old (400x)

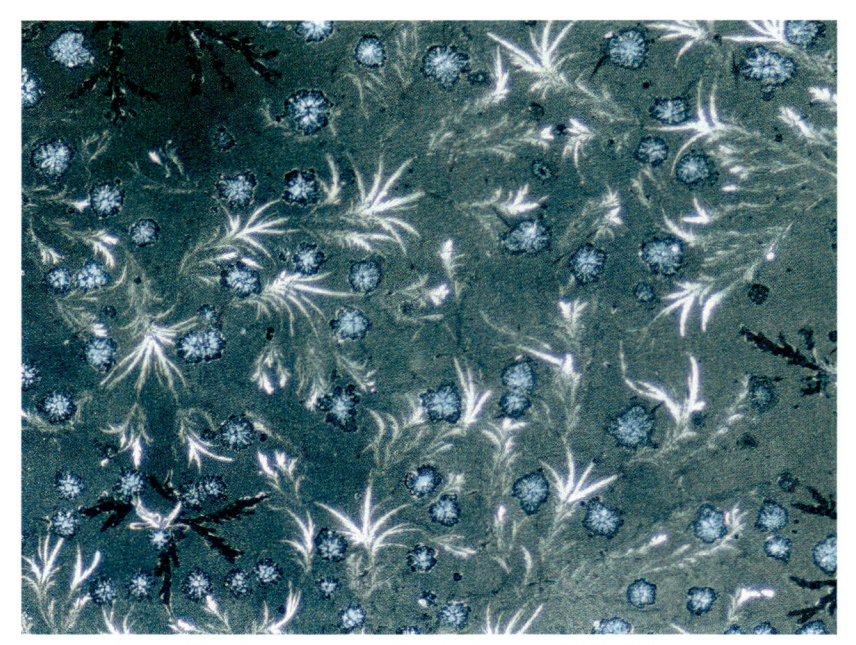

Image 394: Viscum carbo, batch 5 years old (200x)

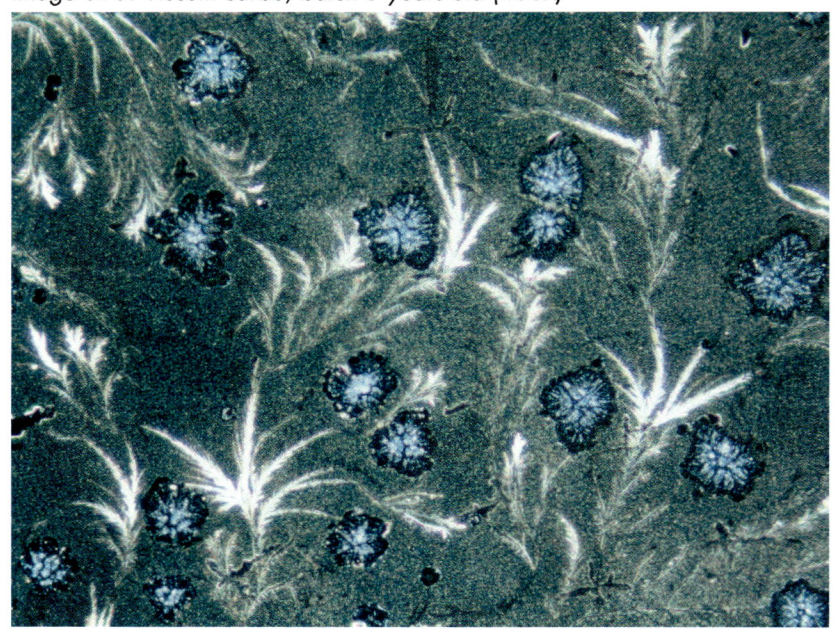

Image 395: Viscum carbo, batch 5 years old (400x)

Plants, Fruits, Seeds, Barks, and Ashes in Water

Image 396: Winter mistletoe (400x)

Image 397: Winter mistletoe (400x)

Four more images of different fermenting mistletoe saps are now shown. One is the fermenting juice of the oak winter mistletoe and one of the oak summer mistletoe. They were dripped on.

Images 396 and 397: Fermenting sap of winter mistletoe (*Quercus*).

Images 398 and 399: Fermenting sap of summer mistletoe (*Quercus*).

Both plants had been prepared and stored by Theodor Schwenk for medicinal purposes in 1984.

The difference between winter and summer mistletoe comes into focus here in that the winter mistletoe shows a crystalline design and the summer mistletoe a delicate, flower-like design. In the case of the summer mistletoe, the centered figures are also surrounded by a shell. We see how, even after twenty-five years, an image of the essence of mistletoe can appear in the water.

It should also be briefly mentioned that Rudolf Steiner had mistletoe prepared in a special way for cancer. In this preparation process, summer mistletoe and winter mistletoe are mixed together by dropping, circling, and potentized for healing purposes.

Image 398: Summer mistletoe (400x)

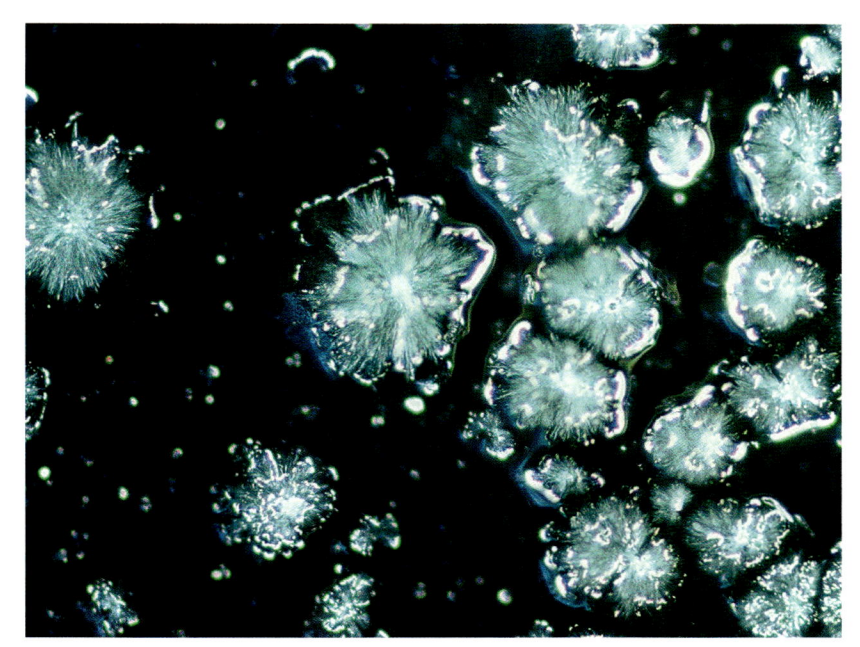

Image 399: Summer mistletoe (200x)

— THE SILENT LANGUAGE OF LIFE —

Image 400: Cinis oxalis foliis (wood sorrel ash) (100x)

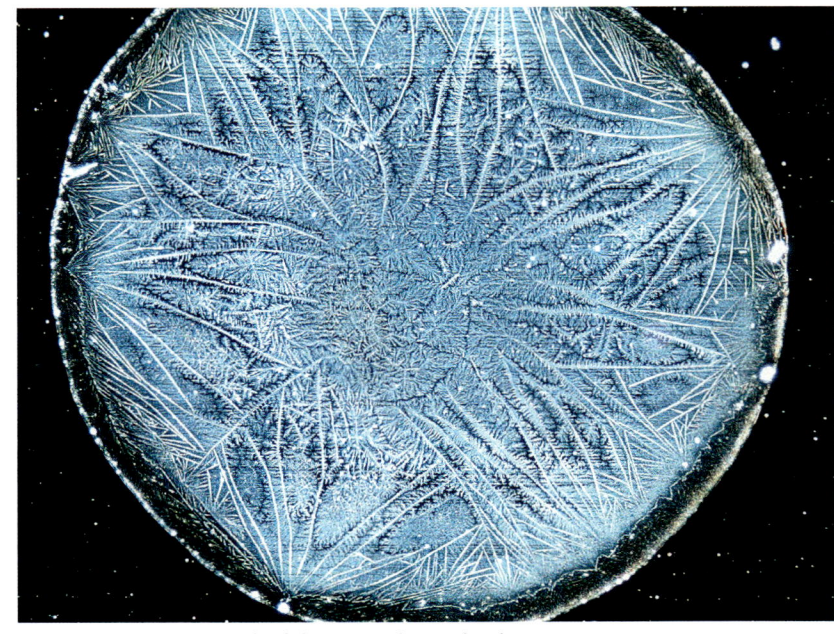

Image 401: Cinis oxalis foliis (wood sorrel ash) (100x)

Plant ash in water

Some of the many ash preparations of the plants are shown here (images 400 to 421). The ashes—with a much lighter materiality and light color—are prepared in test tubes and left to stand for many months. The ash settles on the bottom.

Cinis oxalis foliis, wood sorrel ash (images 400 to 403).

Cinis vitis vinifera foliis, vine leaf ash (images 404 and 405).

Cinis tiliae floribus, lime blossom ash (images 406 and 407).

Cinis equiseti vernale, horsetail harvested and ashed in spring (image 408).

Cinis equiseti vernale, horsetail harvested and ashed in autumn (image 409).

We see differences in the ash structures. They can be recognized as species-specific, even if some can only be distinguished on close examination. In the Equisetum ash assemblages we can see clear differences in the seasons.

Image 402: Cinis oxalis foliis (wood sorrel ash) (400x)

Image 403: Cinis oxalis foliis (wood sorrel ash) (400x)

Plants, Fruits, Seeds, Barks, and Ashes in Water

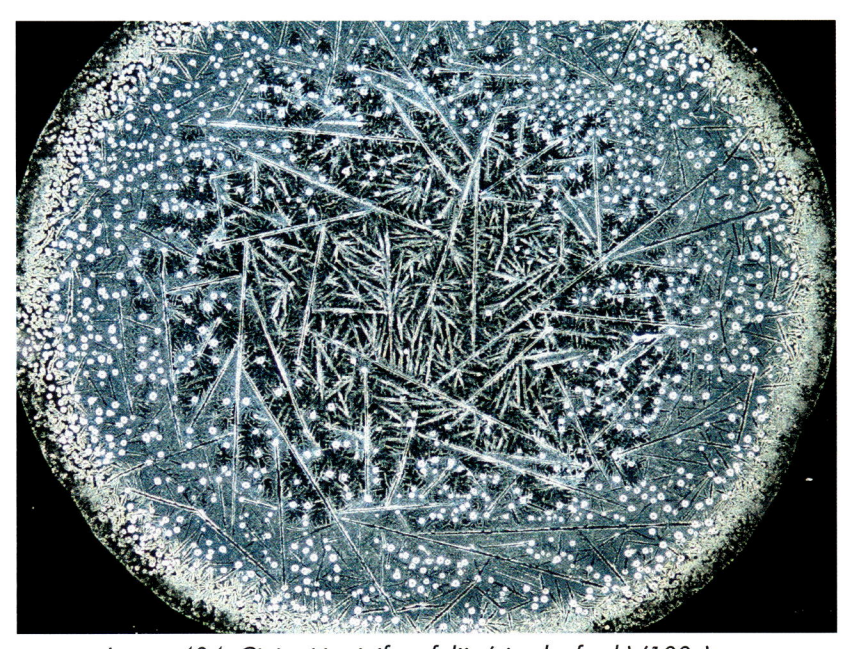

Image 404: Cinis vitis vinifera foliis (vine leaf ash) (100x)

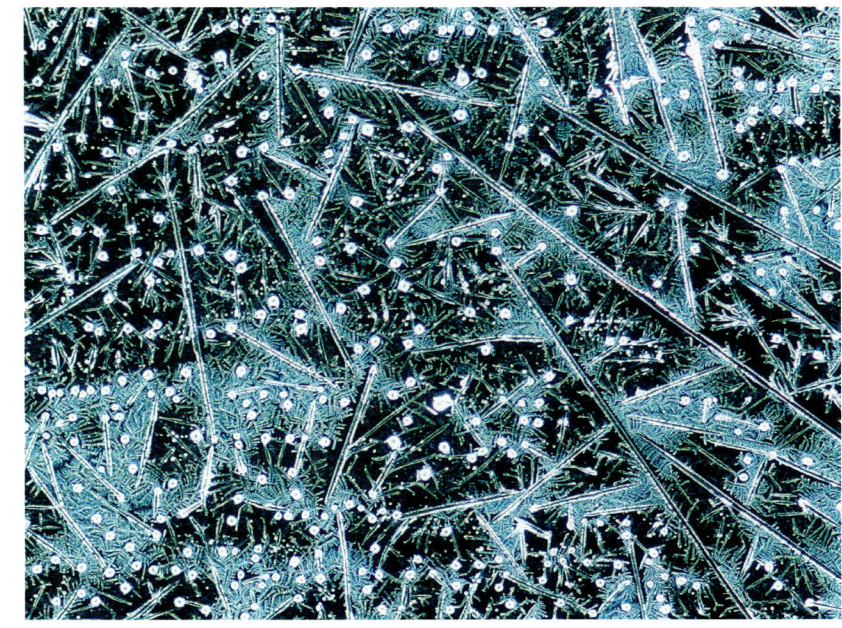

Image 405: Cinis vitis vinifera foliis (vine leaf ash) (200x)

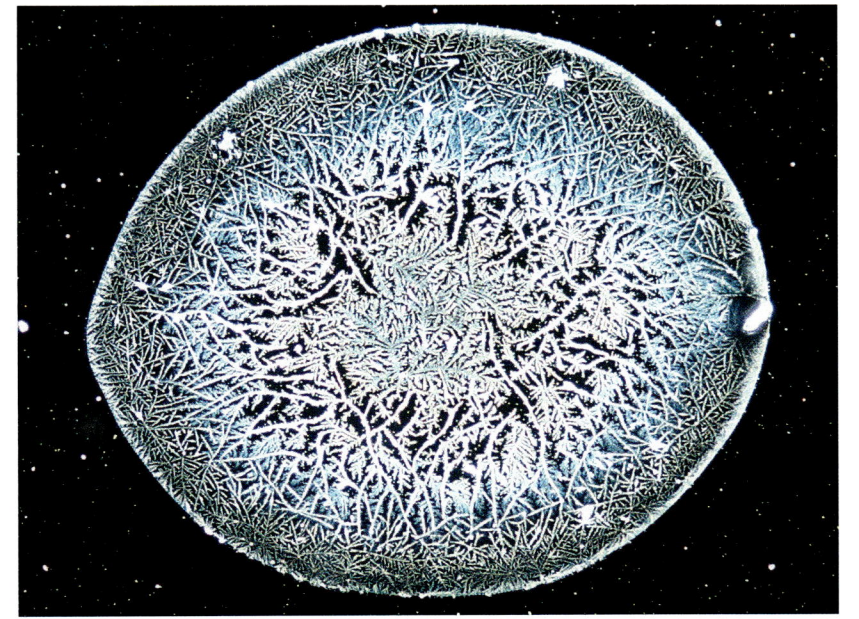

Image 406: Cinis tiliae floribus (lime blossom ash) (100x)

Image 407: Cinis tiliae floribus (lime blossom ash) (400x)

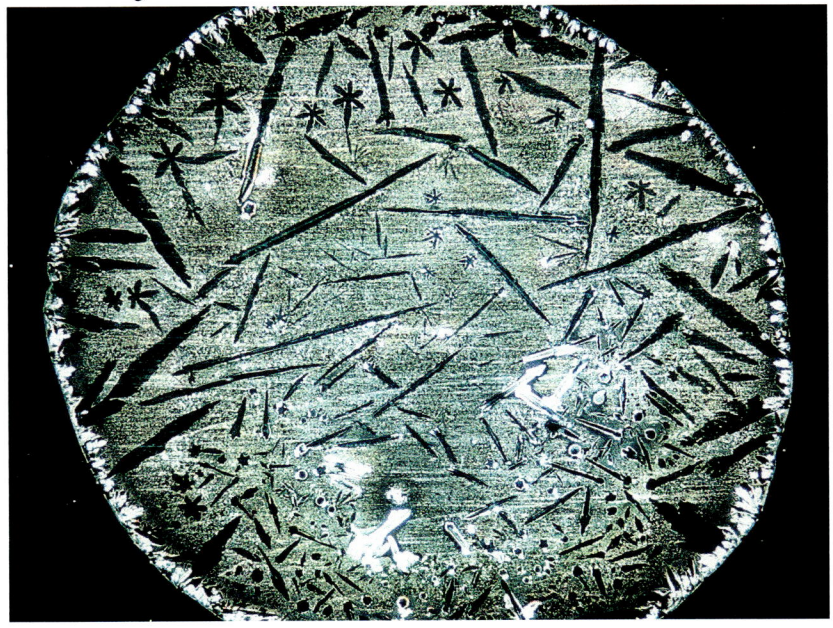

Image 408: Cinis equiseti vernale (horsetail, spring) (100x)

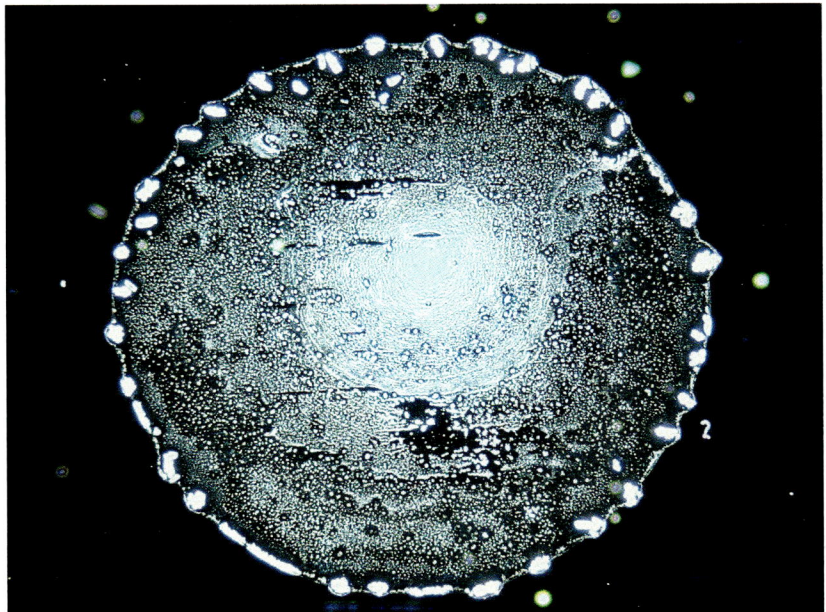

Image 409: Cinis equiseti vernale (horsetail, autumn) (40x)

Image 410: Cinis equiseti vernale (horsetail, spring, very old batch) (100x)

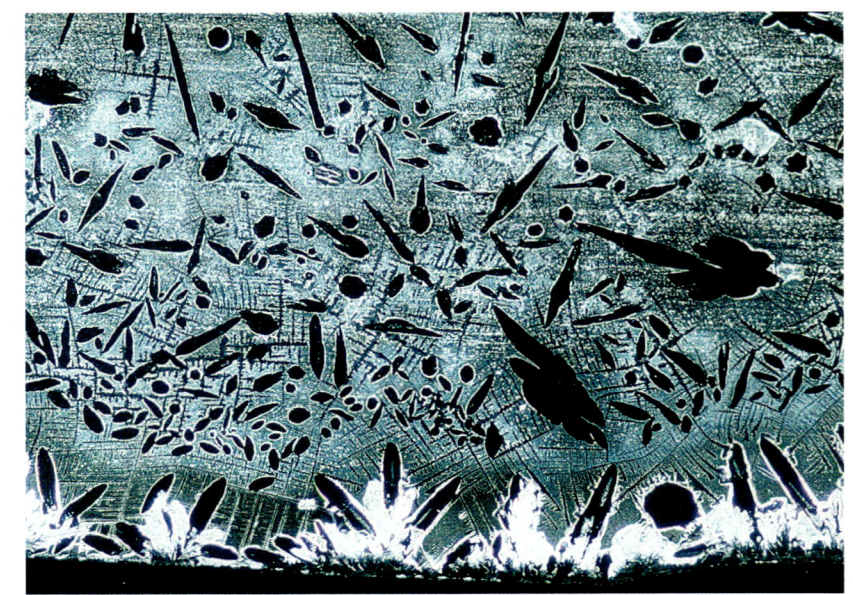

Image 410: Cinis equiseti vernale (horsetail, spring, very old batch) (400x)

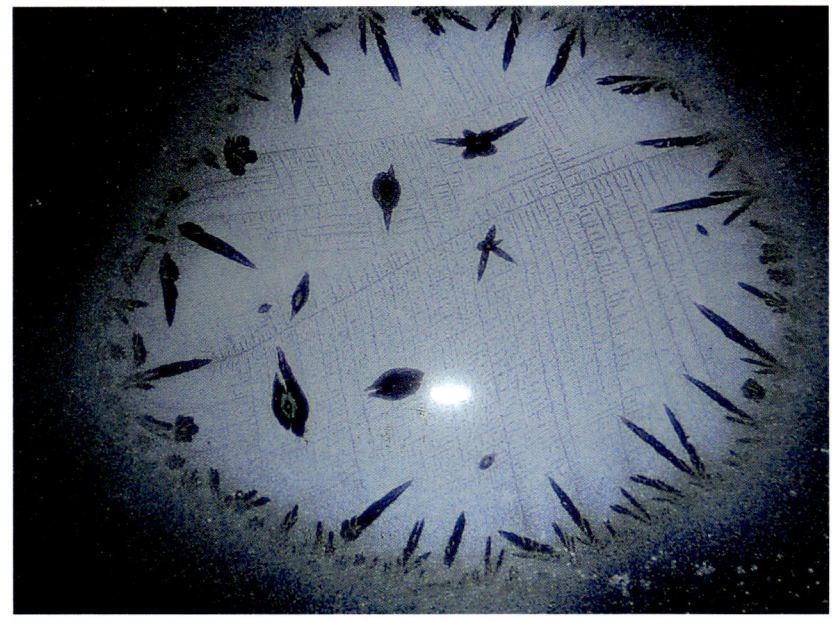

Image 412: Dandelion seeds (many years old batch) (200x)

Image 413: Dandelion seeds (many years old batch) (400x)

In images 410 and 411 we see a preparation of spring horsetail, ashed, but which had stood for many years. Even though the typical ash structure shows up in it, other black figures of a strange shape appear. In a dandelion seed clump (*Taraxacum semen*) that was many years old, these same figurations of a deep black color appeared. Thus they are juxtaposed here as puzzling phenomena (images 412 and 413).

Reference should be made to a pastel sketch by Rudolf Steiner, November 12, 1923, which can be found in the portfolio *Ein malerischer Schulungsweg*[47] (A path of education in painting). Here we encounter similar figures.

Plants, Fruits, Seeds, Barks, and Ashes in Water

Image 414: Wood ash (40x)

The following two images come from a wood ash preparation (images 414 and 415, digital camera). We are amazed at small "tree formations" between long fields of needles.

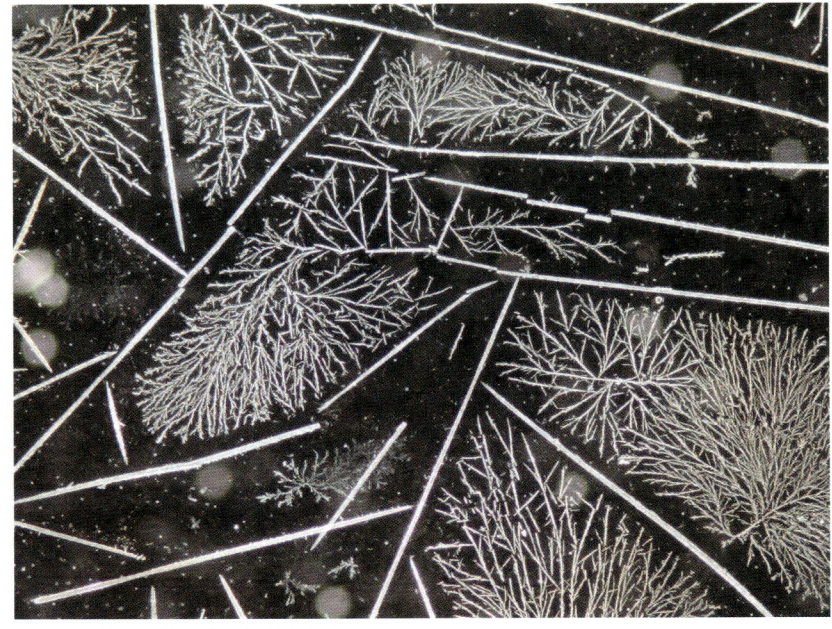

Image 415: Wood ash (200x)

THE SILENT LANGUAGE OF LIFE

Image 416: St. John's wort ash in water (40x)

Image 417: St. John's wort ash in water (40x)

Image 419: St. John's wort ash in water (400x)

Image 418: St. John's wort ash in water (400x)

Image 420: St. John's wort ash in water (400x)

144

Plants, Fruits, Seeds, Barks, and Ashes in Water

Image 421: St. John's wort ash in water (40x)

Further on, images 416 to 421, we see ash images of St. John's wort, Hypericum perforatum (digital camera).

The black radiance of the St. John's wort ash images and the juxtaposition of light and darkness are very impressive and have not been seen in any other ash. The healing effects of the plant, especially for depression, have been sufficiently researched and tested.

Regarding the botany of this plant, the double-edged stem is perhaps noteworthy as it is very rarely seen in the plant kingdom. In the magnifications of the ash images, numerous small dark dots show up in the black delimited fields. As is so often the case, here too the plant—even in the ash—seems to give an image of itself. If you look at the leaf of Hypericum perforatum against the light, it appears perforated. However, these "perforations" are secretion vessels containing a light-colored liquid of essential oils and resin.

In the following three sub-chapters, the question is posed as to whether the plants treated according to spiritual-scientific knowledge reveal something of their newly gained powers through the treatment in the drop image.

Image 422: Mesembryanthemum, salt beads (100x)

Image 423: Mesembryanthemum, 2nd week, exposed (100x)

Image 424: Mesembryanthemum, 2nd week, exposed (400x)

Image 425: Mesembryanthemum, 2nd week, exposed (400x)

8. Exposed and Unexposed Plant

Preparation of remedies and preservation (without alcohol) can be done by (among other things) subjecting a plant preparation in a screw jar to a rhythm of light and darkness over seven weeks. This is done in such a way that the preparations are placed in the light for one hour in the early morning hours (between 6 am and 8 am) and in the evening hours (between 6 pm and 8 pm) and spend the rest of the time in a dark box. Since there is of course no sunlight outside at all during these "exposure times" in the winter time, something else is important here. It was Rudolf Steiner who pointed out the significance of this kind of rhythmic treatment of medicinal plants.[48]

It has to do with the *Pisces-Virgo* forces of the zodiac that form a directional axis each morning and evening. The fact that this is the case throughout the year has been proven and documented by Theodor Schwenk in numerous experiments—e.g., in his brief paper "Vom Einströmen kosmischer Kräfte in irdische Substanzprozesse"[49] (On the influx of cosmic forces into earthly substance processes).

Steiner spoke of these morning and evening forces, which would have to be used in the laboratories of future science—science that will have spiritualized itself in the sense of the good "if humanity does not want to go completely into decadence," and further:

> It will be the task of good, health-giving science to find certain cosmic forces which can arise on earth through the interaction of two cosmic directional currents. These two cosmic directional currents will be Pisces-Virgo.[48]

In the following images we see, using the example of a plant preparation, that there are differences in the drop image when a plant is exposed in the way described above or when the preparation is left untreated.

Plants, Fruits, Seeds, Barks, and Ashes in Water

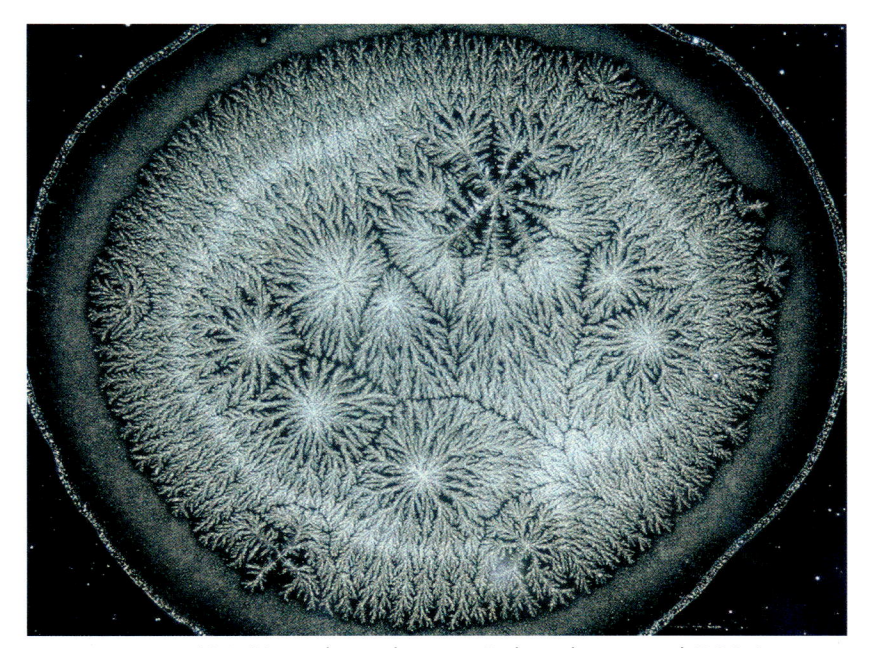

Image 426: Mesembryanthemum, 3rd week, exposed (100x)

Image 427: Mesembryanthemum, 3rd week, exposed (400x)

Image 428: Mesembryanthemum, 5th week, exposed (200x)

Image 429: Mesembryanthemum, 5th week, exposed (100x)

Mesembryanthemum (also called ice plant) grows in the Mediterranean region and on the Canary Islands as a succulent ground cover. The fleshy leaves are very hairy and densely covered with "salt beads." For some years it has been used as a medicinal plant in ointment form for skin diseases.

The plant was harvested on Lanzarote, cut into small pieces and put into two water jars with screw caps.

The first batch was exposed to light for one hour in the morning and one hour in the evening for seven weeks, before which the water was briefly agitated; the rest of the time it stood in the dark. The second batch remained on the laboratory bench.

Image 422: Salt beads from Mesembryanthemum, freshly dripped on.

1st batch of Mesembryanthemum, exposed:

   Images 423 to 425: Mesembryanthemum, 2nd week, exposed.

   Images 426 and 427: Mesembryanthemum, 3rd week, exposed.

   Images 428 and 429: Mesembryanthemum, 5th week, exposed.

After that the designs changed into a barely visible geometric drawing.

147

THE SILENT LANGUAGE OF LIFE

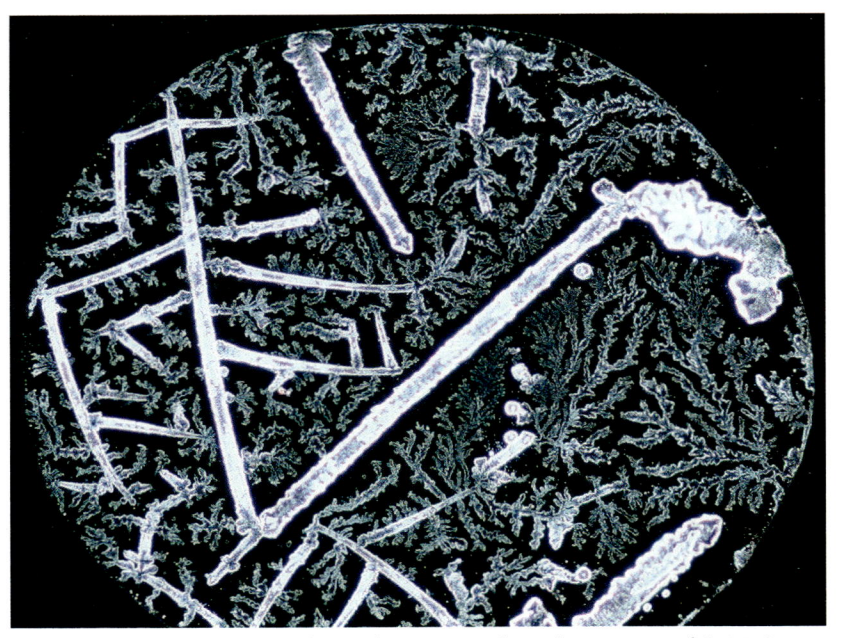

Image 430: Mesembryanthemum, 2nd week, unexposed (100x)

Image 431: Mesembryanthemum, 3rd week, unexposed (200x)

2nd attachment of Mesembryanthemum, unexposed:

Image 430: Salt beads and leaves of Mesembryanthemum, 2nd week unexposed.

Image 431: Mesembryanthemum attachment, 3rd week unexposed.

The drop pattern was constant until the second month.

Images 432 to 435 show us the drop pattern of the soda plant (unexposed) in its truly imaginative, crystalline structuring after more than two months in the water. Here the plant shows an image of itself, for all leaves and stems are densely hairy.

Rhythmically exposed and unexposed plants give an impression of different degrees of unfolding force in the drop image. One could say that rhythmic exposure makes swelling, more lively effects visible.

After the fifth week, the image formations of the exposed preparation change into geometric figures, as can also be seen in homeopathic potency enhancement (not shown here). The unexposed preparation retains its crystalline imprint for months in the water.

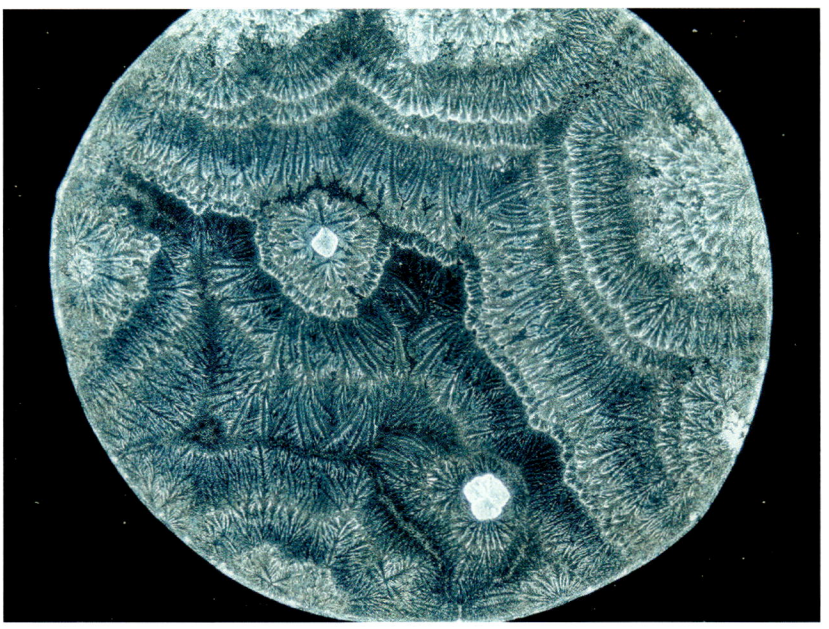

Image 432: Mesembryanthemum, over 2 months, unexposed (100x)

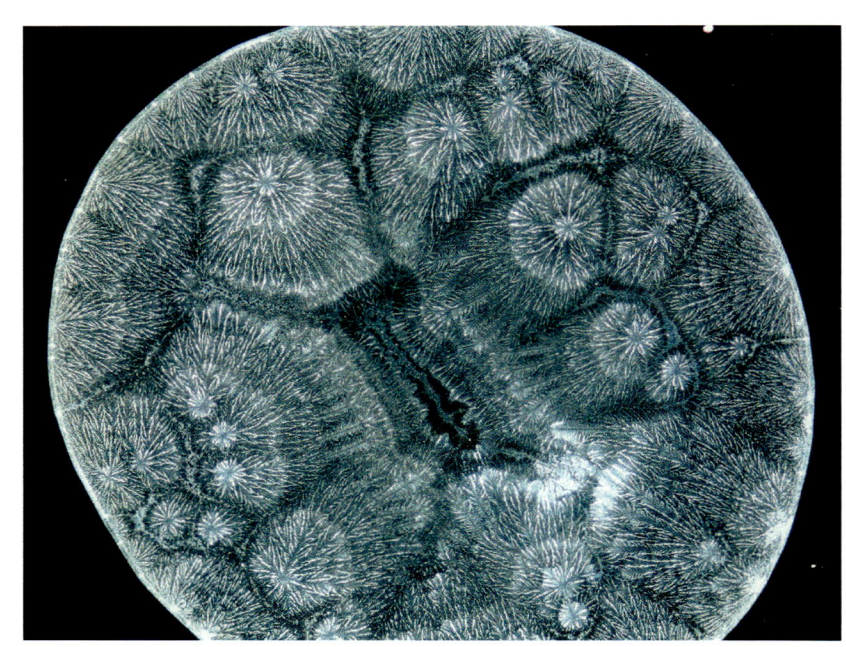

Image 433: Mesembryanthemum, over 2 months, unexposed (100x)

Plants, Fruits, Seeds, Barks, and Ashes in Water

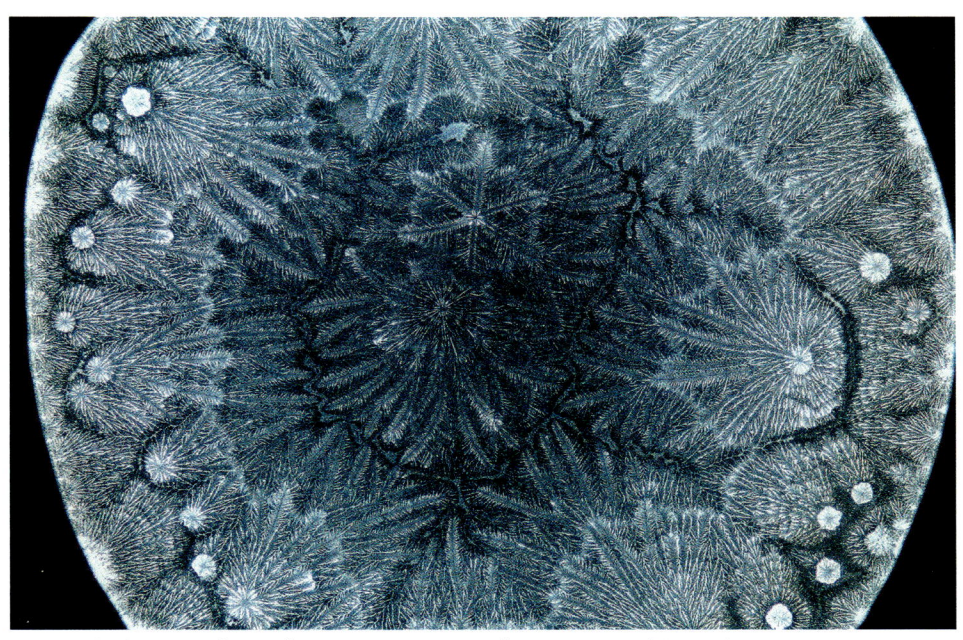

Image 434: Mesembryanthemum, over 2 months, unexposed (100x)

Image 435: Mesembryanthemum, over 2 months, unexposed (400x)

THE SILENT LANGUAGE OF LIFE

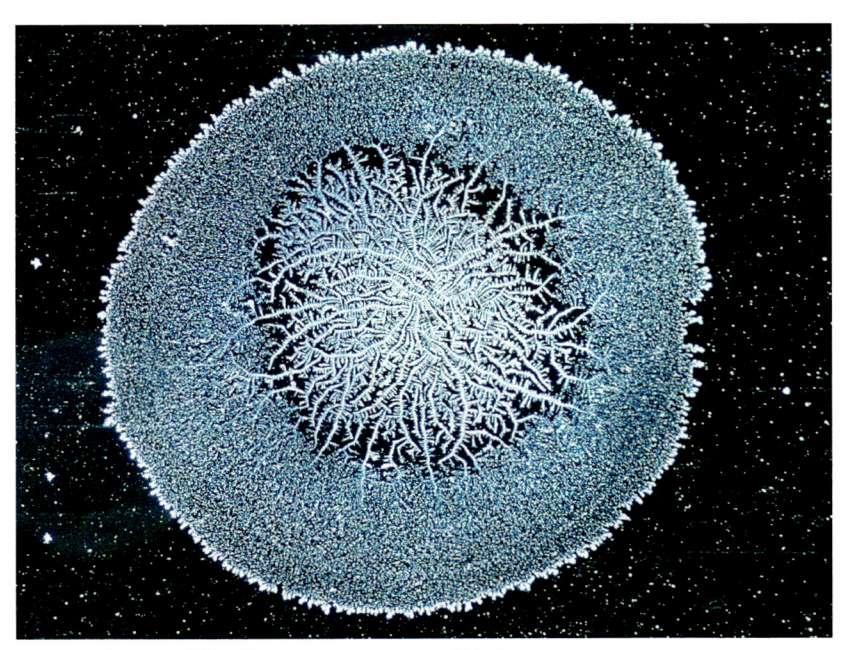

Image 436: Yarrow preparation (40x)

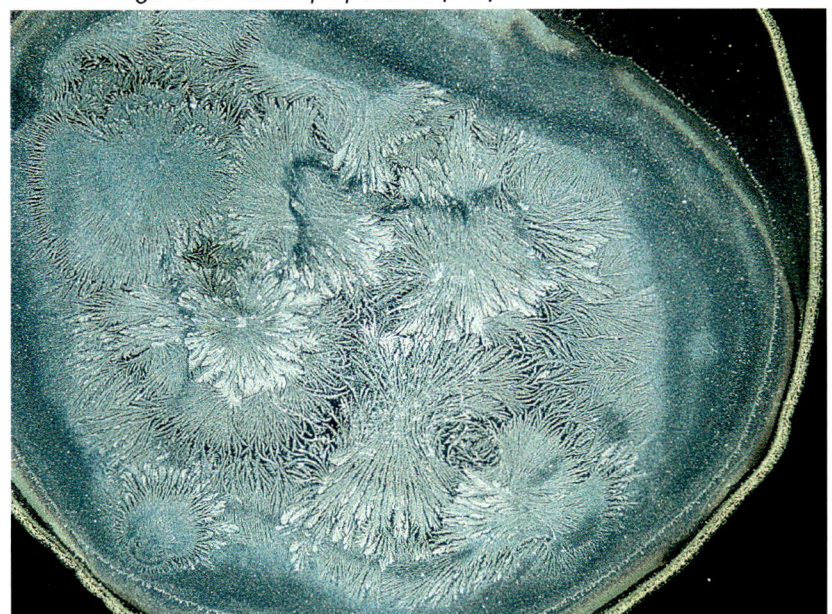

Image 437: Camomile preparation (100x)

Image 438: Camomile preparation (400x)

Image 439: Camomile preparation (100x)

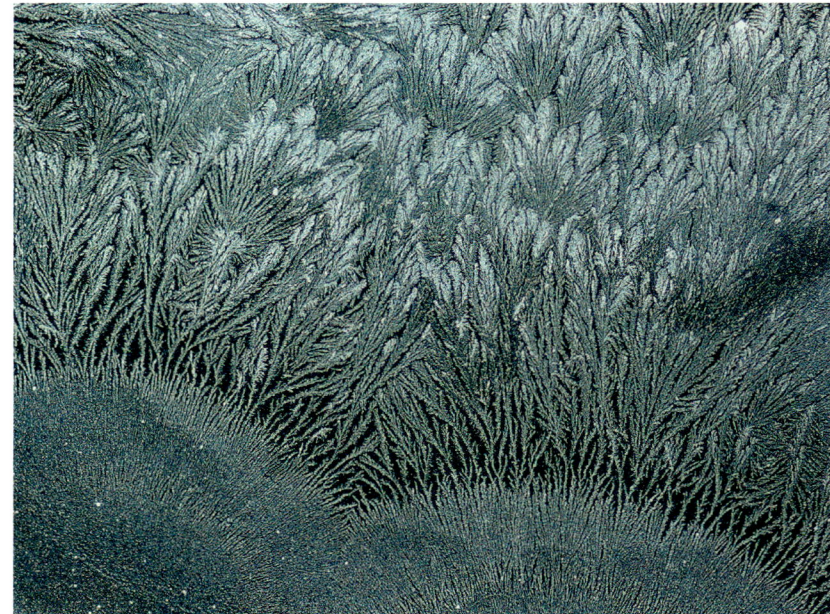

Image 440: Camomile preparation (400x)

9. Four Preparations from Rudolf Steiner's Agricultural Course

In 1924, in Koberwitz near Breslau, Steiner gave farmers completely new, groundbreaking suggestions for cultivating the soil, for raising and caring for food crops and domestic animals through spiritual-scientific insights in his course, *Spiritual Foundations for the Renewal of Agriculture*.[50]

Without cultivation of the soil based on these insights, the food crops would no longer be able to give humans what they need for actually "being human."

This impulse, which carried far into the future, was taken up by the farmers and flowed into practice through the Demeter method of cultivation. At the center of this "biodynamic farming method" are eight preparations, six of which are used to prepare compost and manure using medicinal plants. The complex method of preparation of the medicinal plants is attuned to the cosmic forces, modified by planetary effects. Through the extraordinarily differentiated preparations, the earth and the plant being are strengthened in relation to their physiological forces.

Two preparations (horn manure and horn silica) are added directly to the soil by spraying in drop form, the others via the compost or manure to which the preparations have been introduced.

Four of the total of eight "preparations" were prepared over several weeks in distilled water (A. d.) and are shown here (images 436 to 448).

Image 436: Yarrow preparation in aqua distillata (A. d.).
Images 437 to 440: Camomile preparation in A. d.
Images 441 and 442: Oak bark preparation in A. d.
Images 443 to 448: Horn manure preparation in A. d.
    (digital camera).

Although the images only document a part of the preparations and although the images can reveal something only to those who are familiar with this comprehensive subject, they are nevertheless presented here in this brief form.

The horn manure preparation from the cow horn shows whimsical figurations. We could consider these—as well as the other preparation drop images—in a context of their production and mode of action in order to come to a certain understanding. However, this would go beyond the scope given here.

The horn manure preparation is stirred rhythmically to the right and left before it is used. The direction of rotation is reversed in each case when the vortex has been formed all the way to the bottom of the container, where the vortex is stopped so that chaos is created, and another vortex is created with the opposite direction of rotation.

Image 441: Oak bark preparation (100x)

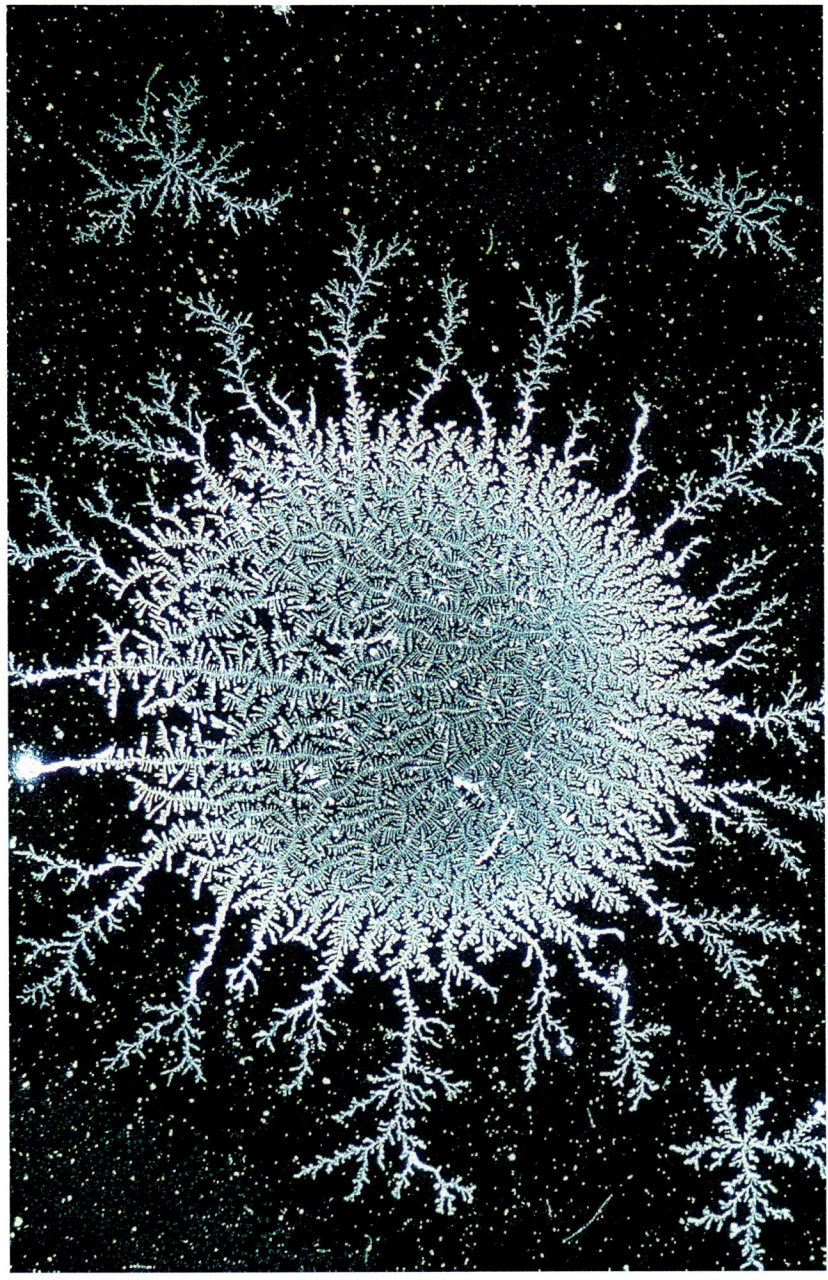

Image 442: Oak bark preparation (400x)

THE SILENT LANGUAGE OF LIFE

Image 443: Horn manure preparation (100x)

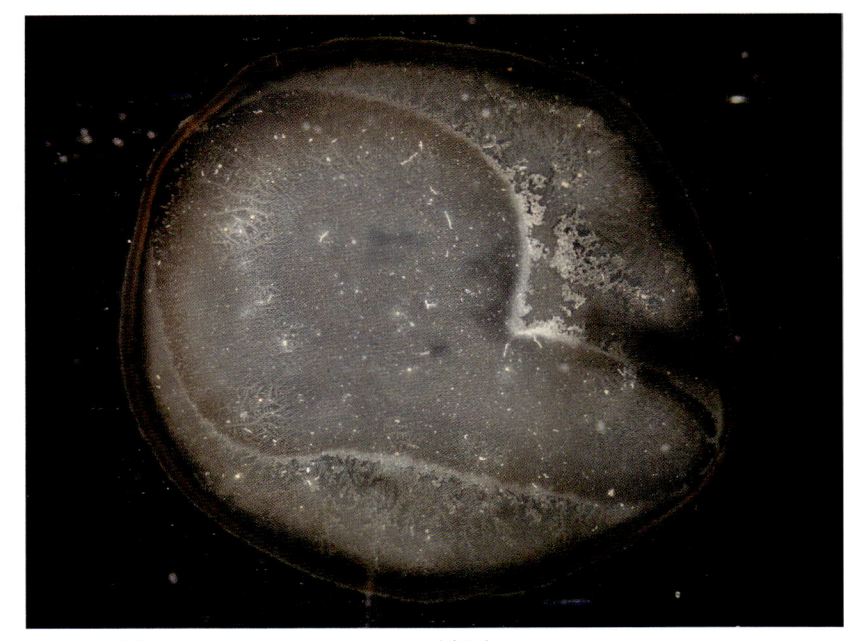

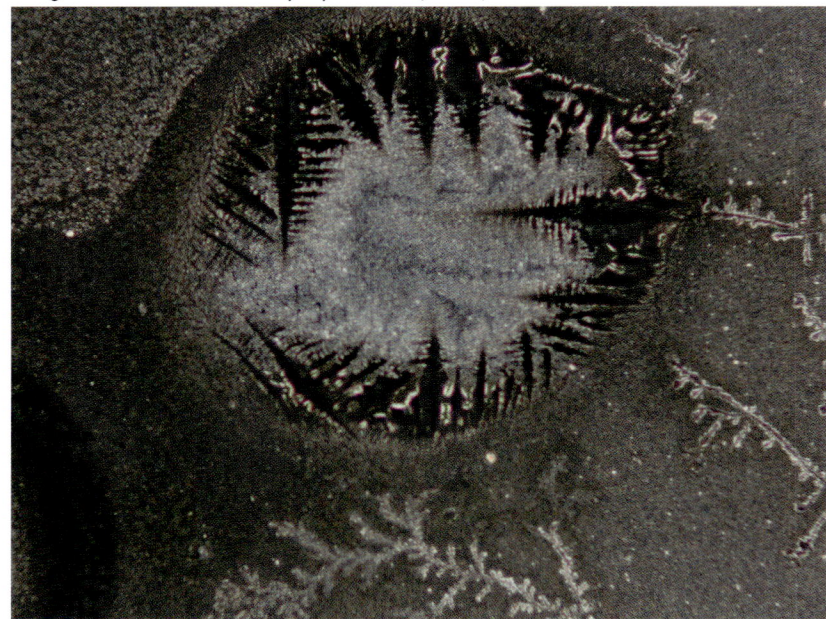

Image 444: Horn manure preparation (40x)

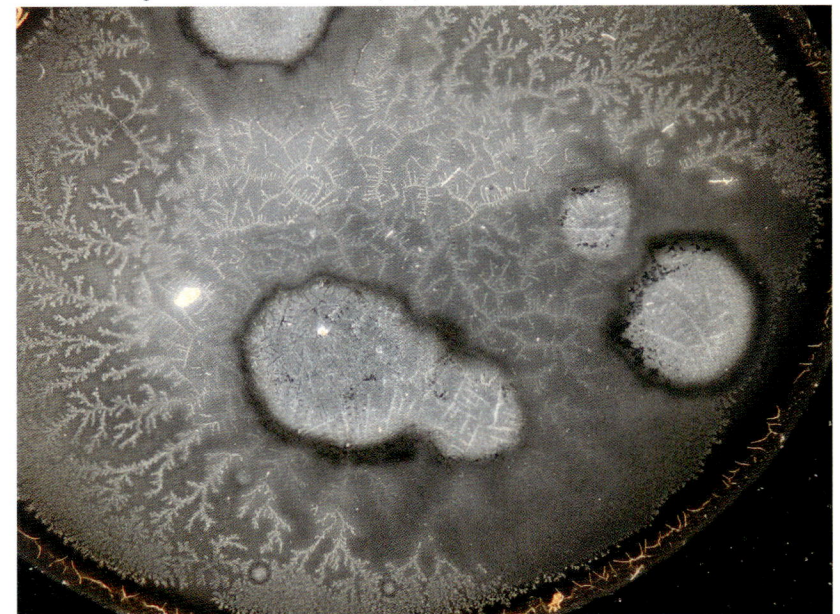

Image 445: Horn manure preparation (100x)

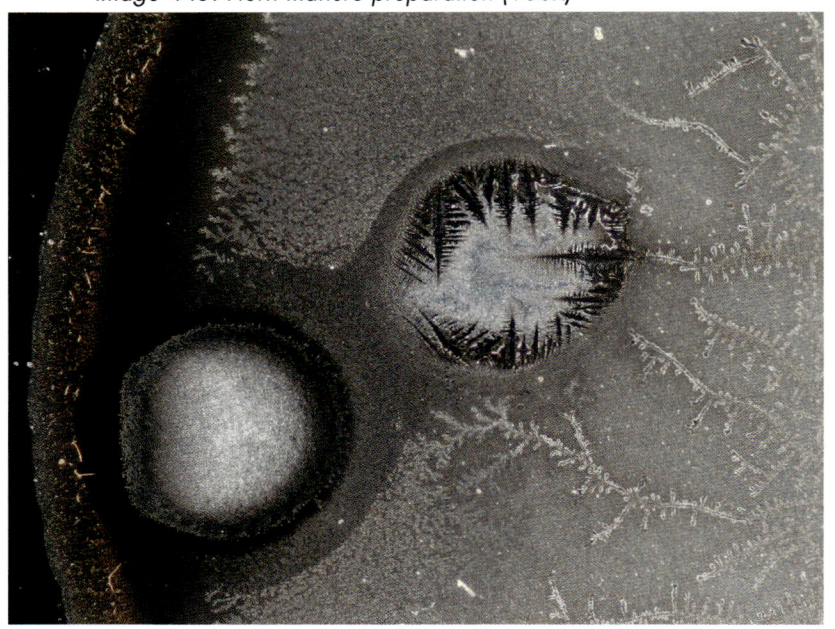

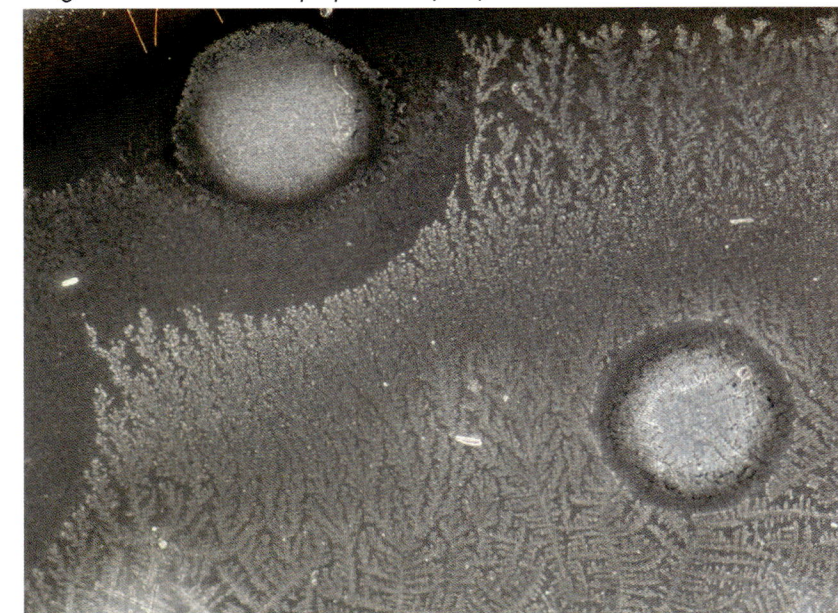

Image 446: Horn manure preparation (200x)

Image 447: Horn manure preparation (200x)

Image 448: Horn manure preparation (400x)

Plants, Fruits, Seeds, Barks, and Ashes in Water

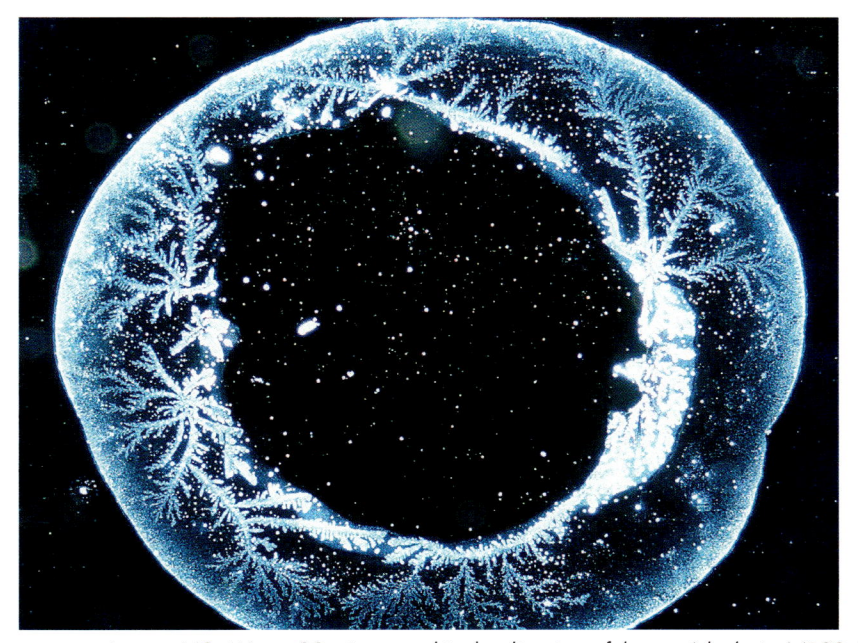

Image 449: Water, 20 min. turned in the direction of the sun (clockwise) (100x)

Image 450: Water, 20 min. rotated in the direction of the moon (counterclockwise) (100x)

Image 451: Water, turned right-left for 20 minutes (100x)

Image 452: Water, turned right-left for 20 minutes (100x)

In the four images above, which always appeared in a similar form after frequent experiments, the stirring process was first examined separately. In the first standing container, water was turned clockwise with a wooden spoon for twenty minutes, in the second standing container, also for twenty minutes, in the opposite direction. Then, in a third standing container, the rotation was alternately clockwise and counterclockwise, as with the horn manure preparation.

Image 449: Water turned to the right (direction of the sun)
Image 450: Water turned to the left (direction of the moon)

Peripheralization and contraction show in the image what happens in the water due to the polar directions of rotation.

Image 451: Rhythmically right- and left-turned water (20 minutes)
Image 452: Rhythmically right- and left-turned water (20 minutes)

Delicate branches sprouting from centers appear, spreading over the entire surface of the drop. The polarities are suspended. New, centered and growing structures appear.

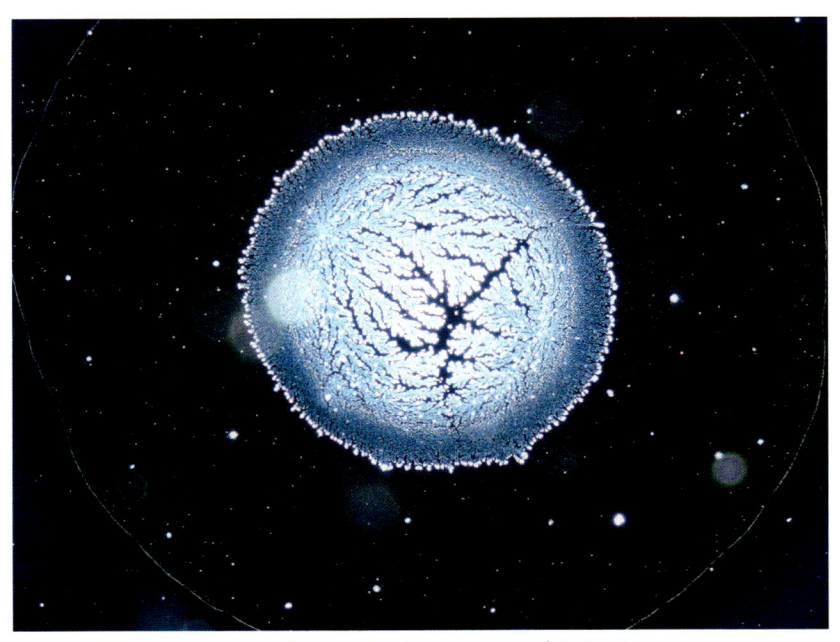

Image 453: Peat from Ireland, unimproved (100x)

Image 454: Peat from Ireland, unimproved (100x)

10. Peat from Ireland and the Swabian Alps, Refined Peat

Peat bogs "grow" for centuries in the northern, temperate zones of our earth. Of the sedges, it is above all the cotton grass, Eryphorum vaginatum—the cutting grass—that piles up the dead, mummified plant layers year after year. This is because sedges do not decay like all other plants, which are "digested" into humus. They constantly mummify and suck the groundwater up to the surface. This is how the raised bogs grow.

This peculiarity of sedges not being absorbed into the normal dying process of nature is of great importance.

We know from Rudolf Steiner that the nature beings are released from the withering and dying plants in the course of the year. Since this does not happen in the upland moors, the elemental beings remain tied to the non-decaying, mummified plant forms. This may be the reason for the uncanny, eerie atmosphere that can grip us in the moor.

Now Rudolf Steiner has given instructions for a refinement of the peat fibers. This rather complicated process has been taken up by some people and leads to the production of spinnable fabric that can be made into garments. The process of peat fiber refinement has great significance not only for human beings, who in our time have to live integrated into an increasing electromagnetic radiation field, but also for the elemental beings themselves.

Rudolf Hauschka reports from Ita Wegman's conversations with Rudolf Steiner, of which he was aware, that the refinement of peat fibers would succeed in "freeing the bound elemental beings, and these would then, out of gratitude, protect man from what is imminent in the foreseeable future, namely that the atmosphere will be so permeated by electricity, magnetic fields, airplanes and much worse that life on earth will become a torment for mankind. Garments made of peat fibers, however, could protect people from these influences."[51]

In the meantime, there are many reports on the use of peat fiber products. Not only garments, but also peat oils, peat ointments, and so on are used. Essentially, a noticeable warming up and strengthening of the vital forces is described, especially when work at the computer is unavoidable. It is permitted here to refer to Wandil, Peter Böhlefeld's peat workshop in Germany, and to Ruth Erne and Anita Borter's "Torf-Faser-Atelier" (peat fiber studio) in Switzerland.

In the following images we look at what the peat in the water reveals of its secrets. Peat from Ireland was placed in water for many weeks.

In images 453 and 454, we see a strongly animated, delimited center in the water drop.

Images 455 to 457 show refined peat fibers in water. The interior of the clearly enlarged center has a similar design as the unimproved peat. However, the outer edges are broken up and branch expressively into the surrounding area. The peat has clearly changed due to grafting. It is peat from Sweden that has been refined by Ruth Erne in Switzerland through a complicated process.[52]

———————— Plants, Fruits, Seeds, Barks, and Ashes in Water ————————

Image 455: Peat from Sweden, refined (100x)

Image 456: Peat from Sweden, refined (100x)

Image 457: Peat from Sweden, refined (200x)

THE SILENT LANGUAGE OF LIFE

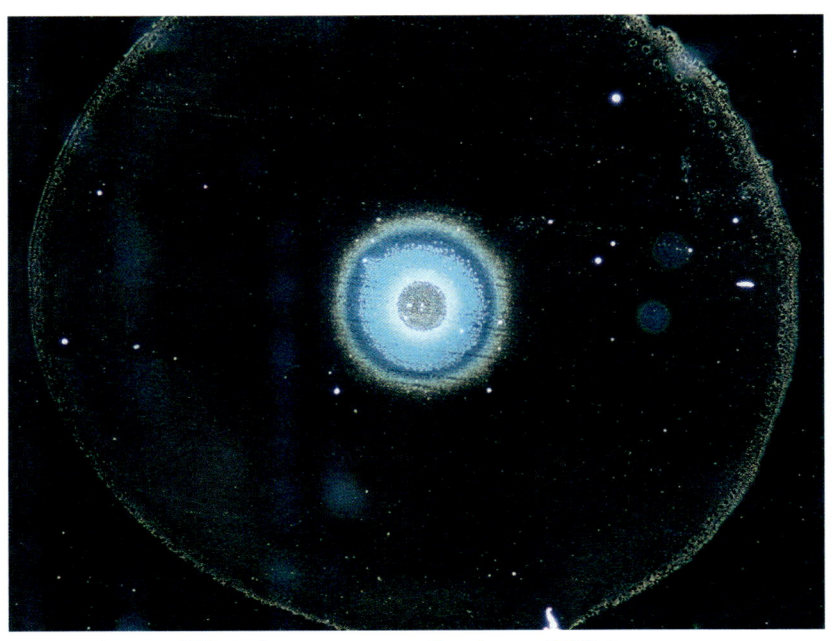

Image 458: Irish peat next to refined peat (100x)

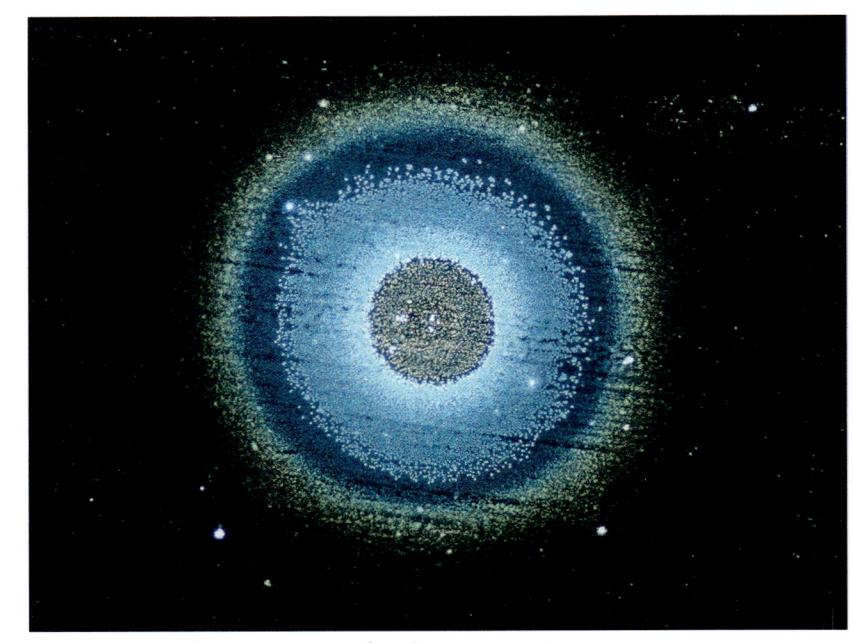

Image 459: Irish peat next to refined peat (200x)

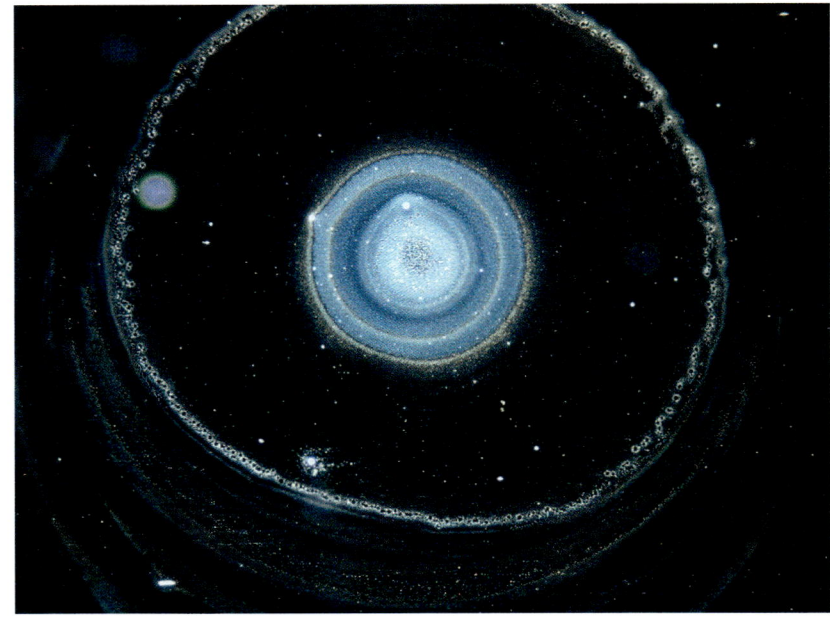

Image 460: Irish peat next to refined peat (100x)

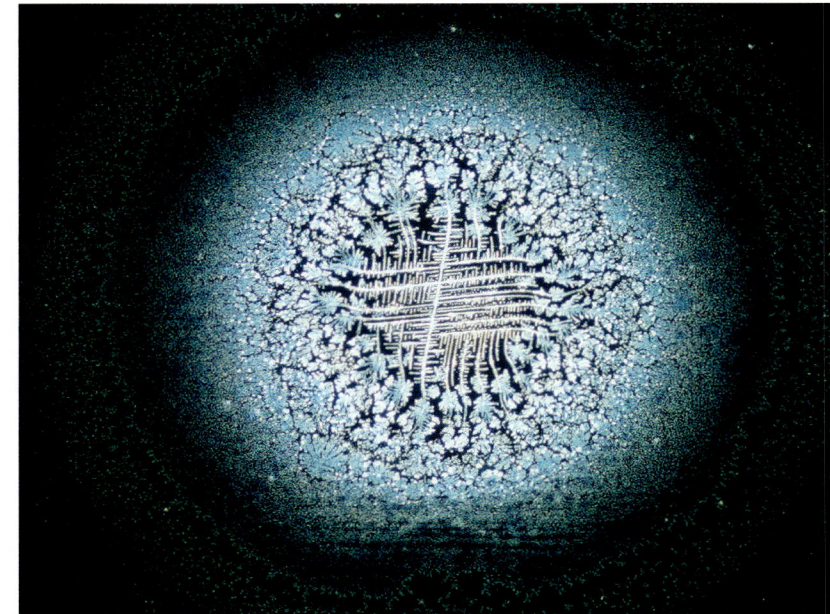

Image 461: Irish peat next to refined peat (400x)

Image 462: Irish peat next to refined peat (400x)

Image 463: Irish peat next to refined peat (400x)

———————— Plants, Fruits, Seeds, Barks, and Ashes in Water ————————

One day the Irish peat got too close to the jars with refined peat. This was supposed to be avoided by keeping the jars at a safe distance from each other.

The images 458 to 463 show what comes to light through the "conversation" of the different preparations. It is the Irish peat that has undergone a tremendous change. The drop center remains but "blossoms" in colors and in the most delicate designs. We see how a radiating power of the refined peat unfolds an extraordinary effect beyond the vessel.

Image 464: Here a small water jar has been wrapped in a coat of refined peat and left to stand. After a few weeks, the water appears "revitalized."

Image 465: There was a surprise when the (unrefined) Irish peat showed a budding rosette. What had happened?

The preparation water of Ireland peat stood on the laboratory table for many months. Flower pollen was also prepared in test tubes during this time. The test tube stand with the yellow pollen water had come too close to the Irish peat. So they had started a "conversation" with each other. The flower pollen transformed the rather crystalline peat structures into a budding rosette. The flower pollen preparations showed no change.

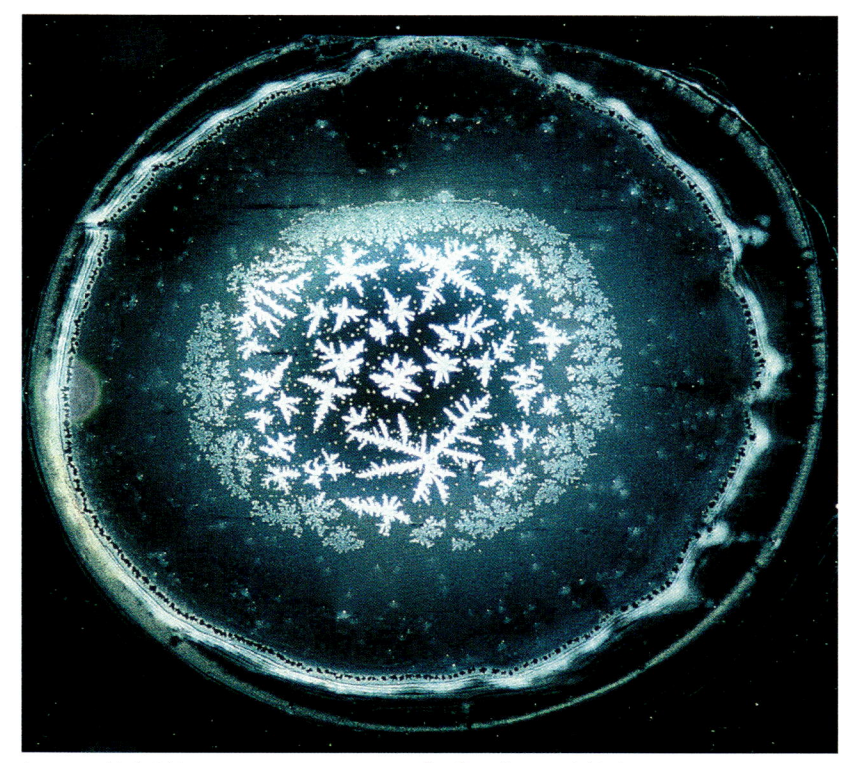

Image 464: Water jars in wrapping of refined peat (40x)

Image 465: Irish peat next to flower pollen preparaton (200x)

157

THE SILENT LANGUAGE OF LIFE

To conclude these observation, we look at peat images from the Swabian Alps.

Images 466 to 468 show peat from the Swabian Alps. Strong crystalline, rythmically ordered, growing structures are formed in the rather large drop center.

Before we come to the last peat images, an intermediate remark should be made. Is it any wonder that during this research, with its exceedingly astonishing results, doubts arise again and again? Can we simply take the richly shaped formations in the water for granted? Can we easily include them in what lives in us as images of sensory experience of nature? If that were the case, there would be no research! Again and again, we have to test and cross the border in order to encounter the truth of the forces at work in water anew. Doubt and not taking for granted what is true are driving moods of the soul in this research.

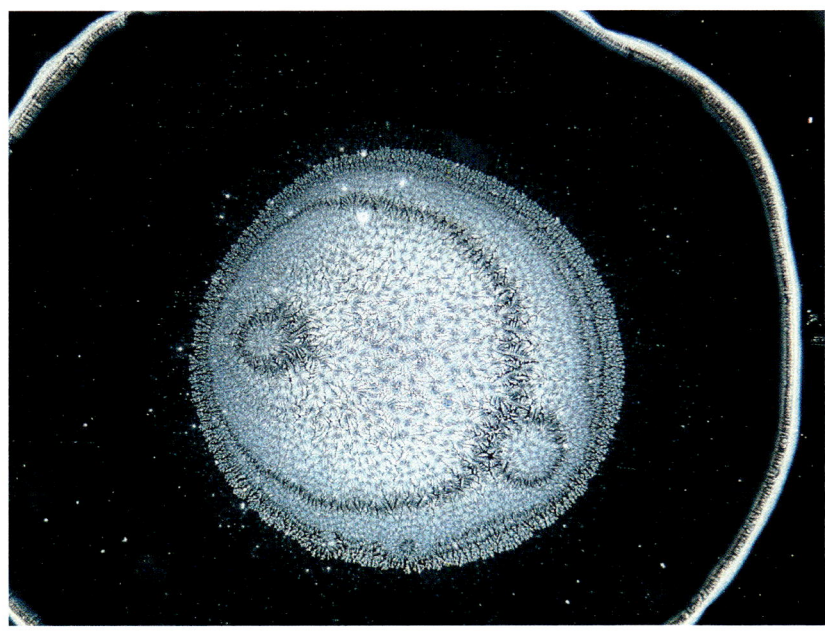

Image 466: Peat from the Swabian Alps (100x)

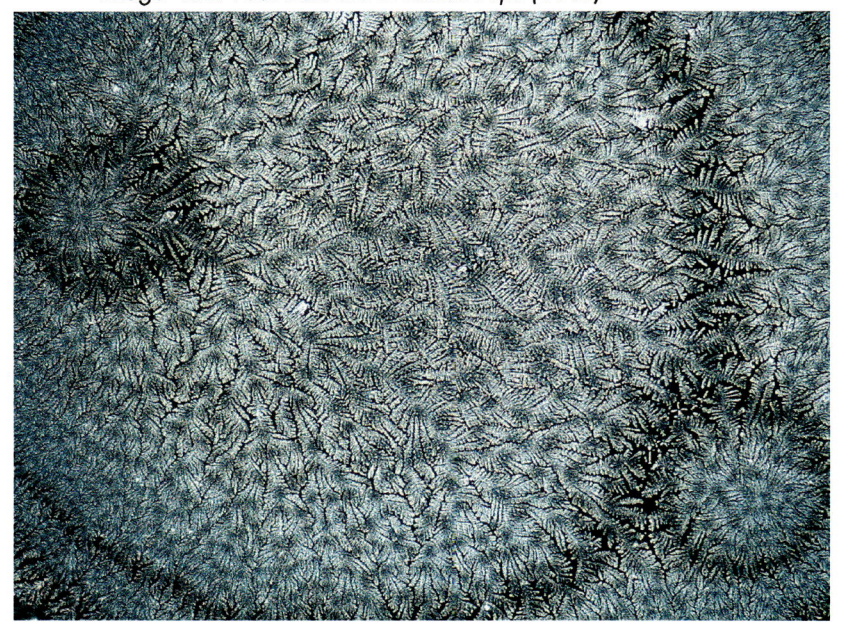

Image 467: Peat from the Swabian Alps (200x)

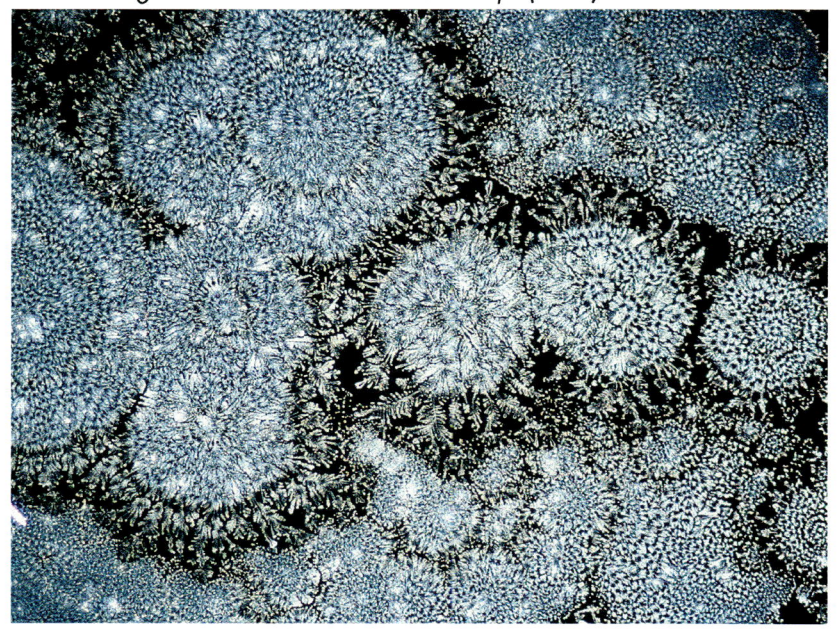

Image 468: Peat from the Swabian Alps (400x)

Plants, Fruits, Seeds, Barks, and Ashes in Water

Images 469 to 475: The following images resulted when the Alps peat was dripped in this inquiring mood. After drying, impressive special formations appeared in the center in seven of the eighteen drops. The astonishment at this was so great that holding one's breath while looking into the microscope was forgotten. (The dried crystallizations could not tolerate the moist exhalation.) In one of the drops, a "creature" of elongated shape immediately transformed—it shrunk into a "curly" ball and thus changed its shape (image 471). Mysterious, moving life in the peat?

With awe and respect one looks at these various, creature-like appearances in the peat images. According to the above, we do not know what happens to them after they have been solidified by the water image.

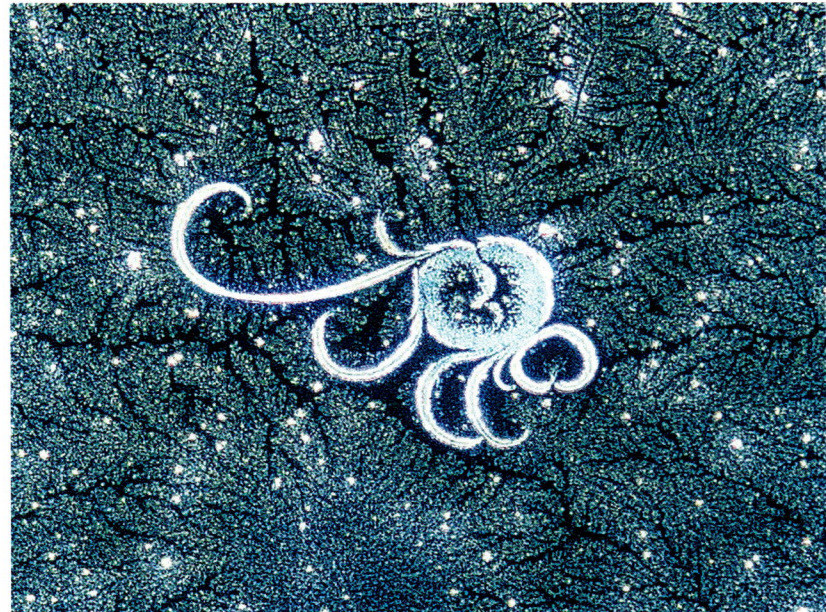

Image 469: Peat from the Swabian Alps (200x), special formation

Image 470: Peat from the Swabian Alps (200x), special formation

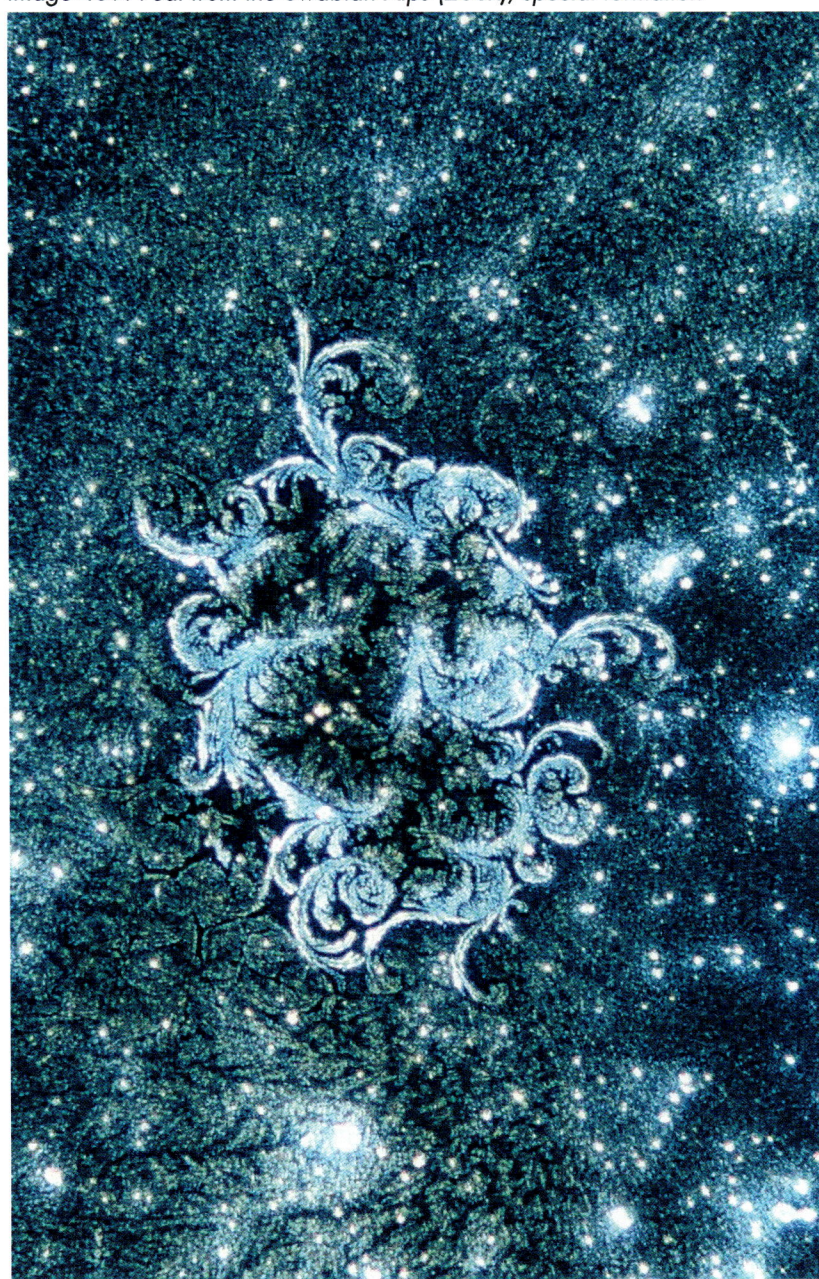

Image 471: Peat from the Swabian Alps (200x), special formation

THE SILENT LANGUAGE OF LIFE

Image 472: Peat from the Swabian Alps (200x), special formation

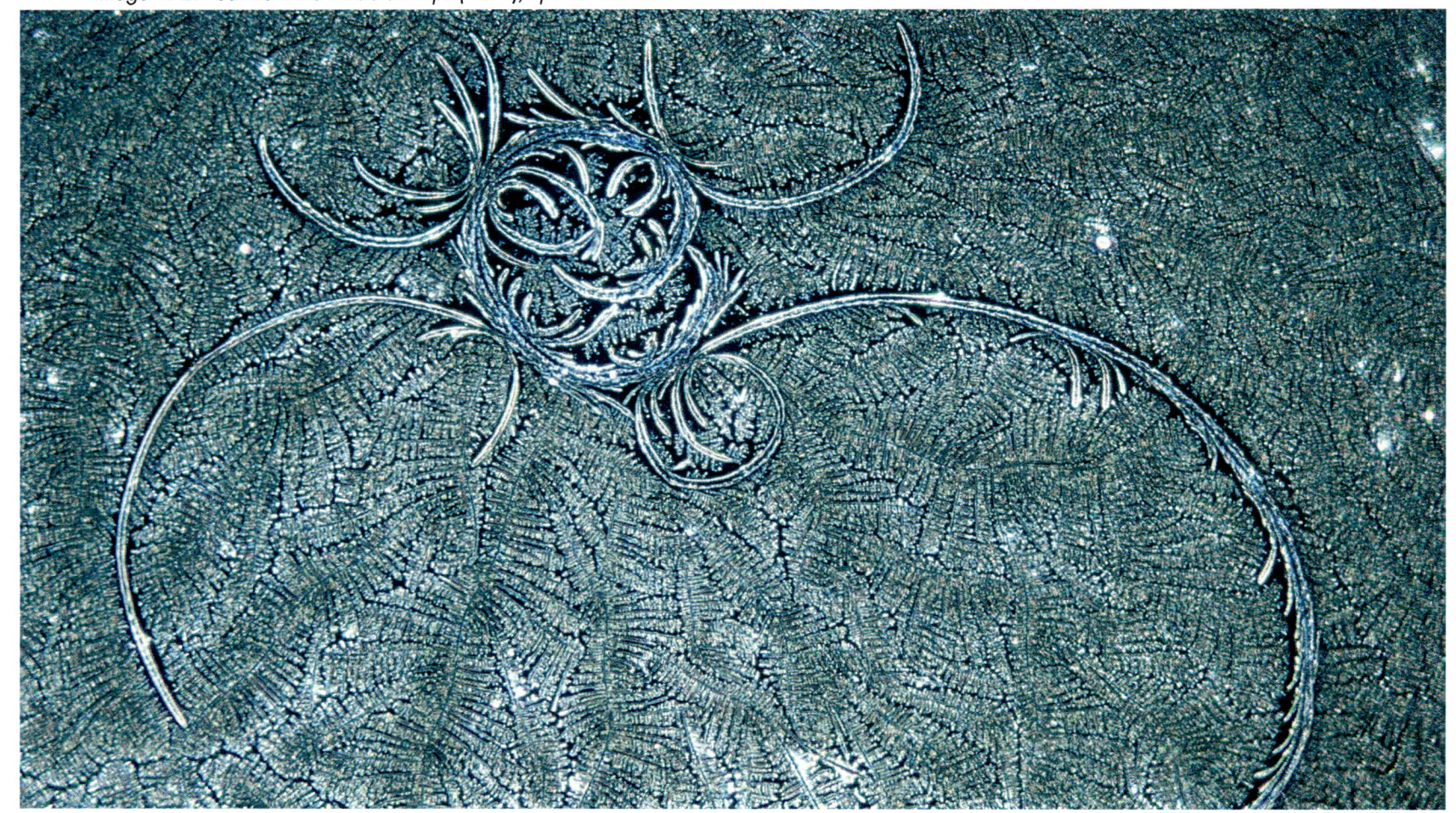

Image 473: Peat from the Swabian Alps (400x), special formation

Plants, Fruits, Seeds, Barks, and Ashes in Water

Image 474: Peat from the Swabian Alps (200x), special formation

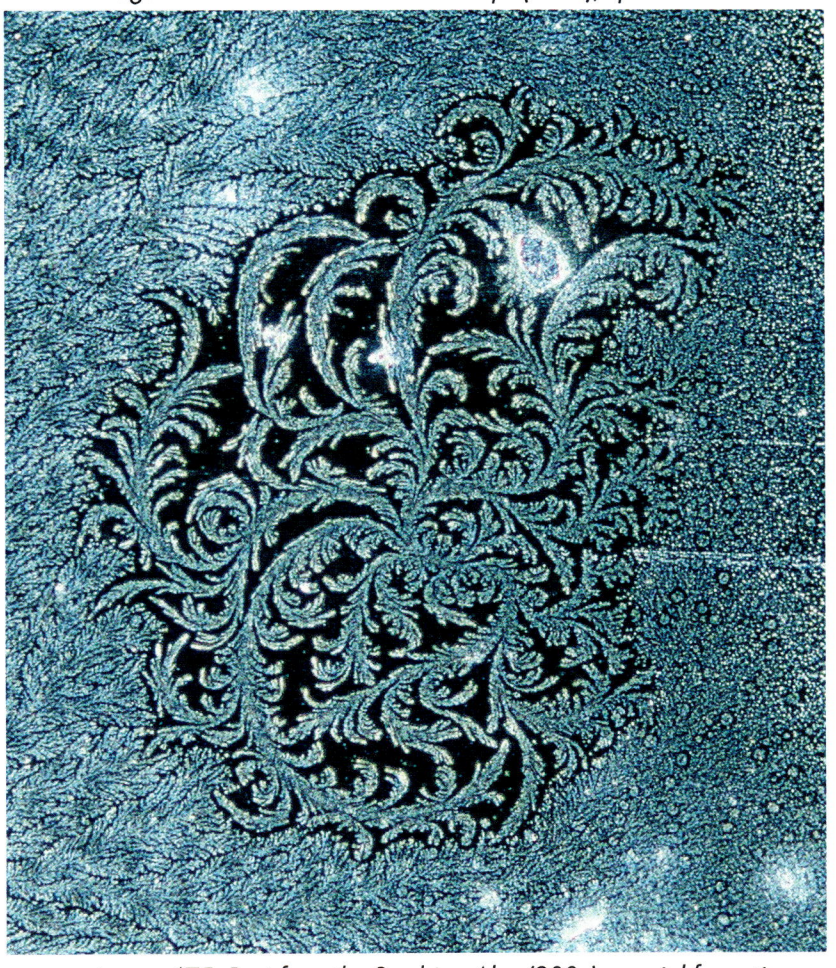

Image 475: Peat from the Swabian Alps (200x), special formation

11. Reflections on Chapter IV, Sub-chapters 8–10

In these last three sub-chapters, it is no longer just individual natural objects that are placed in the water. Man handles the plant in the most diverse ways. He follows the suggestions of the spiritual scientist and brings the plant into a new connection with the cosmos through complicated procedures. The astonishing thing is that here, too, images appear in the water that resonate with the actions, for example, when the exposure of the plant to light—in contrast to the unexposed one—results in a clear revival in the drop image (images 422 to 435).

The four preparations from the Agricultural Course also produce something in the image that can be harmonized with the background of the production and effect of the respective preparation. This could of course be more evident to the Demeter farmer than to those less familiar with this method (images 436 to 452).

Peat, refined and unrefined, appears quite different again. What the drop of water reveals to us here is the breaking up of its central structure through refining and its radiating power to the surrounding area (images 453 to 475).

Of course, these images do not serve to "prove" anything. Rather, they lead to a deeper experience and recognition of what has become a living knowledge for us through the meaningful background of the activities.

What we see on a small scale opens our eyes to the magnitude of the formative forces in the context of nature.

Microscopy becomes macroscopy.

Image 476: Tear fluid (400x)

Chapter V
Human Fluids

In this final chapter, we see some samples from the complex and highly differentiated liquid organism of human beings. Here, there is no longer any need for a "time of exposure" in the liquid—the specific action of forces is revealed immediately after the drops have dripped on and dried.

No diagnostic purposes were pursued, but the question was posed, solely out of human interest, as to how the available fluids of different areas of organ function reveal themselves in the drop image.

Some structures appear, as they can also be seen in the previous chapters. The three basic faculties of water—metabolic, rhythmic, and sensory—also meet us here in a hidden way. They link up in close cooperation.

What could previously be recognized as form-giving forces from the stars, now shifts inward in the human being, the only spiritually gifted being in creation. It is the I of the human being that, as a spiritually shaping force, penetrates and shapes the physical body in a specific way through the warmth of the soul body and the life body. The drop images can be viewed against this (even if described only briefly) background. For example, the transformations of the blood under various conditions are impressive.

The images may essentially speak for themselves, except for necessary explanations of individual procedures and a few questions about the cerebrospinal fluid.

The path leads from the exterior of the human being, the fluids of the sense organs, which also includes the skin, via the spinal fluid and the breath, all the way inward to the blood and serum.

1. Fluids from the Sensory Area

Image 476: Tear fluid.
Image 477: Nasal fluid.
Image 478: Saliva.

Image 477: Nasal fluid (200x)

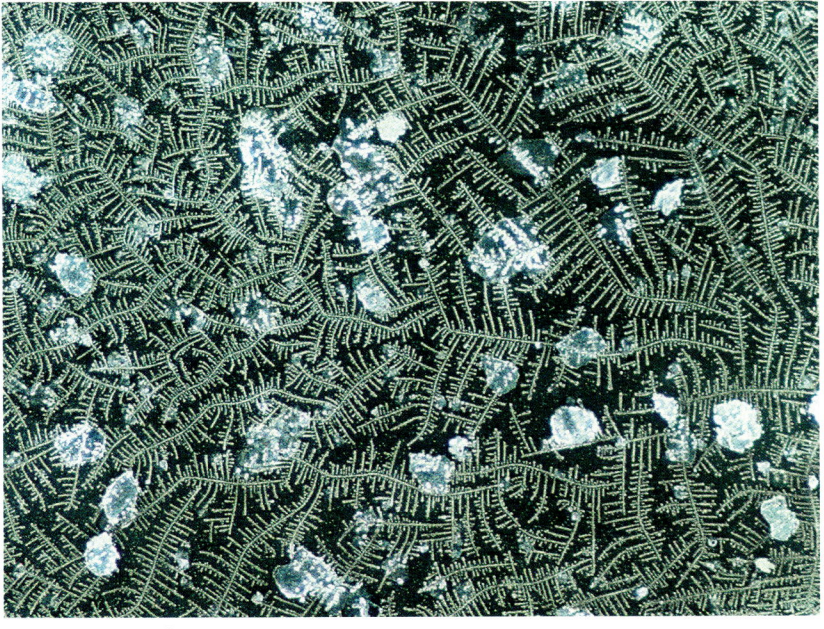

Image 478: Saliva (200x)

163

THE SILENT LANGUAGE OF LIFE

Image 479: Saliva, salt (100x)

Image 480: Saliva, salt (400x)

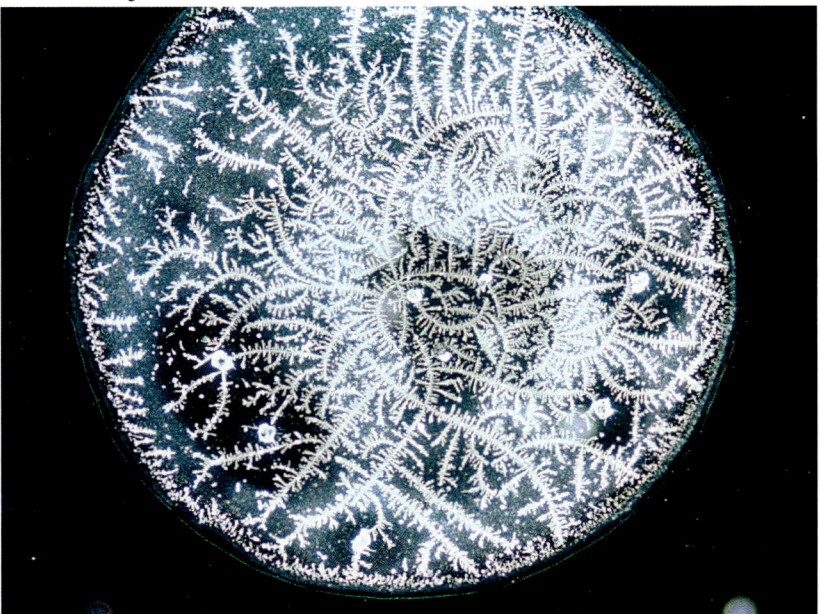

Image 481: Saliva, sugar (100x)

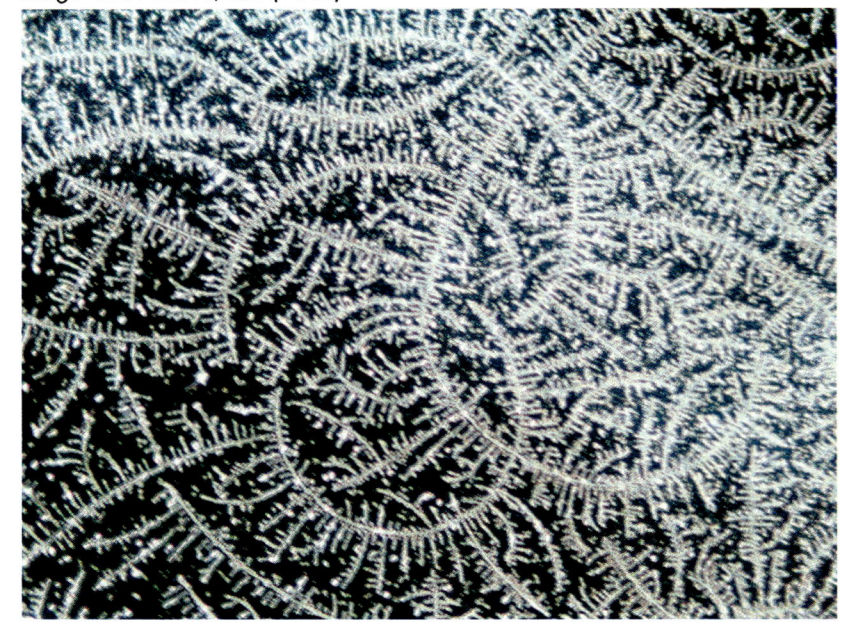

Image 482: Saliva, sugar (400x)

Image 483: Sweat (100x)

Image 484: Sweat (400x)

Now, in an experiment, a few grains of salt are placed on the tip of the tongue. The liquid immediately collected from the gland under the tongue is dripped on. The saliva did not come into contact with the salt!

Images 479 and 480: Saliva and salt—distinct crosses and vegetable structures appear.

The same procedure is done with a few grains of sugar on the tip of the tongue.

Images 481 and 482: Saliva and sugar—large-arched curves appear, but no animation.

Images 483 and 484: Human sweat.

2. Cerebrospinal Fluid (CSF)

The brain and spinal cord are enveloped in a light yellow, clear fluid that is constantly flowing in circles (cerebrospinal fluid). It lifts the brain, which is floating in it, from heaviness to lightness. It is formed from a network of veins that protrude in a special way into the central cerebrospinal fluid chambers of the brain ("blood/cerebrospinal fluid barrier"). The cerebrospinal fluid is absorbed into the venous blood in the periphery—the circumference of the brain. In total, there are four—arranged in a special way—liquid-filled inner chambers that communicate with the spinal fluid.

Images 485 to 487 and 489 to 506: CSF (cerebrospinal fluid), dripped on.

For diagnostic purposes, the cerebrospinal fluid is punctured in the clinic under sterile conditions in the area of the lumbar spine. Here we see CSF with normal findings from a thirty-eight-year-old patient.

There is an abundance of elongated, linear, rhythmic formations, mostly arising from star figures. The leading lines are black! Very occasionally, salt crystals and also crosses appear. The salt content of the CSF corresponds approximately to that of the blood.

If we follow the anatomical and physiological principles of nerve formation, then the comparison of the CSF structures in the drop image with the nerves of humans is quite obvious. It is assumed that the CSF structures of different subjects are similar in appearance because the nervous system has almost come to an end in its physical formation and does not show the same variability as, for example, the blood.

Image 488 shows the schematic representation of a neuron.[53] The neuron shown consists of a nerve cell (soma) from which the dendrites and the elongated nerve emanate. The nerve (axon) is enclosed by the medullary sheath at intervals of a few millimeters. This develops from a cell that is able to form a very thin, flat skin in order to wrap the nerve spirally in it. (The medullary sheaths as surface formers were already mentioned.)

The remarkable rhythmic constrictions, known as "Ranvier's lacing rings," cause interspaces (interruptions of the medullary

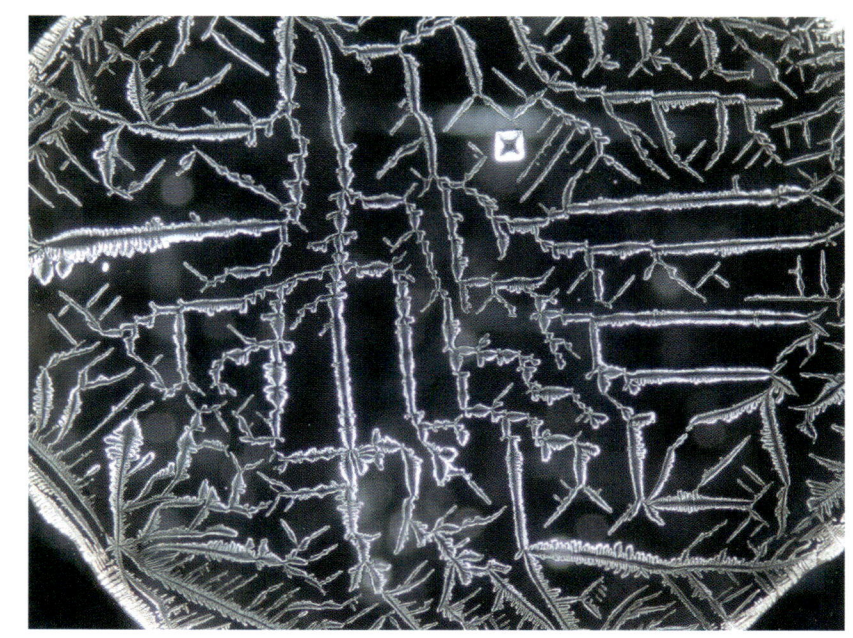

Image 485: Cerebrospinal fluid (100x)

Image 486: Cerebrospinal fluid (100x)

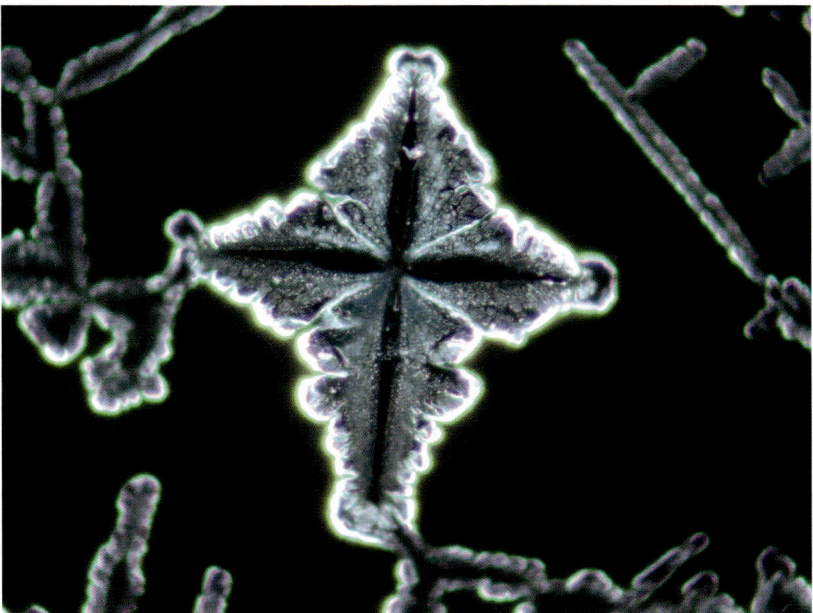

Image 487: Cerebrospinal fluid (400x)

THE SILENT LANGUAGE OF LIFE

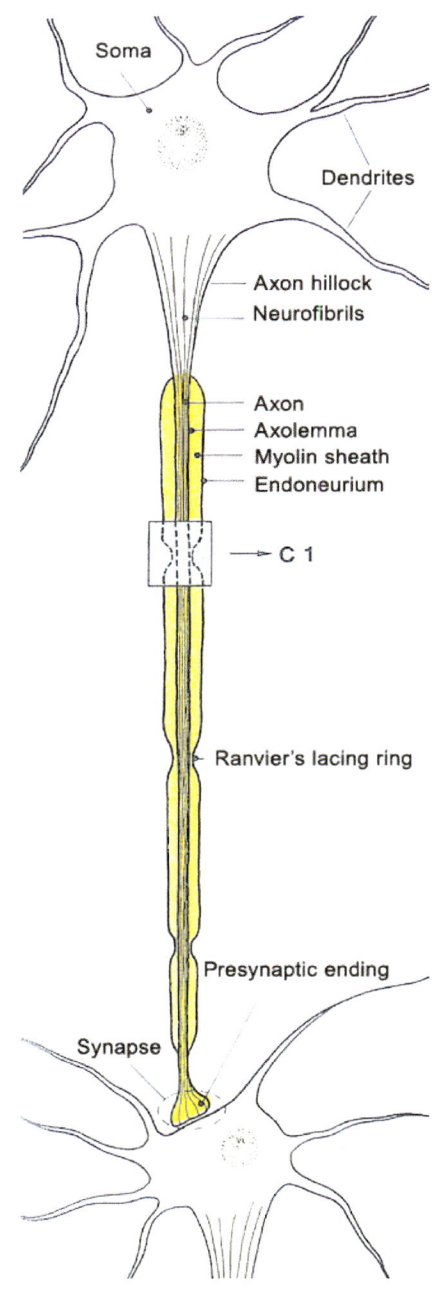

Image 488: Neuron

Image 489: Cerebrospinal fluid (200x)

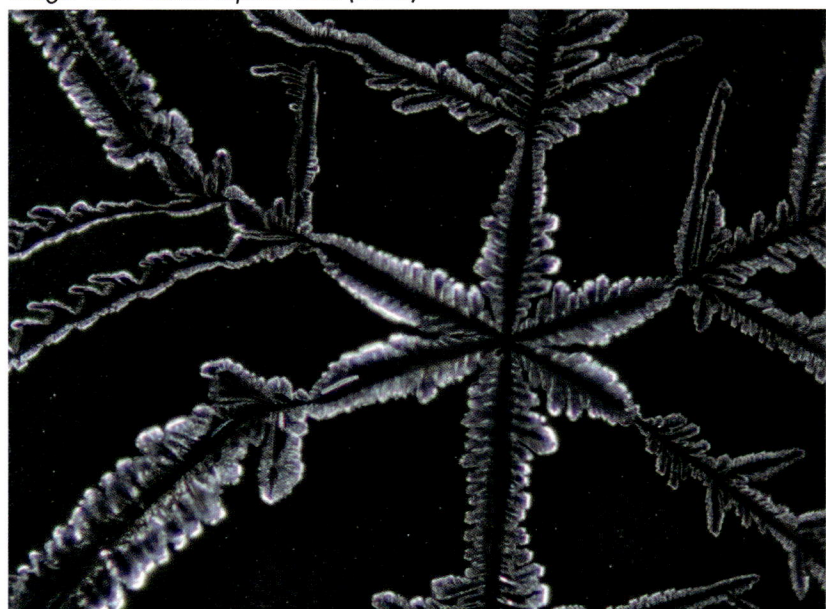

Image 490: Cerebrospinal fluid (200x)

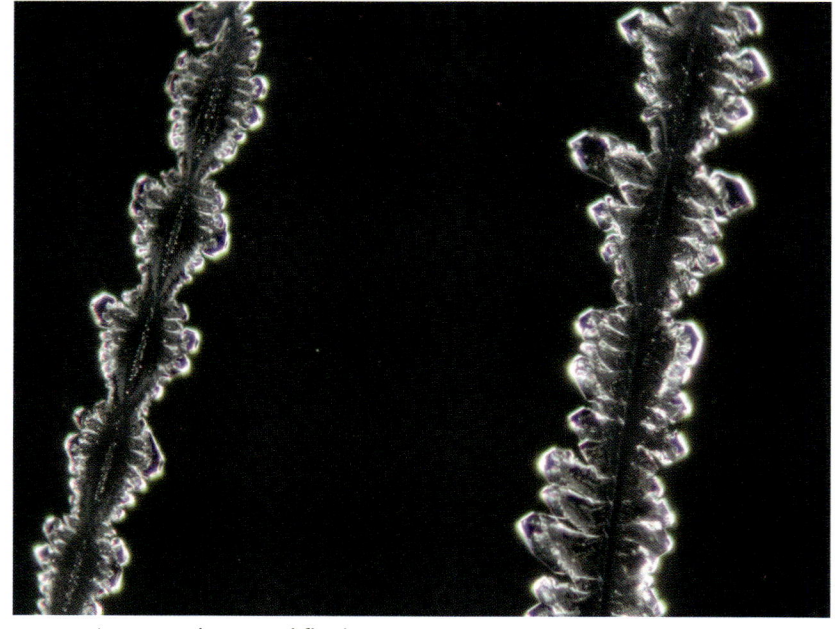

Image 491: Cerebrospinal fluid (400x)

sheaths) so that the axon lies free at these points. "The axon of the central nerve fiber lies completely free in the region of the node of Ranvier."[54]

Image 491 could give us an image of these rhythmic constrictions, especially here too in the magnification (400x).

It should also be briefly noted that this medullary constriction leads to a considerable acceleration of the nerve conduction velocity (120 m/s). In the medullary nerve fibers, nerve conduction takes place much more slowly (up to 2 m/s).[53]

The first impression that the dripped-on cerebrospinal fluid gives is of these long, rhythmic extensions, which often emerge from star-shaped figures.

A word by Novalis from the "Fragments" may be cited: "Nerves are the higher roots of the senses."[55]

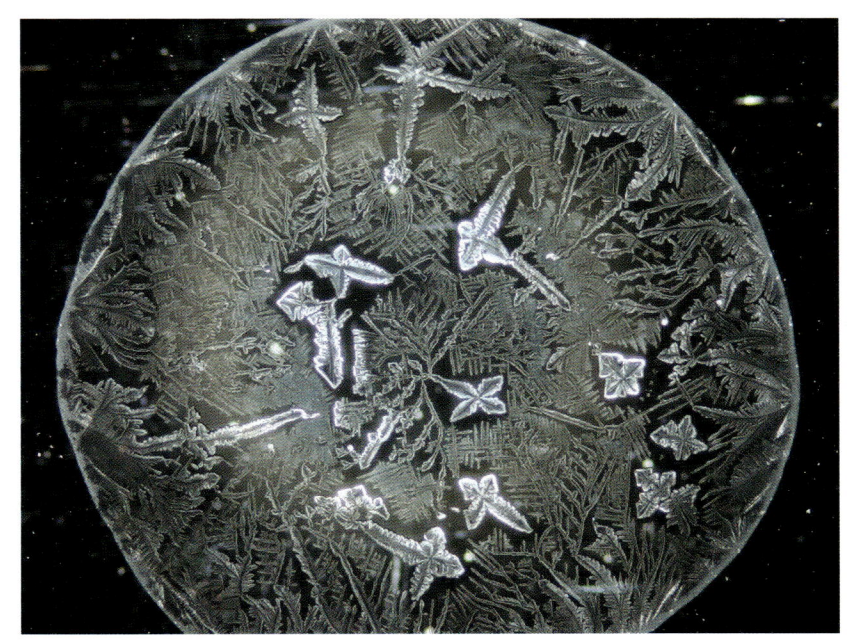

Image 492: Cerebrospinal fluid, 1:1 (100x)

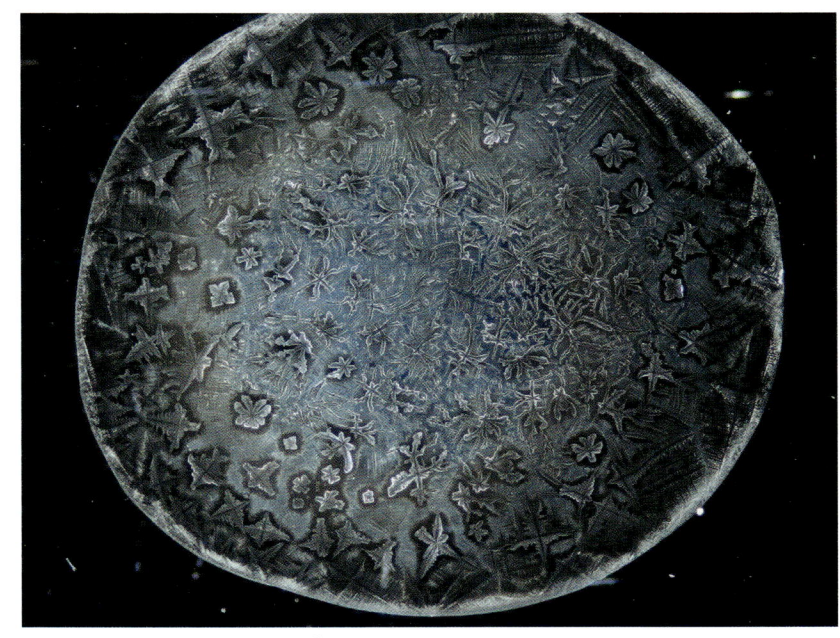

Image 493: Cerebrospinal fluid, 1:1 (100x)

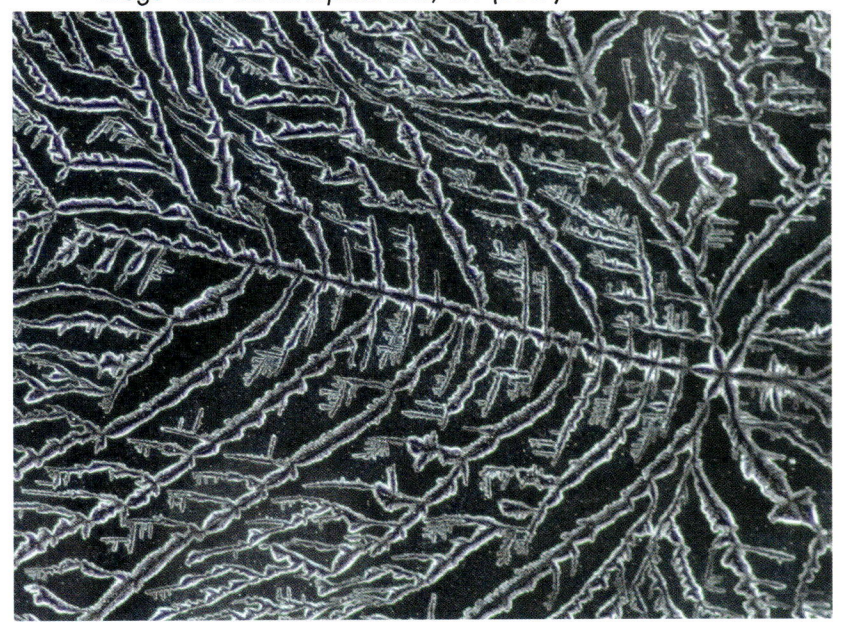

Image 494: Cerebrospinal fluid, 1:1 (200x)

Image 495: Cerebrospinal fluid, 1:1 (400x)

The star formations with their sometimes whimsical figures and spurs seem like root formations. Despite the variety of shapes and variability, mysterious numerical and angular laws seem to underlie them. Around the crosses and salts we see more right-angled branches from the main ray. In the branches of the six-pointed stars we see angles of 60°, which are usually strictly observed in the sometimes very long side branches. We encountered six and four in the salt chapter as an expression of cosmic and earthly structure formation. In the CSF, the "six-pointed stars" (snowflakes!) are predominant, but we find all kinds of variations in the metamorphosis of four-, five-, seven- and eight-pointed figures. We do not encounter vegetative growth structures in the CSF! It would be interesting to check this in a pathological CSF.

If the CSF is loaded—drop by drop—with lauretana water (without SiO_2), the "playfully designed numerical order" reveals itself even more clearly in an imaginative way.

Images 492 to 503: CSF dilution 1:1

The water as a "selfless medium" is here only the "revealer" of what is present and carries no imprint of its own into the dilution process. However, the question also arises as to whether the molecular structure of water discussed above perhaps resonates with the CSF tendency in the formation of six stars.

Of particular importance—possibly also with regard to rhythmization—is the connection of the cerebrospinal fluid with respiration, as described by Rudolf Steiner on August 27, 1915:

———————————— THE SILENT LANGUAGE OF LIFE ————————————

Image 496: CSF, 1:1 (200x)

Image 497: CSF, 1:1 (400x)

Image 498: CSF, 1:1 (200x)

Human Fluids

Image 499: CSF, 1:1 (200x)

Image 500: CSF, 1:1 (200x)

Just as the diaphragm moves up and down during breathing, just as inhalation and exhalation take place in general, so this cerebral water, in which the brain swims and participates in breathing, moves rhythmically in this way. And the whole thought process, in so far as the brain is its tool, has in it its physical connection with the breathing process.[56]

Owing to its complexity, this connection cannot be elaborated further here.

Image 501: CSF, 1:1 (400x)

Image 502: CSF, 1:1 (200x)

Image 503: CSF, 1:1 (400x)

— THE SILENT LANGUAGE OF LIFE —

Image 504: CSF, 1:10 (400x)

Image 505: CSF, 1:10 (200x)

When liquid is diluted with distilled water 1:10 (images 504 to 506), more strictly formed star figures are revealed. Here we quote Rudolf Steiner once again, this time from *Anthroposophical Leading Thoughts*:

> The thinking organization is an organization of the stars. If it lived and expressed itself *as such* alone, man would bear within him not a consciousness of self but a consciousness of the gods. The thinking organization is, however, lifted out of the cosmos of the stars and transplanted into the realm of earthly processes. Man becomes a self-conscious being in that he experiences the world of stars within the earthly realm.[57]

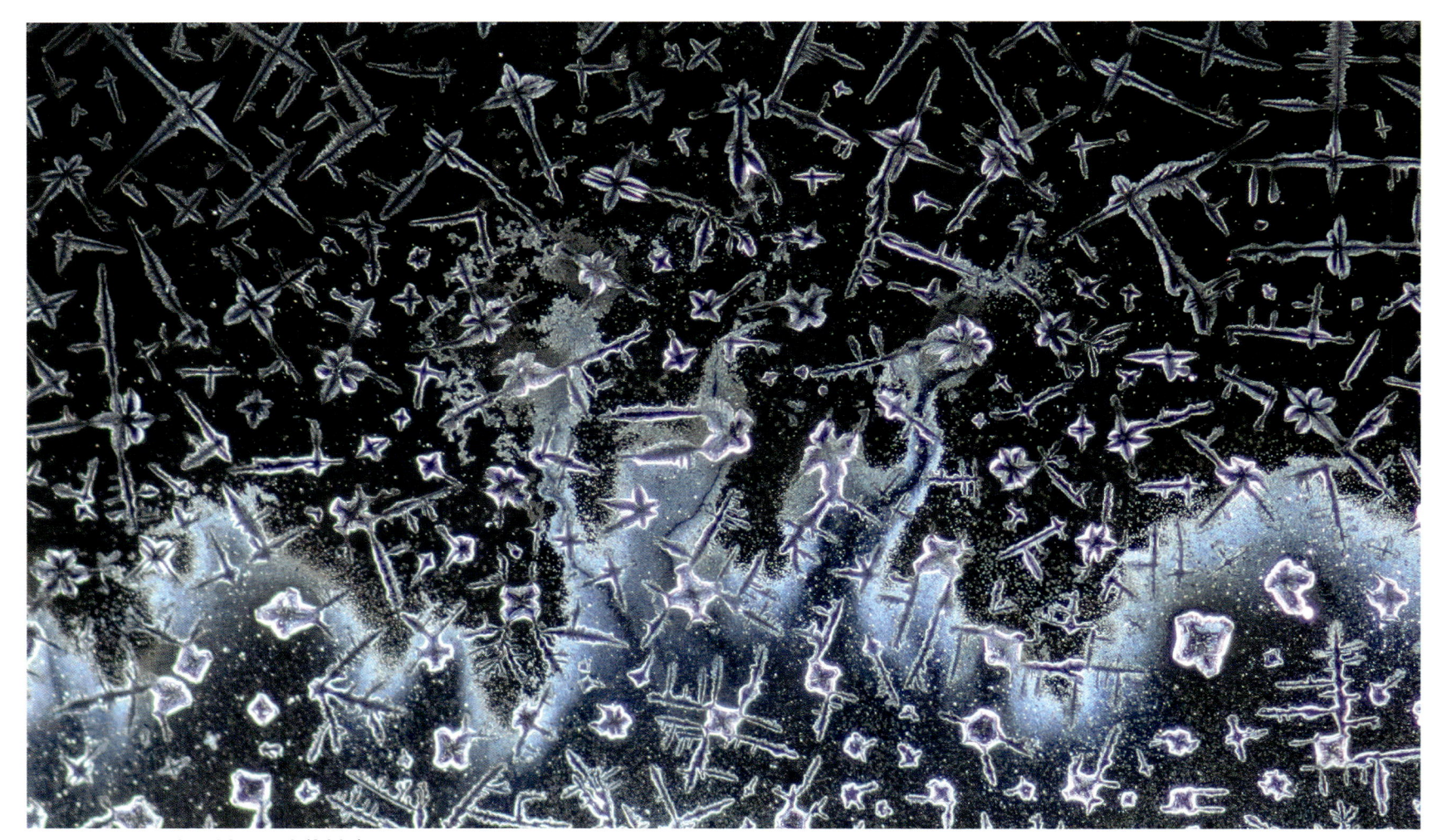

Image 506: CSF, 1:10 (200x)

Human Fluids

Image 507: Pleural exudate (100x)

Image 508: Pleural exudate (400x)

3. Pleural Exudate and Breath

Images 507 and 508 show pleural exudate, the pathological accumulation of fluid between the lungs and the pleura.

Image 509: Breath; the drop structures now change into vegetative shapes. The expiratory flow has met the blood!

Image 509: Breath (400x)

— THE SILENT LANGUAGE OF LIFE —

4. Blood

Image 510: Human native blood, taken from the vein.

Human blood appears much more differentiated when it is collected in a filter bag, as in Ehrenfried Pfeiffer's sensitive blood crystallization. A circular shape is cut out of the dried surface and redissolved in 1 ml aqua distillata of a small petri dish. This soft reddish liquid is dripped on.

Images 511 to 514: We see four drop images of different subjects. Even though no two images are alike, we can speak of an ideal image. The animation of the entire drop surface and the halo of rays shining into the periphery are impressive!

Image 510: Native blood vein (100x)

Image 511: Blood (100x)

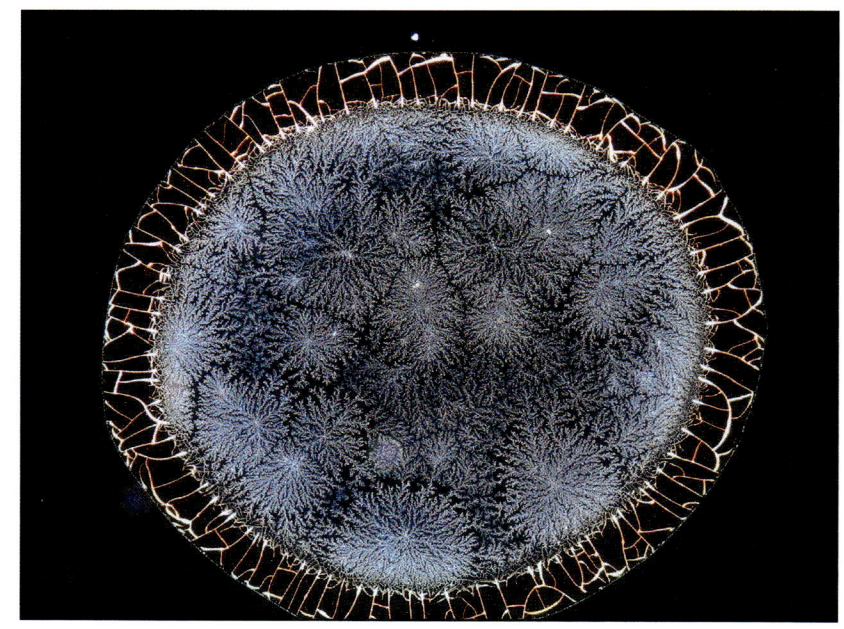

Image 512: Blood (40x)

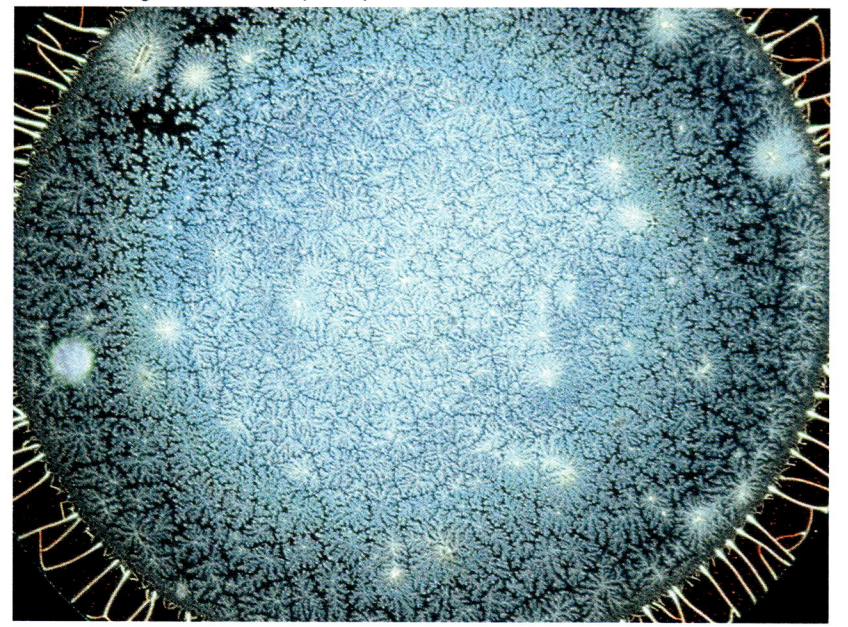

Image 513: Blood (100x)

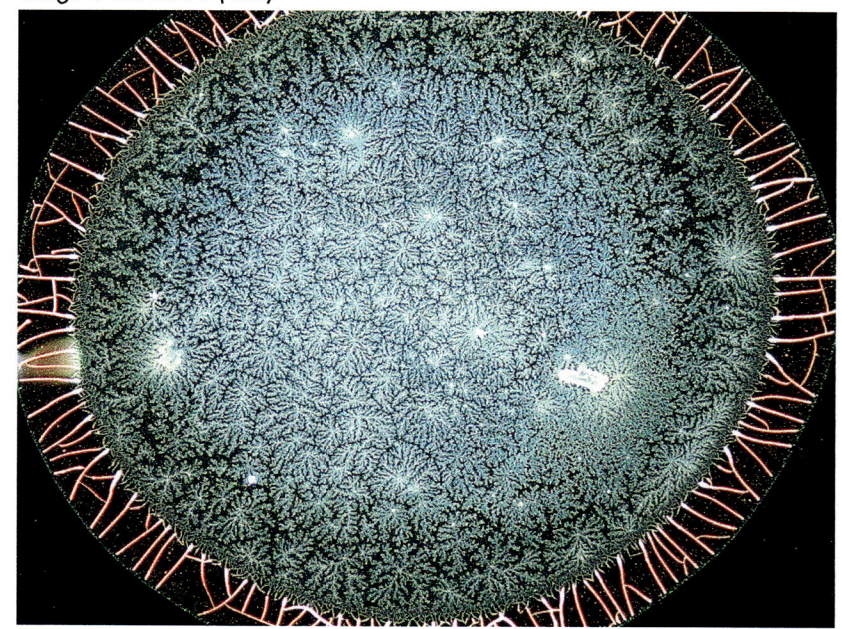

Image 514: Blood (100x)

Human Fluids

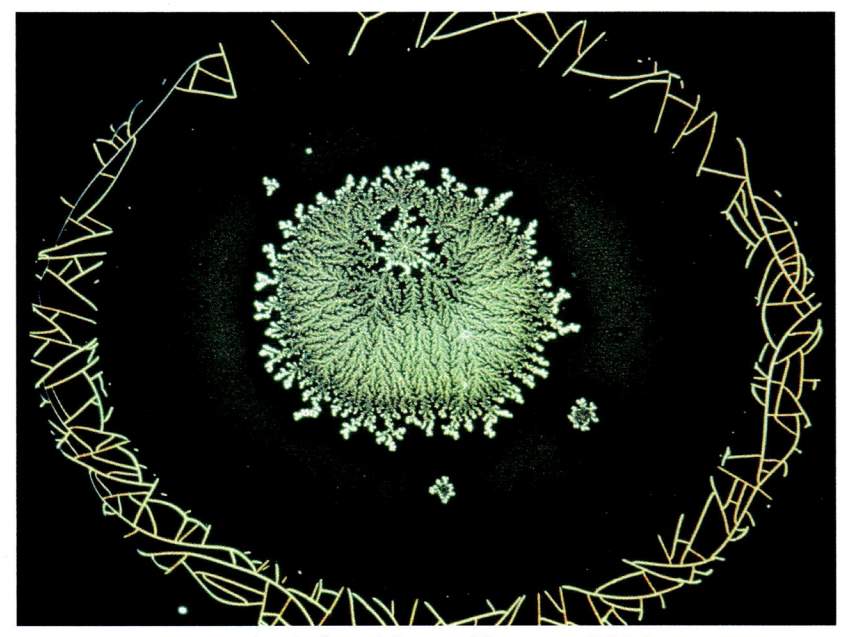

Image 515: Blood of a 104-year-old woman (100x)

Image 516: Blood of a 104-year-old woman (100x)

The variability of human blood is immense and also shows rapid transformations. The most diverse circumstances and conditions are expressed in it. Here are some examples:

Images 515 to 517: Blood of a 104-year-old, very spry woman; there is a certain contraction in the center of the drop next to a halo of rays which is also preserved at this age!

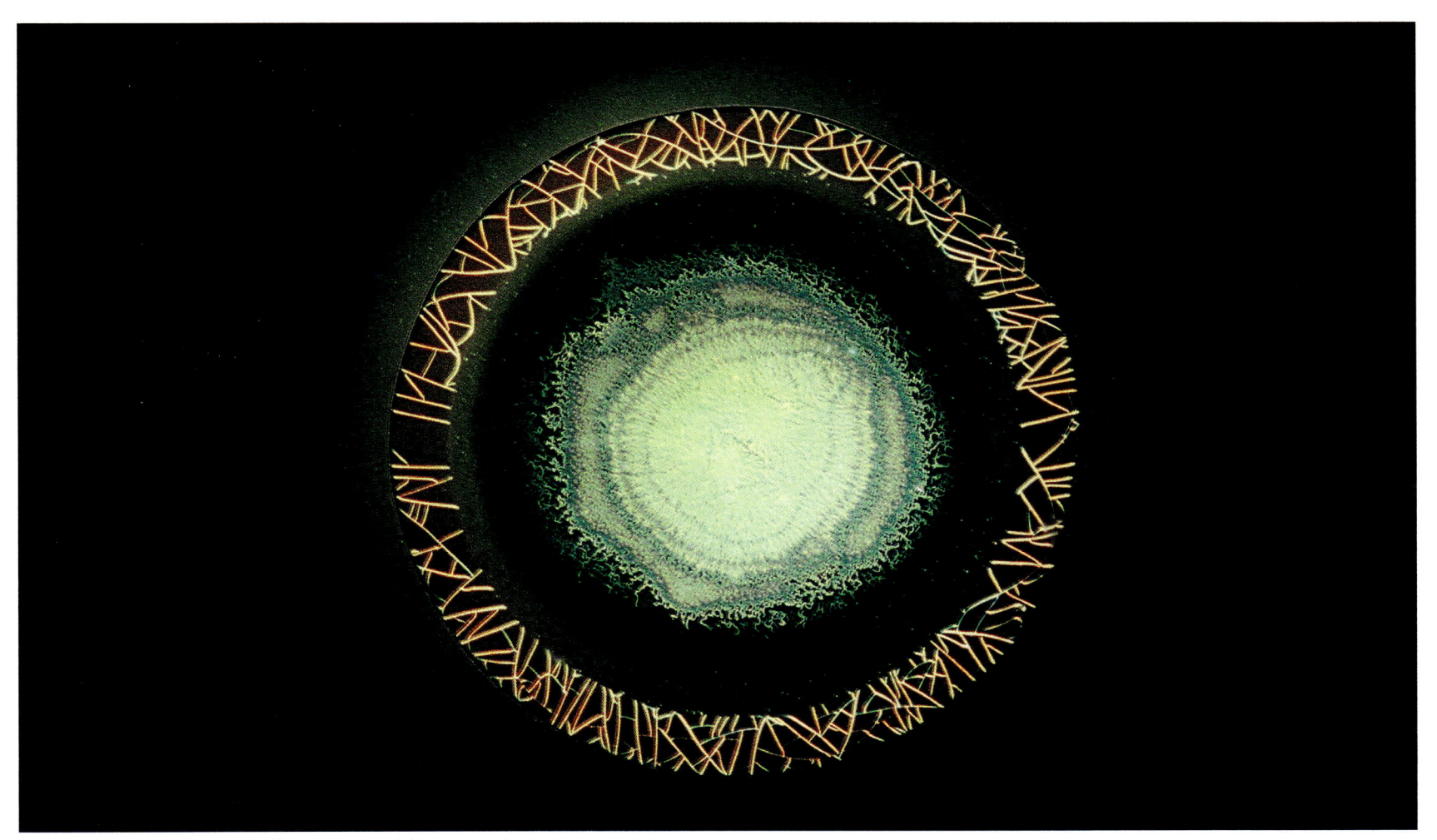

Image 517: Blood of a 104-year-old woman (40x)

173

THE SILENT LANGUAGE OF LIFE

Image 518: Blood in the morning (100x)

Image 519: Blood in the evening (100x)

Image 518: Blood in the morning.
Image 519: Blood in the evening.
Image 520: Blood at night.
Image 521: Blood after taking a sauna.
Image 522: Blood in the morning after taking a sauna.

In the following images we see the changeability of the blood after exertion and subsequent eurythmy on the same day:

Human Fluids

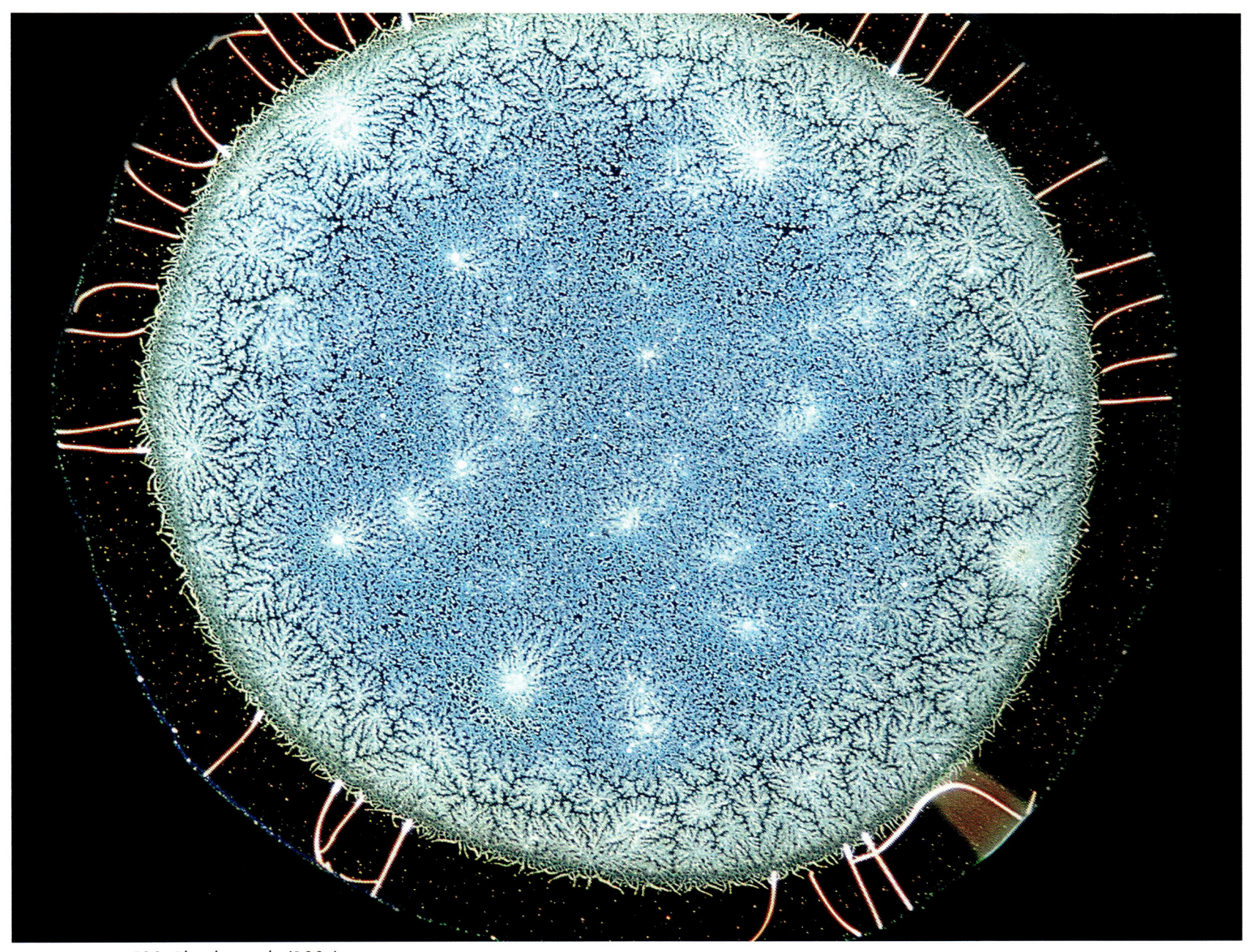

Image 520: Blood at night (100x)

Image 521: Blood after taking a sauna (100x)

Image 522: Blood in the morning after taking a sauna (100x)

THE SILENT LANGUAGE OF LIFE

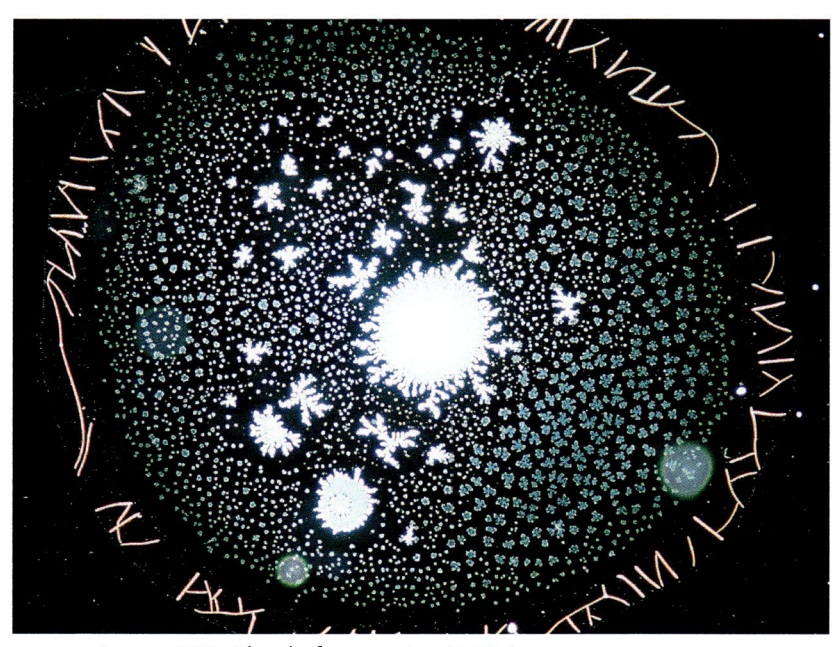

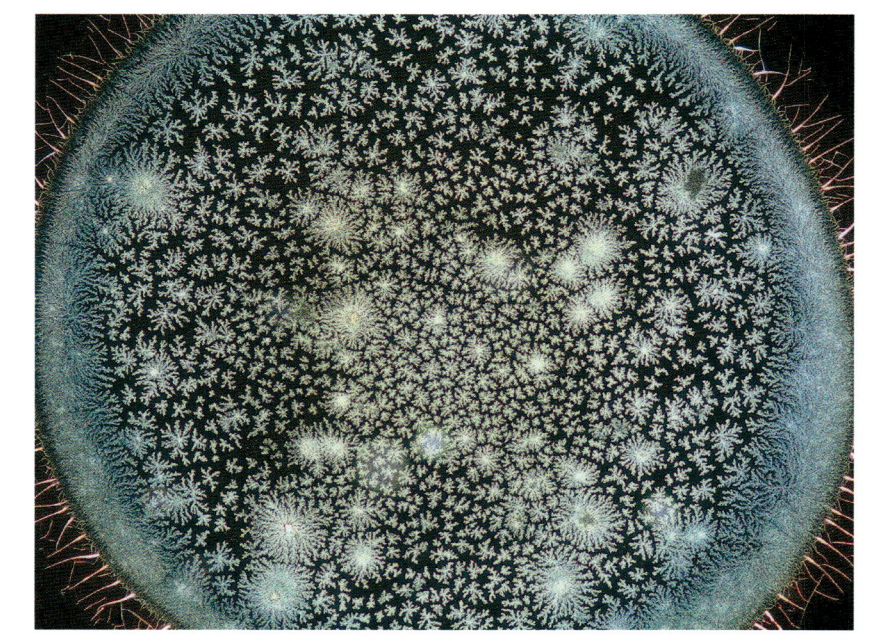

Image 523: Blood after exertion (100x)

Image 524: Blood after eurythmy (100x)

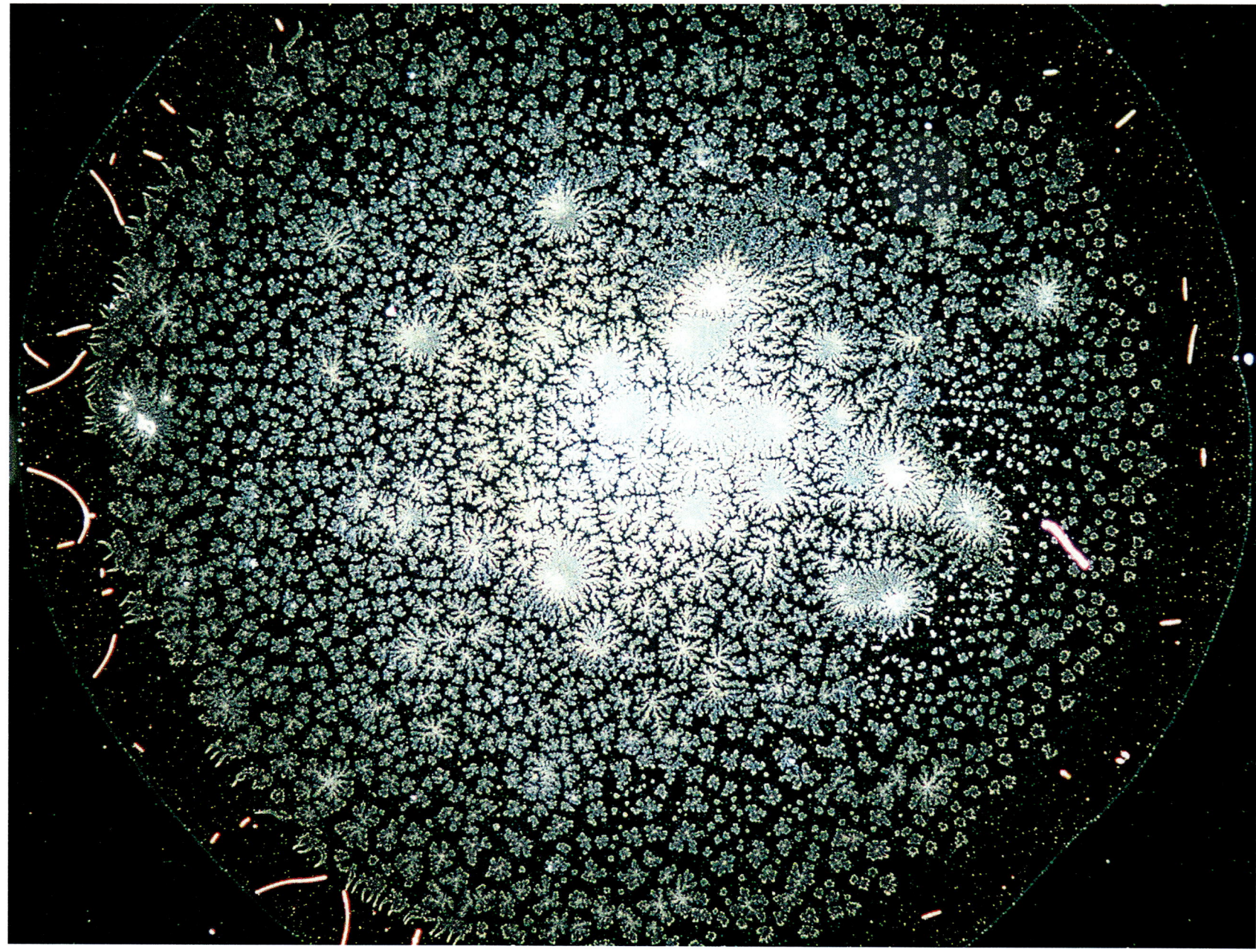

Image 525: Blood after exertion (100x)

Image 523: Blood after exertion.
Image 524: Blood after eurythmy .
Image 525: Blood after exertion, here we see again one of these numerous trials over the years.
Image 526: Blood after eurythmy, the same day.
Image 527: In this drop of blood of an exhausted person, the source stars have strung themselves up like a ball of wool. This is a common phenomenon.
Image 528: Here, too, it rearranges itself, after relaxation and recovery on the same day.

These few images shown here are, as always, not one-off experiments, but a selection of many variations with the same expressive power.

Image 526: Blood after eurythmy (100x)

THE SILENT LANGUAGE OF LIFE

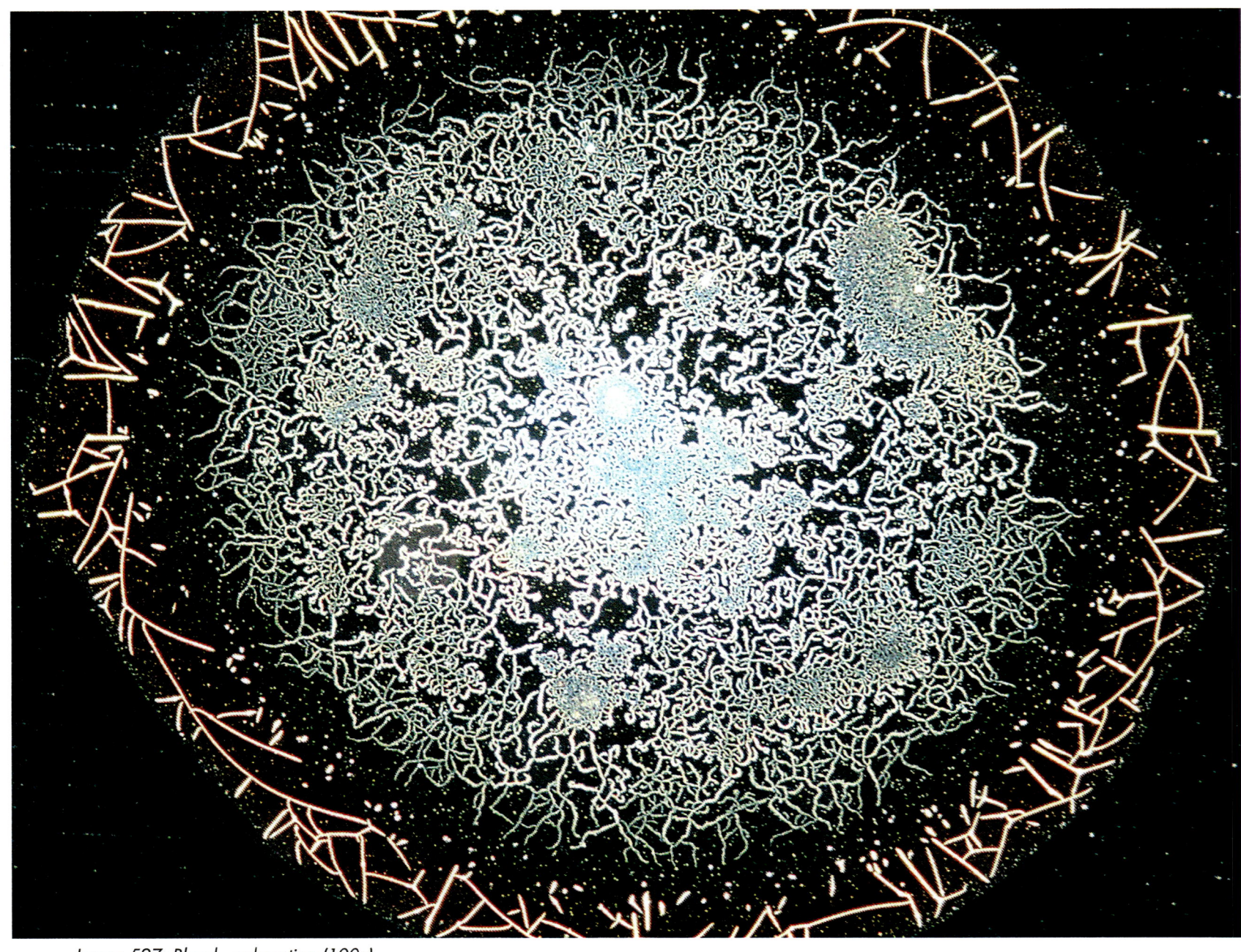

Image 527: Blood—exhaustion (100x)

Human Fluids

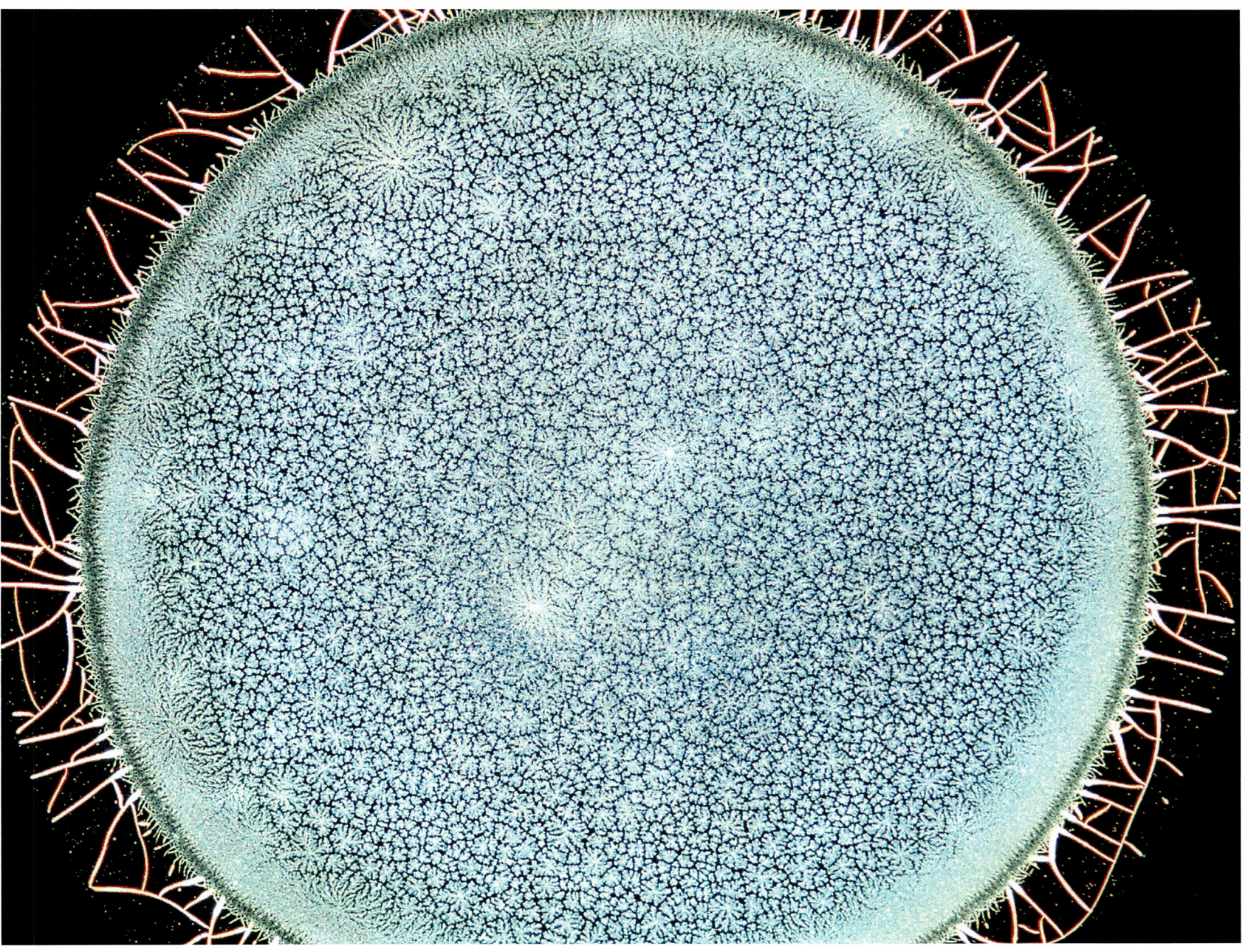

Image 528: Blood—recovery (100x)

THE SILENT LANGUAGE OF LIFE

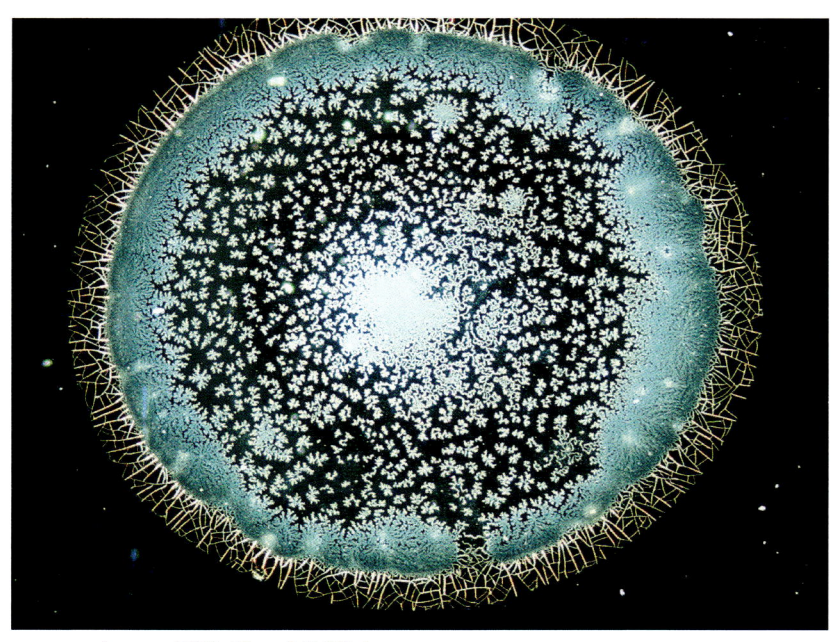

Image 529: Blood (100x)

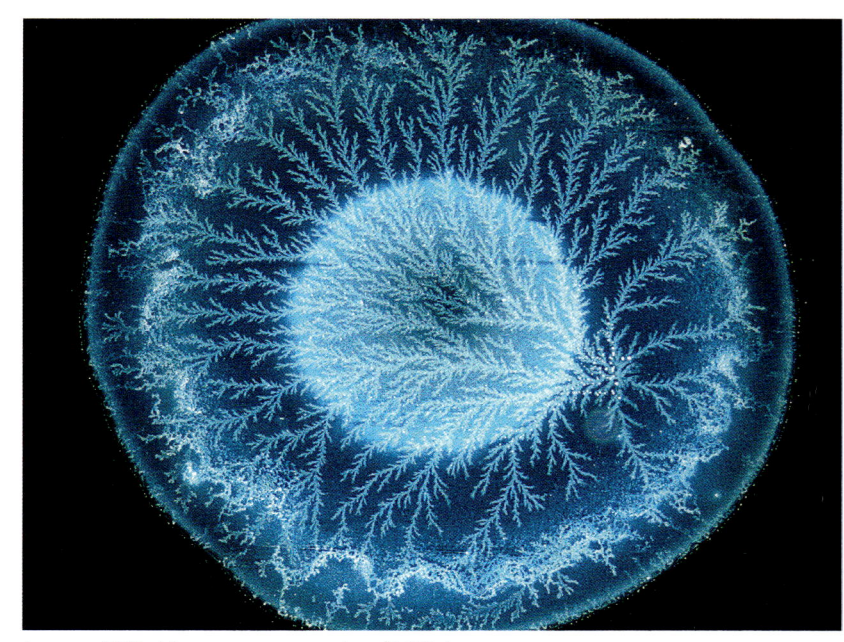

Image 530: Mercury preparation (100x)

Image 531: Blood next to mercury preparation (40x)

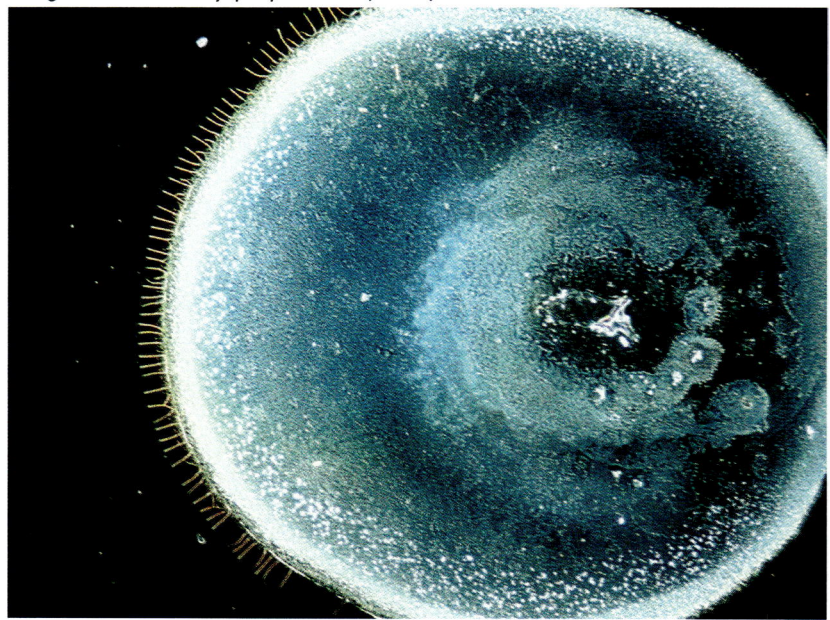

Image 532: Mercury preparation with rays (100x)

5. Drop Communication

When different drops come too close to each other, they often influence their respective designs quite considerably. In the process, the drop edges must be preserved so that they do not flow into each other.

This phenomenon of communicating drops is shown here—as one example of many—with blood.

Image 529: Blood drop

Image 530: Mercury drop

Image 531: Meeting of both at a relatively far distance

Image 532: Meeting at a closer distance; the mercury drop has partially taken over the halo of the blood, but has lost its own structure in the process. The blood drop itself does not come into the image here.

180

Human Fluids

6. Human Serum

The bright and clear serum of the human being, which no longer contains any corpuscular components, is dripped on.

Images 533 and 534 show us drops of serum; here, too, we can speak of an "ideal image" as having an off-center starting point from which delicate branches grow into the periphery. This happens in variations that are specific to the individual sample. One could call this formation a "tree of life."

The following variations do not have to be an expression of a severe disease either: Images 535 and 536.

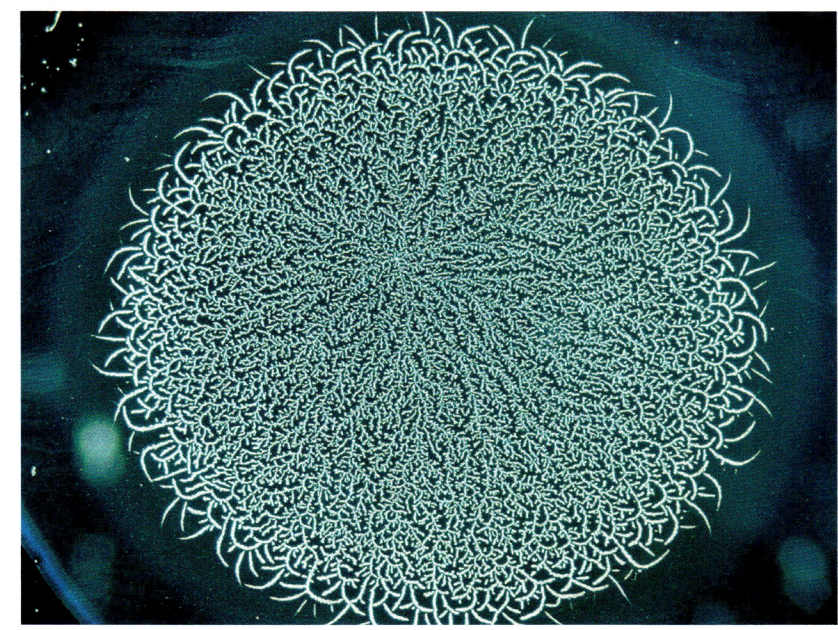

Image 533: Serum (100x)

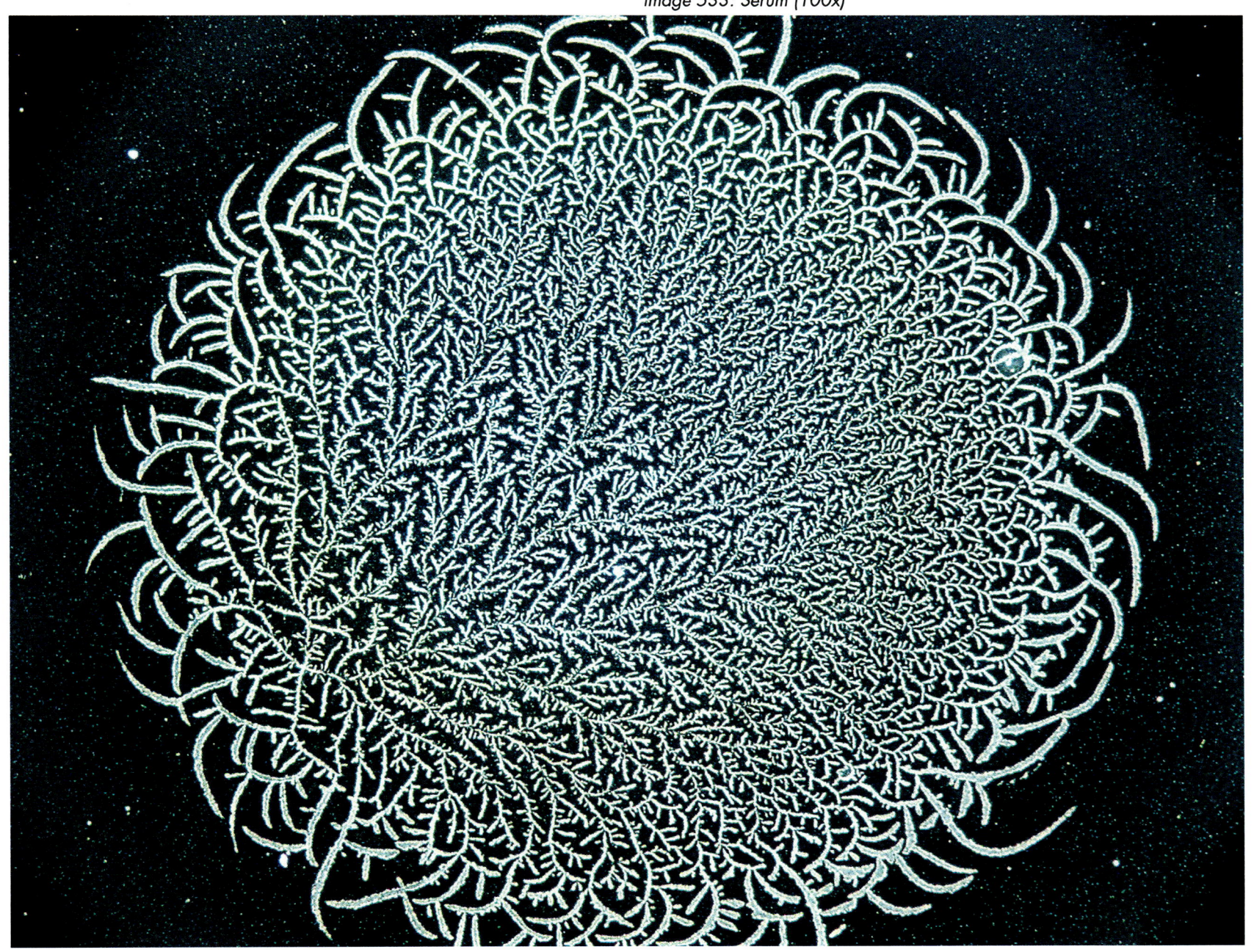

Image 534: Serum (100x)

THE SILENT LANGUAGE OF LIFE

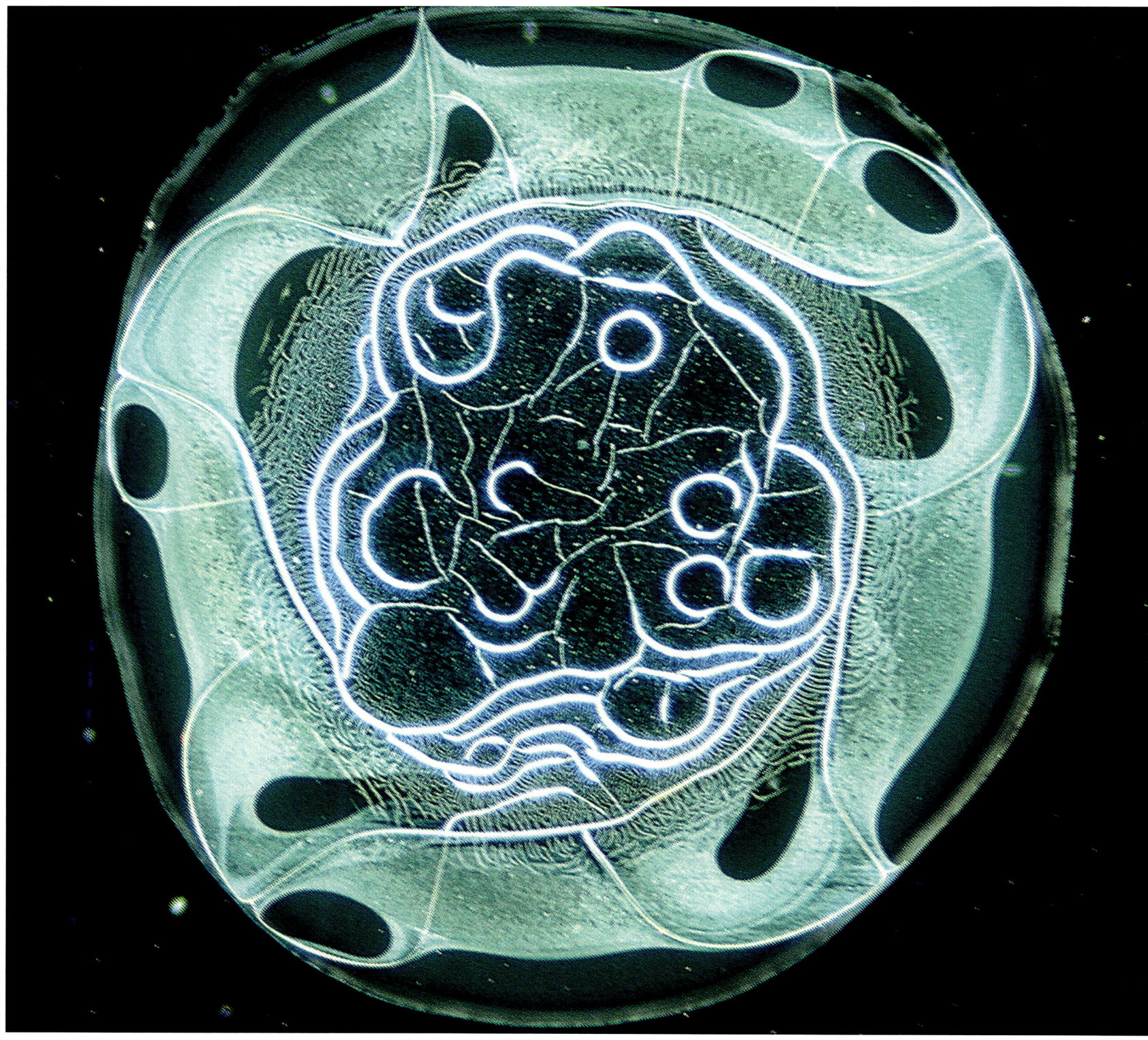

Image 535: Serum (40x)

Human Fluids

Image 536: Serum (40x)

— THE SILENT LANGUAGE OF LIFE —

7. Patient Serum before and after Treatment

In human serum and blood, differentiated effects of homeopathic remedies can be demonstrated with this drop imaging method (in vitro). This is a comprehensive topic in itself and cannot be discussed here. Only the serum of two cancer patients before and after mistletoe treatment is shown.

Image 537: Serum of a cancer patient (62 years old).
Image 538: Serum of the same patient after four months of mistletoe therapy with daily subcutaneous injections. The patient's condition was clearly improved.
Images 539 and 540: Serum of a female cancer patient (67 years old).
Image 541: Serum of the same patient after four weeks of mistletoe therapy with daily injections. The patient feels healthy. In both cases the ideal serum image as a "tree of life" was restored. Neither radiation nor chemo-therapy was used.

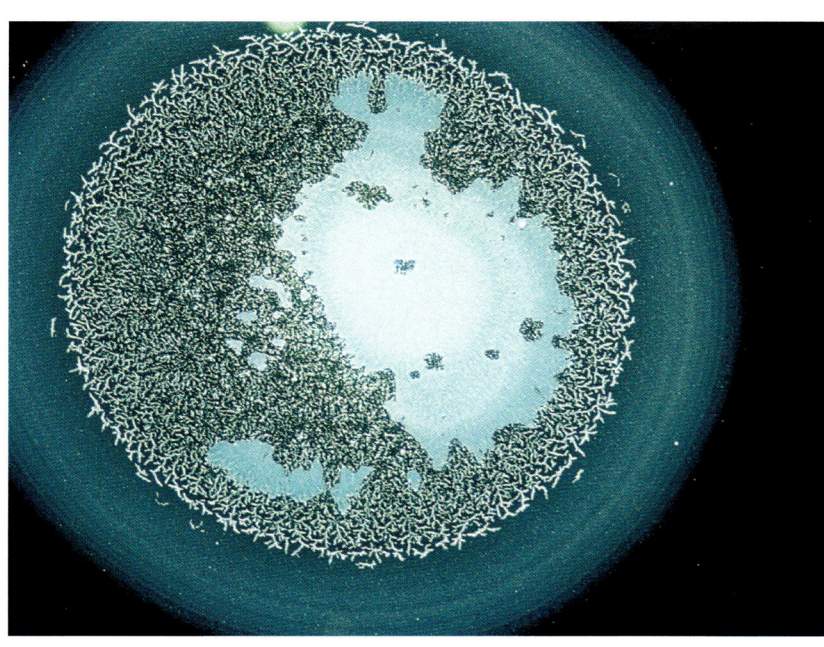

Image 537: Serum of a cancer patient (100x)

Image 538: Serum of a cancer patient after 4 months of mistletoe therapy (100x)

Human Fluids

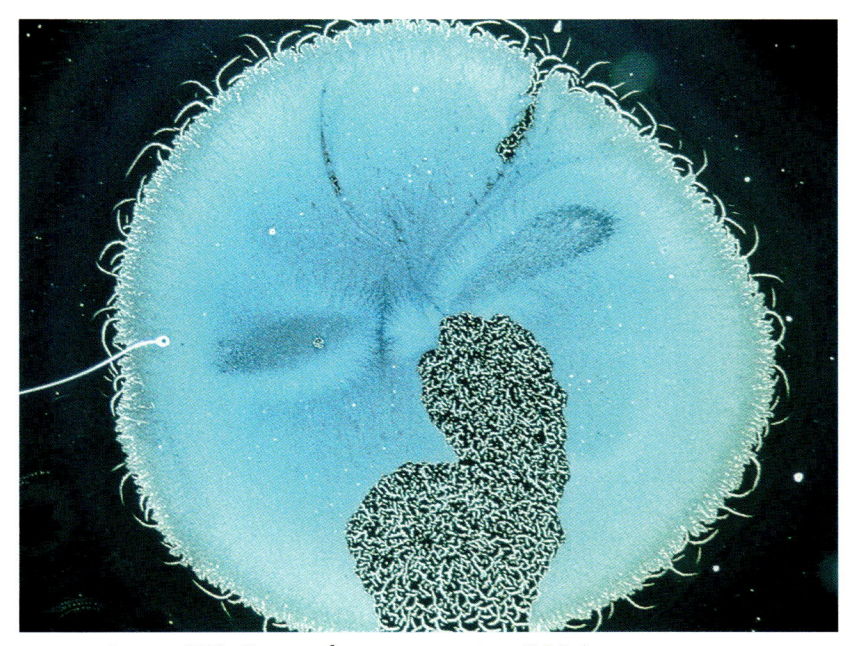

Image 539: Serum of a cancer patient (100x)

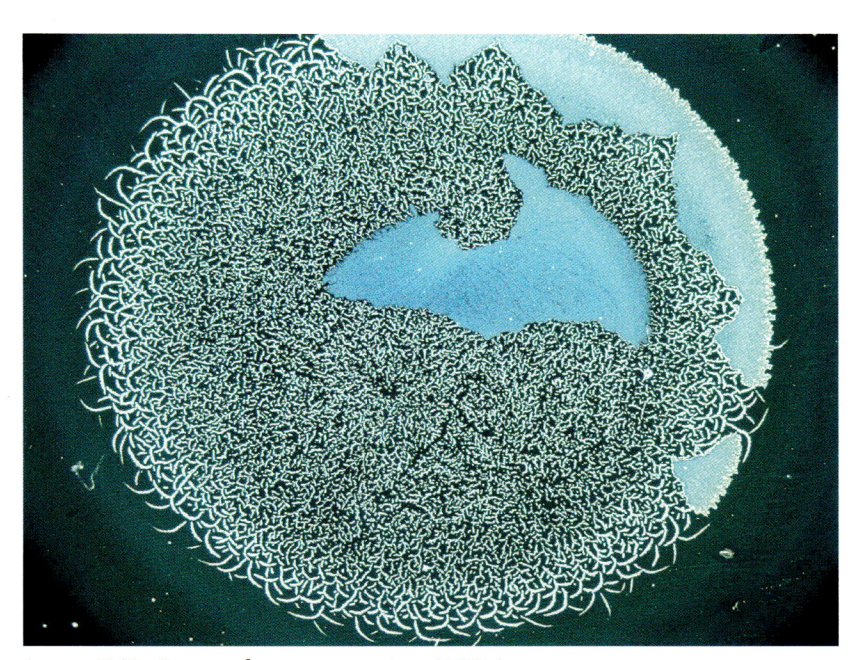

Image 540: Serum of a cancer patient (100x)

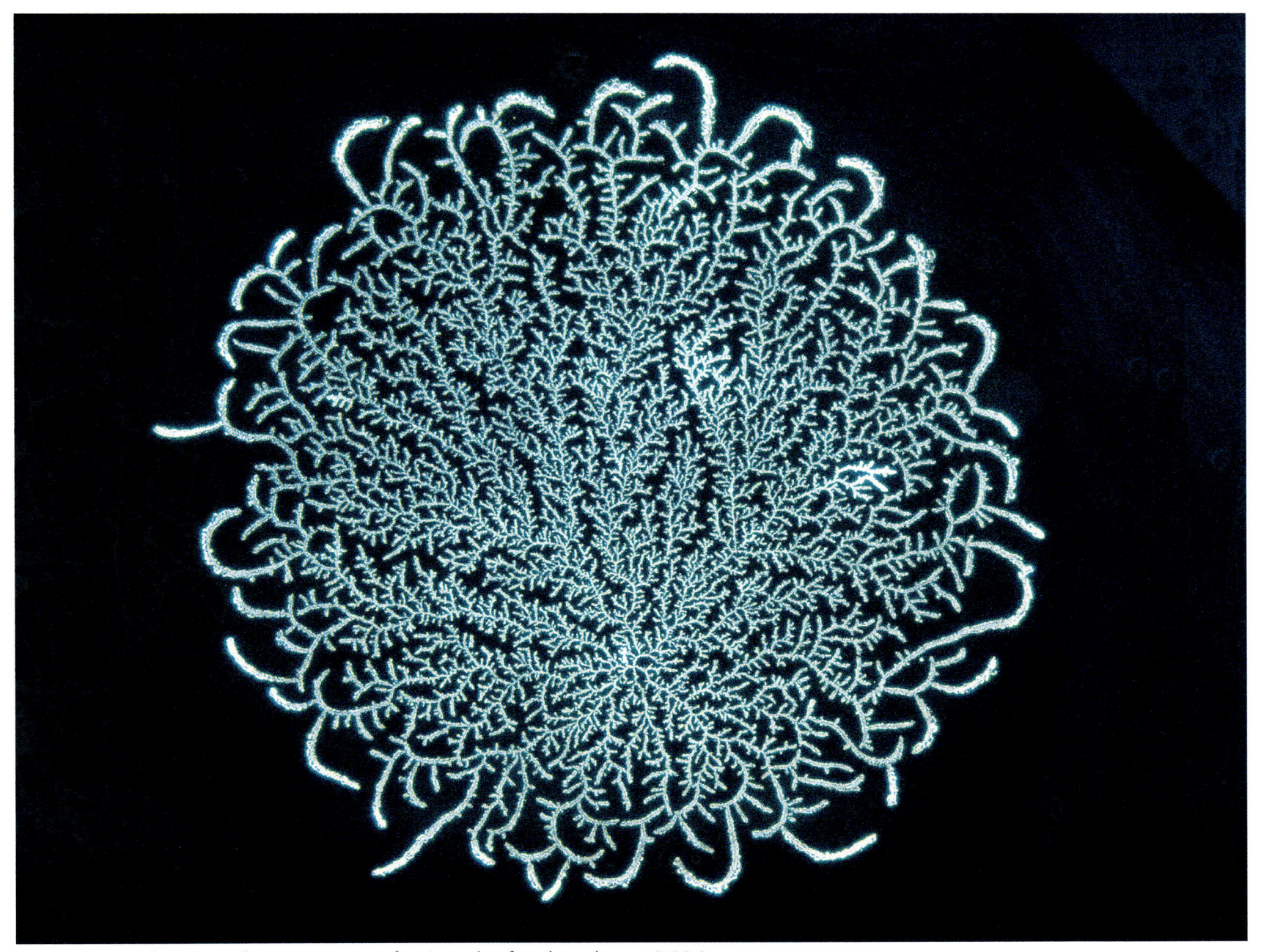

Image 541: Serum of a cancer patient after 4 months of mistletoe therapy (100x)

8. Reflections on Chapter V

What is depicted in this chapter of the human fluids is only a very small insight into individual areas of organ function. Nevertheless, these drop images speak of a polarity between the sensory and nervous spheres on the one hand and blood and serum on the other.

What has the character of salt in the sensory area passes from the breath with its vegetative structures into the transformative and radiating figurations of the blood. One would like to title the chapter "Man between Salt and Sun."

Between the two functional areas, the cerebrospinal fluid appears in the drop image. It may seem bold to see it treated in this way. However, the basis for this can be found in the fact that here, too, the drop image in itself reveals its origin as in a mirror. We do not encounter these elongated, rhythmically organized extensions from the star figures anywhere else. Here, too, one must consider the background. The cerebrospinal fluid is in constant communication with the brain and its nerve cells as well as with the spinal cord and its nerve branches, as this fluid flows rhythmically around the nerve organization. Thus, with reference to the previous explanations, it becomes understandable that an image of the actual event is shown.

The halo of blood is particularly impressive and enigmatic. Isn't it wonderful how it almost disappears in the night (image 520), while the drop itself is very much animated? The red blood cells themselves also send out delicate bright lines into the periphery as they migrate in the liquid native preparation. This becomes visible in the 400x magnification. The connection of the red blood with the iron content of the erythrocytes could certainly be the background here.

The rays of the halo are just as sensitive and vulnerable as the source star structures on the inner part of the drop surface. Rapid transformation in different directions is therefore possible, as can be seen. The human "I" relies entirely on the blood—naturally in the warmth—via the "I"-organization. It is the most flexible organ of the human being. This is how the rapid transformations after exhaustion and subsequent eurythmy can be understood. There is a "hidden germinating power" in the blood of the human being, which can perhaps be compared with the fruits and seeds in chapter IV!

Homeopathic medicines can be shown to have specific effects by the drop-image method, in vitro. As this is a very extensive field of research, it is not presented here.

However, the serum drop changes before and after homeopathic treatment with mistletoe injections are shown from two patients. The clinical situation in both cases corresponded to the serum images shown with a recovered "tree of life."

Human Fluids

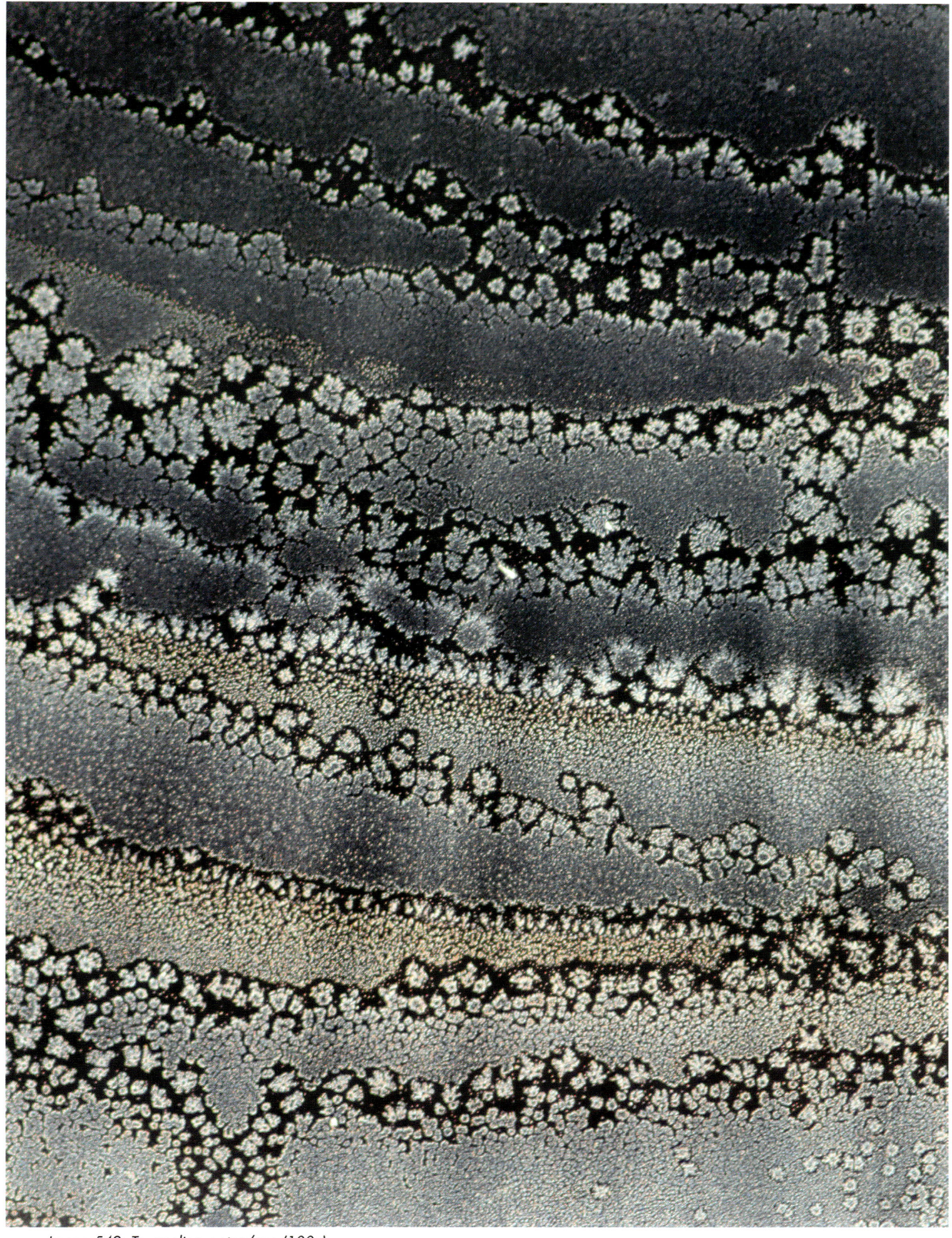

Image 542: Tourmaline water/sun (100x)

Three Phenomena

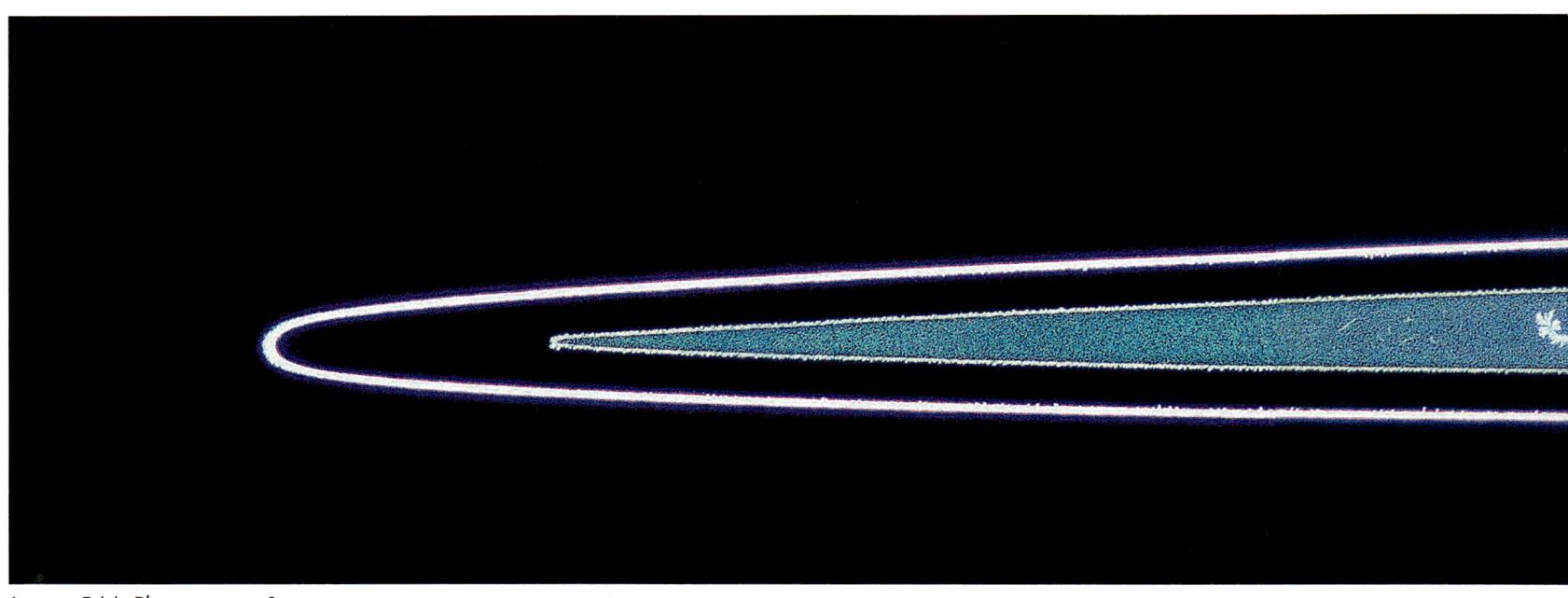

Image 544: Phenomenon 1

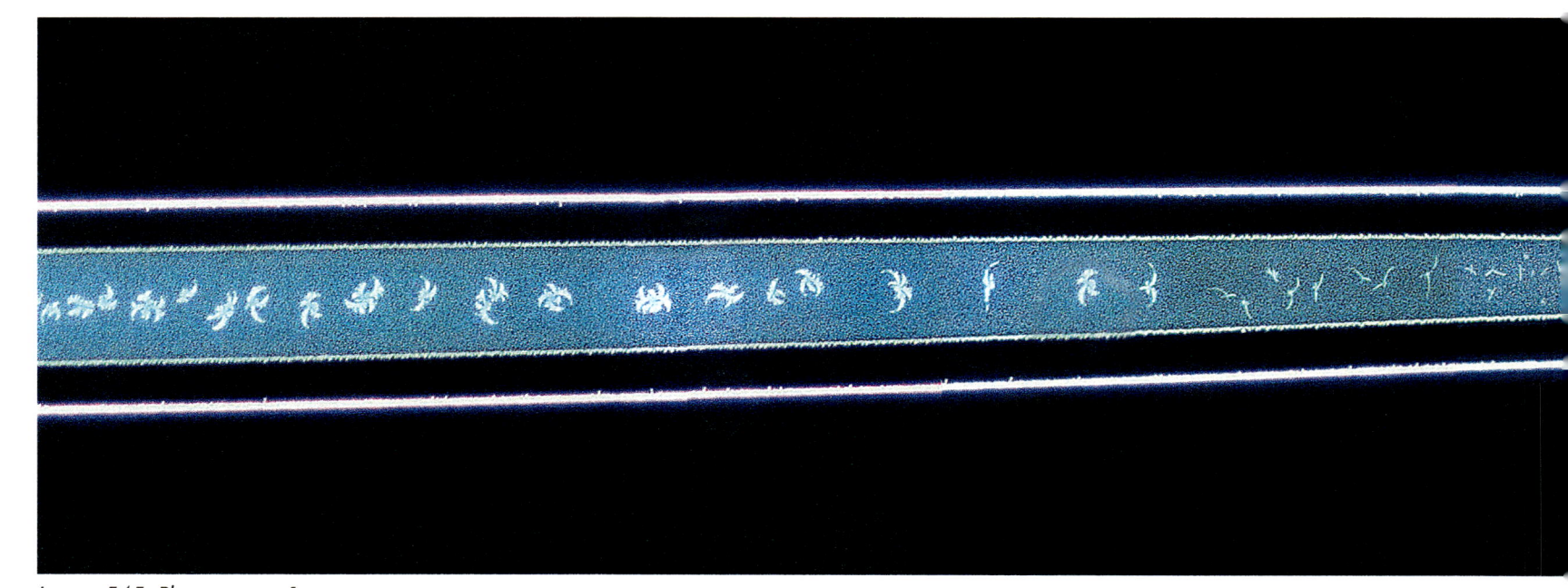

Image 545: Phenomenon 1

Three Phenomena

Three times in the fifteen years of research, these blue, elongated figurations appeared on a slide without reference to the drop image.

Phenomenon 1 was also "labeled." It had formed next to a drop of serum.

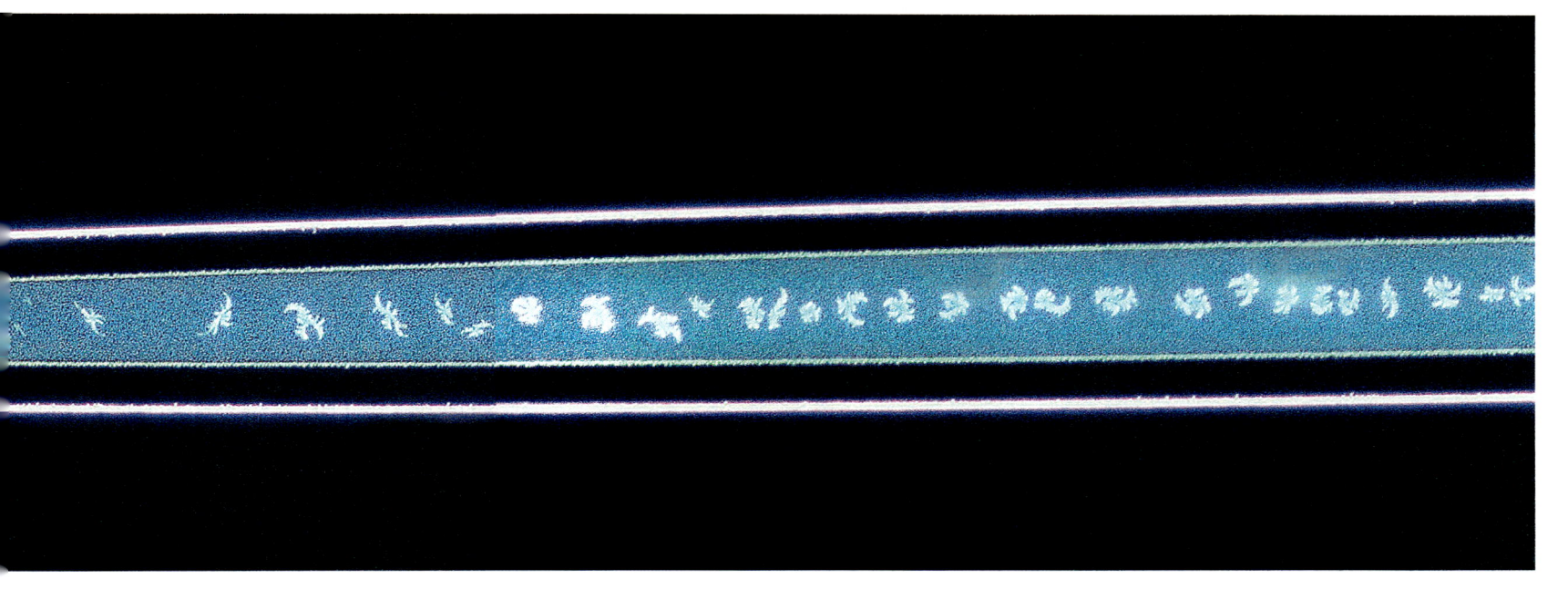

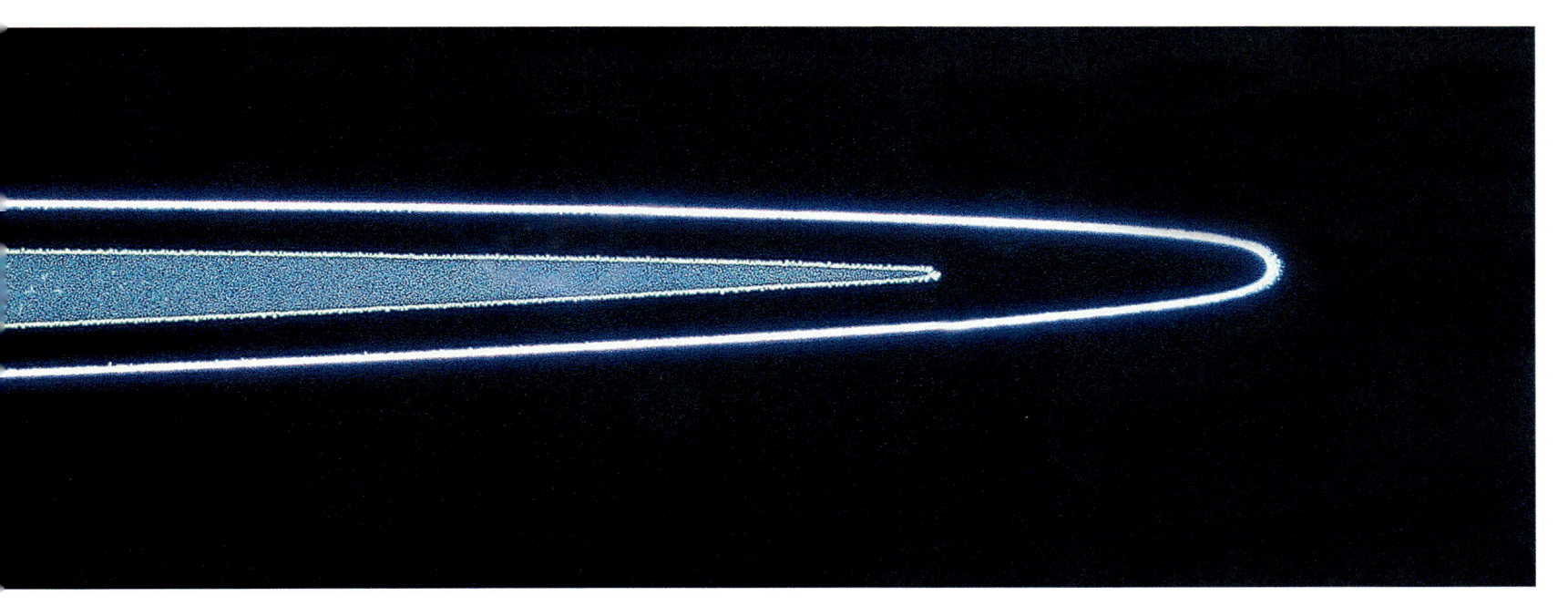

Phenomenon 2 crossed a drop from the Jerusalem artichoke preparation and was "unlabeled."

Phenomenon 3 merely showed its shape in a delicate "line of light."

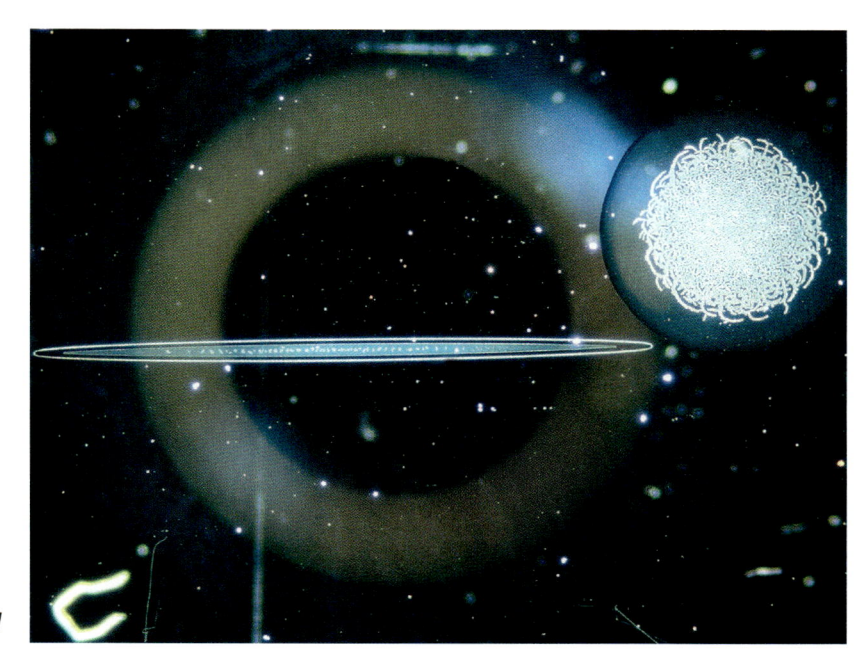

Image 543: Phenomenon 1

THE SILENT LANGUAGE OF LIFE

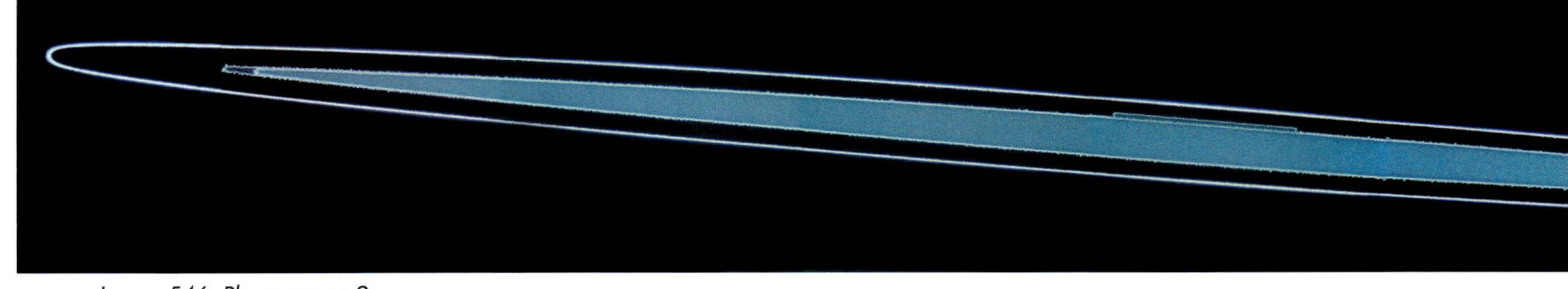

Image 546: Phenomenon 2

Image 548: Phenomenon 3

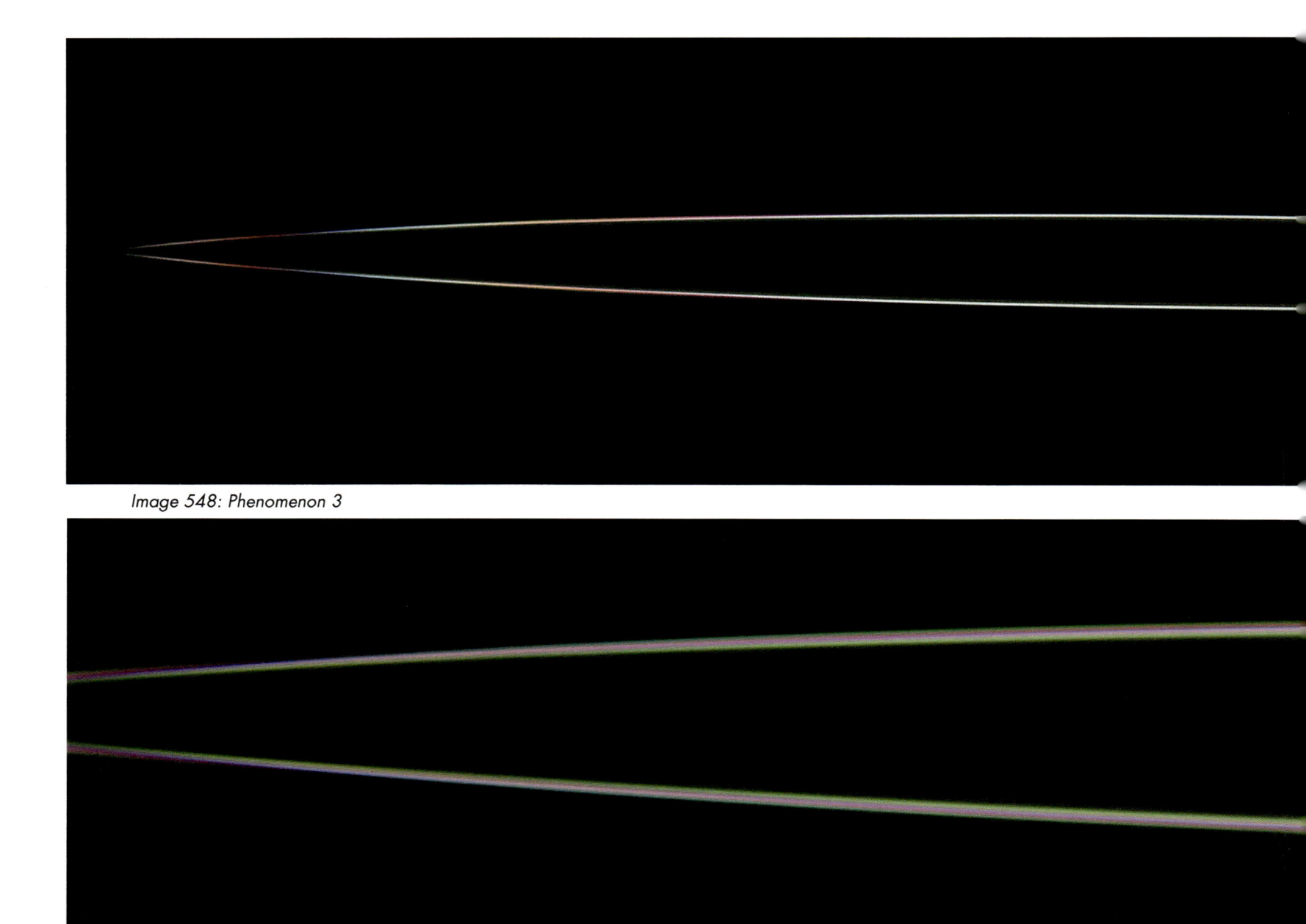

Image 549: Phenomenon 3

Three Phenomena

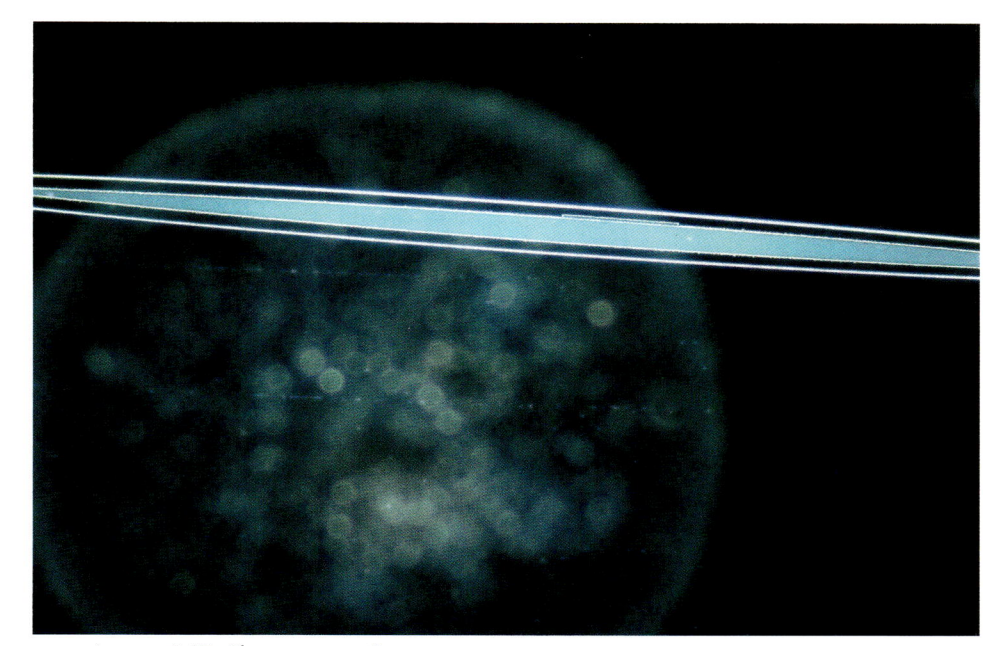

Image 547: Phenomenon 2

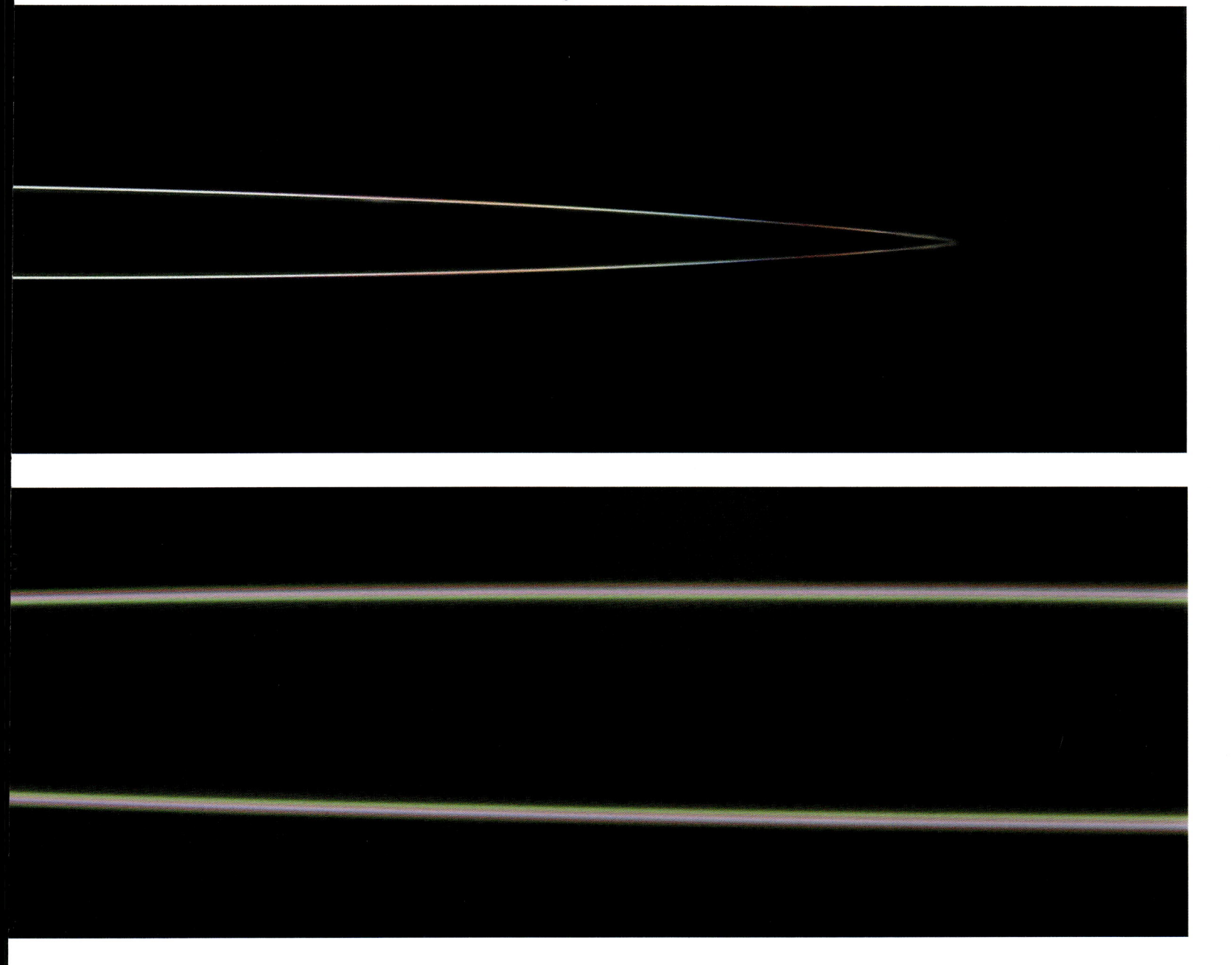

191

Conclusion and Acknowledgments

At the beginning of the fourth chapter is a short quotation by Rudolf Steiner from his book *Goethe's World View*.[35] It says that the natural scientist, in Goethe's view, should not stop at the external appearances of things. His attention should also be directed to the "ideal driving forces" that work in natural phenomena but do not appear. For Goethe, this was a self-evident and at the same time necessary undertaking in order to arrive at a true knowledge of nature. His lifelong training in the perception of natural phenomena allowed him this insight.

Rudolf Steiner's work contains comprehensive descriptions of the living forces in the kingdoms of nature and in the human being. When we speak of the *formative forces,* the etheric forces, or forces of growth and formation, these concepts take on a nuanced meaning that illuminates the respective subject. In our context, the "ideal driving forces" are, as a superordinate concept, contrasted with the events appearing in the differentiated activity of the forces of the drop as described in each case.

In this book with its images, it is water that, as the basis of all life processes, conveys pictorial and reproducible impressions of its conversation with the things of nature. We have attempted to show how this can happen. Of the many peculiarities of this mineral medium, three were highlighted as essential: Numerous inner surfaces ensure functions similar to sensory membranes.

The predisposition of water to rhythmic movement plays a role in this perception, mirroring, and listening. Again, in connection with this ability, water takes hold of its mineral substance and works, through the formation of clusters, on the forms that are generated by the natural objects. The formations produced in this way now reveal the background of their origin and lead us to formative "ideas" from a superordinate, previously invisible realm of nature. Through the water, a threshold of the world of manifestations is crossed, and another world of manifestations appears in the image. Since this cannot be classified immediately in the usual way, questions and justified doubts arise.

A first help in the search for understanding can be the acknowledgment of this other reality. The images are there—they reveal scientifically observable results. The comprehensive potency of water itself guarantees that these results will be produced.

The interpretations presented here can naturally be expanded upon in many directions and are also open to other points of view.

The first impression when looking at the images leads to wonder. The soul encounters order, harmony, and beauty. One can experience how the wonder never goes away—even when observing image phenomena that are many years old. Wonder really was at the beginning and was the impetus for numerous attempts. When we approach a phenomenon with wonder, we gain a much more immediate relationship to it than we do through reason and intelligence, for these are more concerned with the individual parts. "The fact that a person can be amazed by a thing makes it possible to connect with the object that lies below the threshold of consciousness."[58]

In his lecture cycle *The World of the Senses and the World of the Spirit* (CW 134), held in Hanover in 1911, Rudolf Steiner describes four basic prerequisites in the soul life of the human being who wants to attain a real knowledge of nature.

The first prerequisite is the state of wonder from which thinking must emerge as from a seed in order not to become a mere play of thoughts.

> All human inquiry must proceed from amazement...for all real knowledge that wants to have a prospect of having anything at all to do with the riddles of the world must come from the seed of wonder.[59]
>
> All knowledge begins with wonder, and only those who start from wonder are on the way to right knowledge.[60]

What lies hidden in wonder? We find an answer in a lecture Rudolf Steiner gave in Breslau on February 3, 1912. It is about the fact that we can be amazed at all. The reason for this lies in the perceptions and encounters we had in the spiritual world before our existence on earth. By facing the "rising sun," the "roaring sea," the "sprouting plant" here on earth, we are amazed because we feel darkly that we have encountered all this before—but now it looks quite different! If man did not come into this life from a prenatal, spiritual world, there would be no explanation for wonder and the knowledge it brings. "Only through initiation can it be brought as a clear memory."[60]

Wonder grew in the soul and led to a reverent mood toward the creative happening in the water, both during the experiments and when looking at the images. What Rudolf Steiner stipulated as a precondition of all scientific research actually proved in this research to be a way of tracing the unknown. Wonder and awe became accompanying moods of the soul, especially since the images are not the result of one's own activity—forces of will, creative forces from another world have created them! We can even watch this creative process in the drying drop!

In the beginning, there were only images, hieroglyphs, riddle. They had to be brought into context with their origin and creation. It was described how what had come to rest in the image was to be brought into harmony with its process of becoming. This is an active and passive activity of the soul at the same time. If, for example, lavender blossoms, placed alone in water, do not show a structured image—like other parts of the same plant (see chapter IV)—the laws of the plant being, which is subject to a devitalization process from root to blossom, must be taken into account in the search for understanding, and the process depicted in the water is illuminated. If I methodically include the background of the creation of the image in my contemplation, all abstract speculation disappears. Thanks to the mobility of my imagination, the meaningful interaction of the plant parts from root to flower unfolds before my inner eye—and the result of the devitalization process appearing in the image becomes tangible.

In this way, the third prerequisite was fulfilled: "To feel oneself in wisdom-filled harmony with the laws of the world-all."[59] Comprehending what was given in the image was possible only if thinking was prepared to immerse itself in the living impulses of movement in the creation of the image. In this way it could encounter the reality of the wisdom-filled connections.

It is also the fourth stage of the soul's constitution that Rudolf Steiner speaks of: it is the mood of devotion, "which does not seek to research truth for its own purposes but rather awaits all truth from the revelation that flows out of things and can wait until it is ripe to receive this or that revelation."[59]

The water images set us in motion and place us on the path to feeling and willing *with* the world of nature. The spiritual creative power of the originator of all phenomena is at work here. This power is also active in us—otherwise we would be unable to attach any significance to the phenomena.

Thus, in this new territory, a path opens through vision, allowing us to encounter hidden forces—the previously mentioned "ideal driving forces"—in the phenomena of nature.

It goes without saying that this book does not claim to be the only valid interpretation or to be complete.

Many unanswered questions and a mood of wonder can inspire further work in this field.

The most beautiful thing we can experience is the mysterious. It is the basic feeling that stands at the cradle of true art and science.
—Albert Einstein (1879–1955)

Acknowledgements

From the bottom of my heart I am grateful to the originator of this research, Prof. Dr. Bernd Kröplin. Without him, this book would not exist. He encouraged me in the early years of my research to present his and my images to a larger audience in lectures. He was always open to questions and new discoveries and immediately ready for mutual exchange.

Christine Rasch had followed the ever-new images with enthusiasm and understanding over the years, and so she was happy to help with the text correction. An overabundance could be tamed and brought to the point. We thank her from the bottom of our hearts for the many "sessions"!

Melina Elmali did the work on a computer. It is thanks to her technical expertise that text and images could be brought into a condition suitable for publishing.

My thanks also go to the many colleagues and lecture participants who have tirelessly expressed their great interest in the publication of the book over the years.

References

1. Prof. Dr.-Ing. Bernd Kröplin, *Welt im Tropfen*, GutesBuchVerlag, Stuttgart, 2001; http://www.weltimtropfen.de; Bernd Kröplin and Regine C. Henschel, *Die Geheimnisse des Wassers*, AT Verlag, Aarau, Switzerland, and Munich, 2016.
2. Theodor Schwenk, *Sensitive Chaos: The Creation of Flowing Forms in Water and Air*, Forest Row, UK: Rudolf Steiner Press, 2014.
3. Heinz-Michael Peter, *Sensibles Wasser*, in booklet 4, Verein für Bewegungsforschung e.V., Institut für Strömungswissenschaften, Herrischried, 1994.
4. Masaru Emoto, MD, *Wasserkristalle*, KOHA-Verlag GmbH, 1999/2002.
5. Alexander Lauterwasser, *Wasser Klang Bilder*, AT Verlag, Aarau/Switzerland, 2002; Alexander Lauterwasser, *Wasser Musik*, AT Verlag, Baden and Munich, 2005.
6. Lili Kolisko, *Sternenwirken in Erdenstoffen*, Mar. 30, 1927, Orient-Occident Verlag, Stuttgart, Den Haag, London.
7. Thomas Meyer, *Ein Leben für den Geist: Ehrenfried Pfeiffer*, (Pfeiffers autobiografische Erinnerungen...), Perseus Verlag, Basel, 1999.
8. Alla Selawry, MD, *Metall-Funktionstypen in Psychologie und Medizin*, Haug Verlag, Heidelberg, 1985; also, *Ehrenfried Pfeiffer: Pioneer of Spiritual Research and Practice*, Spring Valley, NY: Mercury Press, 1995.
9. Claus Biegert and Georg Gaupp-Berghausen (Ed.), *Vom Wesen des Wassers*, Frederking&Thaler Verlag GmbH, Munich, 2006.
10. Novalis, *Die Lehrlinge zu Sais*, Novalis Schriften 4. Band, published by Eugen Diederichs, Jena, 1923.
11. http://www.volvic.de/unsere-quelle/volvic-von-a-bis-z.html, accessed Mar. 2, 2017.
12. Helmut Knauer, *Erdenantlitz und Erdenstoffe*, Philosophisch-Anthroposophischer Verlag am Goetheanum, Dornach, 1961.
13. Rudolf Hauschka, *The Nature of Substance: Spirit and Matter*, Forest Row, UK: Rudolf Steiner Press, 2003.
14. Kaspar Appenzeller, MD, *Die Quadratur des Zirkels*, Zbinden Verlag, Basel, 1979.
15. Rudolf Steiner, *Warmth Course: The Theory of Heat: Second Scientific Lecture Course*, CW 321, Mar. 11, 1920. Steiner's complete works are in the *Gesamtausgabe* (GA), Rudolf Steiner Verlag, Dornach; in English, The Collected Works (CW), SteinerBooks and Rudolf Steiner Press.
16. Rudolf Steiner, *Eurythmy Therapy*, CW 315, Forest Row, UK: Rudolf Steiner Press, Apr. 15, 1921.
17. Rudolf Steiner, *Anthroposophy and the Inner Life: An Esoteric Introduction*, CW 234, Forest Row, UK: Rudolf Steiner Press, 2015, Feb. 2, 1924.
18. Rudolf Steiner, *Understanding Healing: Meditative Reflections on Deepening Medicine through Spiritual Science*, CW 316, Forest Row, UK: Rudolf Steiner Press, 2013, Apr. 24, 1924, p.m.
19. Rudolf Steiner, *Die menschliche Seele in ihrem Zusammenhang mit göttlich-geistigen Individualitäten* (The human soul in connection with divine spiritual beings), CW 224, Jul. 11, 1923.
20. Ernst Marti, MD, *Die vier Äther. Zu Rudolf Steiners Ätherlehre. Elemente—Äther—Bildekräfte*, Verlag Freies Geistesleben, Stuttgart, 4th ed., 1990; expanded ed. by Irmgard Rossmann, MD: Ernst Marti, MD, *Das Ätherische*, Rudolf Steiner Verlag, Dornach, Switzerland, 1989; 2nd ed., 1994 (ibid.).
21. Rudolf Steiner, *Good and Evil Spirits and their Influence on Humanity*, CW 102, Forest Row, UK: Rudolf Steiner Press, 2014, Jan. 6–Jun. 11, 1908; also, *Harmony of the Creative Word: The Human Being and the Elemental, Animal, Plant, and Mineral Kingdoms*, CW 230, Rudolf Steiner Press, 2002, Oct. 19–Nov. 11, 1923.
22. Rudolf Steiner, *Vorträge und Kurse über christlich-religiöses Wirken*, II. Priester-Zyklus (Lectures and courses on Christian religious work), CW 343, Oct. 5, 1921, a.m.
23. Rudolf Steiner, *Die Erkenntnis des Übersinnlichen in unserer Zeit* (Knowledge of the suprasensory in our time), CW 55, Oct. 25, 1906
24. Steinbachs Naturführer, *Mineralien*, Mosaik Verlag, Munich, 1982.
25. Rudolf Steiner, *Esoteric Lessons 1904–1909: From the Esoteric School vol. 1*, CW 266/1, Great Barrington, MA: SteinerBooks, 2007, Nov. 29, 1907.
26. Rudolf Steiner and Ita Wegman, *Extending Practical Medicine: Fundamental Principles Based on the Science of the Spirit*, CW 27, London: Rudolf Steiner Press, 1996, p 6.
27. Matt. 5:13; Matt. 5:14; John 1:4; Luke, 14:34–35; Col. 4:6; Mark 9:49; Rev. 21:16–17.
28. Rudolf Steiner, *The Apocalypse of St. John: Lectures on the Book of Revelation*, CW 104, Hudson, NY: Anthroposophic Press, 1985, Sep. 16, 1907.
29. Rudolf Steiner, *Rosicrucianism Renewed: The Unity of Art, Science, and Religion*, CW 284, Great Barrington, MA: SteinerBooks, 2006, Sep. 16, 1907.
30. Rudolf Steiner, *Eurythmy as Visible Speech*, CW 279, Jul. 12, 1924.
31. Agrippa von Nettesheim, *Die magischen Werke und weitere Renaissancetraktate*, edited and introduced by Marco Frenschkowski, Marix Verlag GmbH, Wiesbaden, 2008, p. 254.
32. Rudolf Steiner, *Esoteric Christianity and the Mission of Christian Rosenkreutz*, CW 130, London: Rudolf Steiner Press, 2001, Sep. 28, 1911.
33. Friedrich Benesch, *The Tourmaline: A Monograph*, Verlag Urachhaus, Stuttgart, 2003.
34. Rudolf Steiner, *Goethe's World View*, CW 6, Spring Valley, NY: Mercury Press, 1994.
35. J. W. v. Goethe, *Gedichte in zeitlicher Folge*, Insel Verlag, Frankfurt on the Main, 1982, p. 1071.
36. Rudolf Steiner, *Broken Vessels: The Spiritual Structure of Human Frailty*, CW 318, Great Barrington, MA: SteinerBooks, 2002, Sep. 15, 1924.

37 Rudolf Steiner, *Harmony of the Creative Word: The Human Being and the Elemental, Animal, Plant, and Mineral Kingdoms*, CW 230, Forest Row, UK: Rudolf Steiner Press, 2002, Oct. 27, 1923.

38 Rudolf Steiner, *Good and Evil Spirits and their Influence on Humanity*, CW 102, Forest Row, UK: Rudolf Steiner Press, 2014, Jun. 1, 1908.

39 Rudolf Steiner, *Physiology and Healing: Treatment, Therapy, and Hygiene*, CW 314, Forest Row, UK: Rudolf Steiner Press, 2013, Oct. 28, 1923.

40 Wilhelm Pelikan, *Healing Plants Volume III: Insights through Spiritual Science*, Spring Valley, NY: Mercury Press, 2021.

41 Rudolf Steiner, *Truth-Wrought-Words*, CW 40, Spring Valley, NY: Anthroposophic Press, 1979, p. 63.

42 Rudolf Steiner, *The Healing Process: Spirit, Nature and Our Bodies*, CW 319, Aug. 29, 1924

43 Rudolf Steiner, *Physiology and Healing: Treatment, Therapy, and Hygiene*, CW 314, Oct. 27, 1922; and *From Crystals to Crocodiles...: Answers to Questions* CW 347, Sep. 27, 1923.

44 Rudolf Steiner, *Mensch und Welt, das Wirken des Geistes in der Natur, über das Wesen der Bienen* (Man and the world, the work of the spirit in nature; on the nature of bees), CW 351, Rudolf Steiner Verlag, 1995, Oct. 10, 1923.

45 Ernst Lehrs, *Mensch Und Materie: Ein Beitrag Zur Erweiterung Der Naturerkenntnis Nach Der Methode Goethes*, Vittorio Klostermann, Frankfurt, 1987, p. 217 (in English: *Man or Matter: An Introduction to a Spiritual Understanding of Nature on the Basis of Goethe's Method of Training Observation and Thought*, 3rd ed., Forest Row, UK: Rudolf Steiner Press, 2014).

46 Rudolf Steiner, *Das Johannes-Evangelium im Verhältnis zu den drei anderen Evangelien, besonders zu dem Lukas-Evangelium*, CW 112, Jul. 7, 1909 (in English: *The Gospel of St. John: And Its Relation to the Other Gospels*. Hudson, NY: Anthroposophic Press, 1994).

47 Rudolf Steiner, *Ein malerischer Schulungsweg: Pastellskizzen und Aquarelle* (A picturesque path of education: Pastel sketches and watercolors), Nov. 12, 1923, Rudolf Steiner Verlag, Dornach.

48 Rudolf Steiner, *Secret Brotherhoods and the Mystery of the Human Double*, CW 178, London: Rudolf Steiner Press, 2004, Nov. 25, 1917.

49 Theodor Schwenk, *Vom Einströmen kosmischer Kräfte in irdische Substanzprozesse* (On the influx of cosmic forces into earthly substance processes), Forschungslabor der Weleda AG, Schwäbisch-Gmünd, 1961.

50 Rudolf Steiner, *Agriculture: Spiritual Foundations for the Renewal of Agriculture*, CW 327, Biodynamic Farming and Gardening Association, 1993, Jun. 7–16, 1924.

51 *Beiträge zur Rudolf Steiner Gesamtausgabe* (Contributions to the Rudolf Steiner complete edition), no. 122, p. 60.

52 Ruth Erne, *Der Torffaser-Impuls: Eine Forschungsaufgabe von Rudolf Steiner* (The peat fiber impulse: A research task by Rudolf Steiner), Verlagsbuchhandlung, Beer/Zurich, 2017.

53 *Taschenbuch der Physiologie*, S. Silbernagl and A. Despopoulos, Georg Thieme Verlag, Stuttgart, 1979.

54 *Taschenatlas der Anatomie Band III, Nervensystem und Sinnesorgane*, W. Kahle, H. Leonhardt, W. Platzer, Georg Thieme Verlag, Stuttgart, 1979.

55 Novalis, *Fragmente vermischten Inhalts*, Novalis Schriften 2. Band, p. 223, published by Eugen Diederichs, Jena, 1923.

56 Rudolf Steiner, *Chance, Providence, and Necessity*, CW 163, Hudson, NY: Anthroposophic Press, 1988, Aug. 27, 1915.

57 Rudolf Steiner, *Anthroposophical Leading Thoughts: Anthroposophy as a Path of Knowledge: The Michael Mystery*, CW 26, London: Rudolf Steiner Press, 1973, Mar. 22, 1925, p. 201.

58 Rudolf Steiner, *The Mission of the New Spirit of Revelation: The Pivotal Nature of the Christ Event in Earth Evolution*, CW 127, Forest Row, UK: Rudolf Steiner Press, 2021, Jan. 7, 1911.

59 Rudolf Steiner, *The World of the Senses and the World of the Spirit*, CW 134, Forest Row, UK: Rudolf Steiner Press, 2014, Dec. 27, 1911–Jan. 1, 1912.

60 Rudolf Steiner, *Three Paths to Christ: Experiencing the Supersensible*, CW 143, Forest Row, UK: Rudolf Steiner Press, 2023, Feb. 3, 1912.

About the Author

Inge Just-Nastansky, general practitioner, Stuttgart, born 1942 in Cologne. After graduating in medicine in Hamburg, three years of research work at the UKE/Hamburg. Then assistant doctor in paediatric and adult surgery for six years. In 1981 she took over a large rural practice in the northern Black Forest for eight years. From 1989 eurythmy studies in Stuttgart, diploma in 1993, then one year eurythmy therapy training in Stuttgart. After working as a school doctor for two years, she set up her own practice again in Stuttgart Bad-Cannstatt for seventeen years until 2014. Since 2001 water drop research with lecturing activities.

ALSO BY INGE JUST-NASTANSKY FROM PORTAL BOOKS

Corona Blood Phenomena

Microscopic Examinations of Blood, Serum, and Cerebrospinal Fluid
Healthy – Vaccinated – Recovered

This book was occasioned by the many patients who sought medical treatment for Covid or after receiving a Covid "vaccine." The blood of these patients was studied using the method of drop-image microscopy.

The results of this research—including examinations of the blood of "vaccinated" and "unvaccinated" as well as healthy and recovered patients—is documented here, along with normal cerebrospinal fluid (CSF) and the CSF of a patient after three "vaccinations."

The drop-image method has been used for more than twenty years. The phenomena in the drop-images documented here have never before been encountered. Covid illness and "vaccination" bring about an unprecedented change in the blood—in some cases, a complete destruction of the protein structures. This raises many questions. Blood, as the central organ of the human organism, not only maintains our life but is also the foundation of human freedom in perception, thinking, feeling, and action.

This book by no means represents a finished scientific study but rather an ongoing documentation that demonstrates the importance of the integrity and wholeness of the human body. The task of the physician is to help protect and maintain this integrity; the task of science is to address the many unanswered questions that arise through such research.

ISBN 9781938685453 | 9.5 x 12.5 inches | 108 pgs
ILLUSTRATED IN COLOR

"The impetus for this [book] came from the many patients who sought medical treatment for COVID or after COVID 'vaccination.' In collaboration with medical colleagues, therapies were developed for those affected, and blood tests using the drop-imaging method were started over a year ago. This revealed an unprecedented type of change in human blood compared to what had been seen for twenty years of blood imaging research. A representative sample of well over fifty COVID-"vaccinated," COVID-diseased, and COVID-recovered people is documented here, showing a wide spectrum of blood thread alterations in the first two groups. The overall coherence of the blood picture was lost, and in many cases numerous mutually similar, recurring phenomena appeared that could not be explained." (from the foreword)